SIGNIFICANT CHANGES TO THE

INTERNATIONAL BUILDING CODE®

2018 EDITION

CENGAGE

Australia • Brazil • Mexico • Singapore • United Kingdom • United States

Significant Changes to the International Building Code® 2018 Edition

International Code Council

Douglas W. Thornburg, AIA;
Sandra Hyde, P.E.

Cengage Staff:

SVP, GM Skills & Global Product Management: Jonathan Lau

Product Director: Matthew Seeley

Senior Product Manager: Vanessa Myers

Senior Director, Development: Marah Bellegarde

Senior Product Development Manager: Larry Main

Associate Content Developer: Jenn Alverson

Product Assistant: Jason Koumourdas

Vice President, Marketing Services: Jennifer Ann Baker

Marketing Manager: Scott Chrysler

Senior Production Director: Wendy Troeger

Production Director: Andrew Crouth

Senior Content Project Manager: Glenn Castle

Design Director: Jack Pendleton

Designer: Angela Sheehan

ICC Staff:

Executive Vice President and Director of Business Development: Mark A. Johnson

Senior Vice President of Product Development: Hamid Naderi

Vice President and Technical Director, Products and Services: Doug Thornburg

Senior Marketing Specialist: Dianna Hallmark

Cover images courtesy of:
iStockPhoto.com/scanrail
Pavel Ganchev-Paf/Shutterstock.com
Gilmanshin/Shutterstock.com

© 2018 International Code Council

ALL RIGHTS RESERVED. No part of this work covered by the copyright herein may be reproduced or distributed in any form or by any means, except as permitted by U.S. copyright law, without the prior written permission of the copyright owner.

For product information and technology assistance, contact us at
Cengage Customer & Sales Support, 1-800-354-9706

For permission to use material from this text or product, submit all requests online at **www.cengage.com/permissions.**
Further permissions questions can be emailed to
permissionrequest@cengage.com

Library of Congress Control Number: 2017949301

ISBN: 978-1-337-27120-2

ICC World Headquarters
500 New Jersey Avenue, NW
6th Floor
Washington, D.C. 20001-2070
Telephone: 1-888-ICC-SAFE (422-7233)
Website: **http://www.iccsafe.org**

Cengage
20 Channel Center Street
Boston, MA 02210
USA

Cengage is a leading provider of customized learning solutions with employees residing in nearly 40 different countries and sales in more than 125 countries around the world. Find your local representative at **www.cengage.com**

Cengage products are represented in Canada by Nelson Education, Ltd.

Visit us at **www.ConstructionEdge.cengage.com**
For more learning solutions, please visit our corporate website at **www.cengage.com**

Notice to the Reader
Publisher does not warrant or guarantee any of the products described herein or perform any independent analysis in connection with any of the product information contained herein. Publisher does not assume, and expressly disclaims, any obligation to obtain and include information other than that provided to it by the manufacturer. The reader is expressly warned to consider and adopt all safety precautions that might be indicated by the activities described herein and to avoid all potential hazards. By following the instructions contained herein, the reader willingly assumes all risks in connection with such instructions. The publisher makes no representations or warranties of any kind, including but not limited to, the warranties of fitness for particular purpose or merchantability, nor are any such representations implied with respect to the material set forth herein, and the publisher takes no responsibility with respect to such material. The publisher shall not be liable for any special, consequential, or exemplary damages resulting, in whole or part, from the readers' use of, or reliance upon, this material.

Printed in United States of America
Print Number: 01   Print Year: 2017

Contents

PART 1
Administration
Chapters 1 and 2 — 1

- 202
  Definition of Greenhouse — 2

- 202
  Definition of Repair Garage — 4

- 202
  Definition of Sleeping Unit — 5

- Chapter 2
  Removal of Definition References — 7

PART 2
Building Planning
Chapters 3 through 6 — 9

- 302.1
  Classification of Outdoor Areas — 11

- 303.4
  Assembly Use of Greenhouses Classification — 13

- 309.1
  Mercantile Use of Greenhouses Classification — 15

- 310.3, 310.4
  Classification of Congregate Living Facilities — 17

- 310.4.2
  Owner-Occupied Lodging Houses — 19

- 311.1.1
  Classification of Accessory Storage Spaces — 21

- 311.2
  Classification of Self-Service Storage Facilities — 23

- 312.1
  Classification of Communication Equipment Structures — 25

- 312.1.1
  Classification of Agricultural Greenhouses — 26

- 403.2.1.1
  Type of Construction in High-Rise Buildings — 28

- 404.6
  Enclosure of Atriums — 30

- 406.1
  Motor Vehicle-Related Occupancies — 32

- 406.3
  Regulation of Private Garages — 34

- 406.6.2
  Ventilation of Enclosed Parking Garages — 35

- 407.5
  Maximum Smoke Compartment Size — 36

- 407.5.4
  Required Egress from Smoke Compartments — 38

- 420.7
  Corridor Protection in Assisted Living Units — 39

- **420.8**
  Group I-1 Cooking Facilities — 41
- **420.10**
  Dormitory Cooking Facilities — 43
- **422.6**
  Electrical Systems in Ambulatory Care Facilities — 45
- **424.1**
  Children's Play Structures — 47
- **427**
  Medical Gas Systems — 49
- **428**
  Higher Education Laboratories — 52
- **503.1, 706.1**
  Scope of Fire Wall Use — 56
- **503.1.4**
  Allowable Height and Area of Occupied Roofs — 58
- **505.2.1.1**
  Mezzanine and Equipment Platform Area Limitations — 60
- **Table 506.2, Note i**
  Allowable Area of Type VB Greenhouses — 62
- **507.4**
  Sprinklers in Unlimited Area Group A-4 Buildings — 64
- **508.3.1.2**
  Group I-2, Condition 2 Nonseparated Occupancies — 66
- **508.4.1, Table 508.4**
  Separated Occupancies vs. Fire Area Separations — 68
- **Table 509**
  Incidental Uses — 70
- **510.2**
  Horizontal Building Separation — 72
- **Table 601, Note b**
  Fire Protection of Structural Roof Members — 74
- **Table 602, Note i**
  Group R-3 Fire Separation Distance — 76
- **602.3, 602.4.1**
  FRT Wood Sheathing in Exterior Wall Assemblies — 78

PART 3
Fire Protection
Chapters 7 through 9 — 80

- **704.2, 704.4.1**
  Column Protection in Light-Frame Construction — 82
- **Table 705.2**
  Extent of Projections — 84
- **705.2.3, 705.2.3.1**
  Combustible Balconies, Projections, and Bay Windows — 86
- **705.8.1**
  Measurement of Fire Separation Distance for Opening Protection — 88
- **706.1.1**
  Party Walls Not Constructed as Fire Walls — 90
- **706.2**
  Structural Continuity of Double Fire Walls — 92
- **708.4**
  Continuity of Fire Partitions — 94
- **708.4.2**
  Fireblocking and Draftstopping at Fire Partitions — 98
- **713.8.1**
  Membrane Penetrations of Shaft Enclosures — 102
- **716.2.6.5**
  Delayed-Action Self-Closing Doors — 103
- **803.1.1, 803.1.2**
  Interior Wall and Ceiling Finish Testing — 105
- **803.3**
  Interior Finish Requirements for Heavy Timber Construction — 108
- **803.11, 803.12**
  Flame Spread Testing of Laminates and Veneers — 109
- **901.6.2**
  Integrated Fire Protection System Testing — 111
- **902**
  Fire Pump and Fire Sprinkler Riser Rooms — 113
- **903.2.1**
  Sprinklers Required in Group A Occupancies — 114
- **903.2.3**
  Sprinklers in Group E Occupancies — 117

- **903.3.1.1.2**
Omission of Sprinklers in Group R-4 Bathrooms — 119

- **903.3.1.2.1**
Sprinkler Protection at Balconies and Decks — 120

- **903.3.1.2.3**
Protection of Attics in Group R Occupancies — 122

- **904.12**
Commercial Cooking Operations — 125

- **904.13**
Domestic Cooking Protection in Institutional and Residential Occupancies — 127

- **904.14**
Aerosol Fire Extinguishing Systems — 129

- **905.3.1**
Class III Standpipes — 130

- **905.4**
Class I Standpipe Connection Locations — 133

- **907.2.1**
Fire Alarms in Group A Occupancies — 135

- **907.2.10**
Group R-4 Fire Alarm Systems — 137

PART 4
Means of Egress
Chapter 10 — 139

- **Table 1004.5, 1004.8**
Occupant Load Calculation in Business Use Areas — 141

- **1006.2.1, Table 1006.2.1**
Group R Spaces with One Exit or Exit Access Doorway — 143

- **1006.3, 1006.3.1**
Egress through Adjacent Stories — 146

- **1008.2.3**
Illumination of the Exit Discharge — 148

- **1008.3.5, 1008.2.2**
Emergency Illumination in Group I-2 — 150

- **1009.7.2**
Protection of Exterior Areas of Assisted Rescue — 152

- **1010.1.1**
Size of Doors — 154

- **1010.1.4.4**
Locking Arrangements in Educational Occupancies — 157

- **1010.1.9.8**
Use of Delayed Egress Locking Systems in Group E Classrooms — 159

- **1010.1.9.12**
Locks on Stairway Doors — 162

- **1010.3.2**
Security Access Turnstiles — 164

- **1013.2**
Floor Level Exit Sign Location — 167

- **1015.6, 1015.7**
Fall Arrest for Rooftop Equipment — 168

- **1017.3, 202**
Measurement of Egress Travel — 170

- **1023.3.1**
Stairway Extensions — 172

- **1023.5, 1024.6**
Exit Stairway and Exit Passageway Penetrations — 174

- **1025.1**
Luminous Egress Path Marking in Group I Occupancies — 176

- **1026.4, 1026.4.1**
Refuge Areas for Horizontal Exits — 177

- **1029.6, 1029.6.3, 202**
Open-Air Assembly Seating — 179

- **1030.1**
Required Emergency Escape and Rescue Openings — 181

PART 5
Accessibility
Chapter 11 — 183

- **1103.2.14**
Access to Walk-In Coolers and Freezers — 184

- **1109.2.1.2**
Fixtures in Family or Assisted-Use Toilet Rooms — 186

- **1109.15**
Access to Gaming Machines and Gaming Tables — 188

- **1110.4.13**
Access to Play Areas for Children — 191

PART 6
Building Envelope, Structural Systems, and Construction Materials
Chapters 12 through 26 — 192

- **1206.2, 1206.3** Engineering Analysis of Sound Transmission — 194
- **Table 1404.2** Weather Covering Minimum Thickness — 196
- **1404.18** Polypropylene Siding — 198
- **1504.3.3** Metal Roof Shingles — 199
- **1507.1** Underlayment — 200
- **1507.18** Building-Integrated Photovoltaic Panels — 206
- **1603.1** Construction Documents — 209
- **1604.3.7** Deflection of Glass Framing — 212
- **1604.5.1** Multiple Occupancies — 213
- **1604.10** Storm Shelters — 215
- **Table 1607.1** Deck Live Load — 217
- **Table 1607.1** Live Load Reduction — 218
- **1607.15.2** Minimum Live Load for Fire Walls — 221
- **1609** Wind Loads — 222
- **1613** Earthquake Loads — 227
- **1613.2.1** Seismic Maps — 230
- **1615, 1604.5** Tsunami Loads — 236
- **1704.6** Structural Observations — 238
- **1705.5.2** Metal-Plate-Connected Wood Trusses — 240
- **1705.12.1, 1705.13.1** Seismic Force-Resisting Systems — 241
- **1705.12.6** Fire Sprinkler Clearance — 243
- **1804.4** Site Grading — 245
- **1807.2** Retaining Walls — 247
- **1810.3.8.3** Precast Prestressed Piles — 248
- **1901.2** Seismic Loads for Precast Concrete Diaphragms — 251
- **2207.1** SJI Standard — 253
- **2209.2** Cantilevered Steel Storage Racks — 255
- **2211** Cold-Formed Steel Light-Frame Construction — 256
- **2303.2.2** Fire-Retardant-Treated Wood — 260
- **2303.6** Nails and Staples — 262
- **Table 2304.9.3.2** Mechanically Laminated Decking — 264
- **Table 2304.10.1** Ring Shank Nails — 266
- **2304.10.5** Fasteners in Treated Wood — 268
- **2304.11** Heavy Timber Construction — 269
- **2304.12.2.5, 2304.12.2.6** Supporting Members for Permeable Floors and Roofs — 274
- **Table 2308.4.1.1(1)** Header and Girder Spans—Exterior Walls — 276
- **2308.4.1.1(2)** Header and Girder Spans—Interior Walls — 278
- **2308.5.5.1** Openings in Exterior Bearing Walls — 280
- **2407.1** Structural Glass Baluster Panels — 282

- **2510.6**
  Water-Resistive Barrier — 283

- **2603.13**
  Cladding Attachment over Foam Sheathing to Wood Framing — 285

PART 7
Building Services, Special Devices, and Special Conditions
Chapters 27 through 33 — 289

- **3001.2**
  Emergency Elevator Communication Systems — 290

- **3006.2.1**
  Corridors Adjacent to Elevator Hoistway Openings — 292

- **3007.1**
  Extent of Fire Service Access Elevator Travel — 294

- **3008.1.1**
  Required Number of Occupant Evacuation Elevators — 296

- **3113**
  Relocatable Buildings — 298

- **3310.1**
  Stairways in Buildings under Construction — 301

- **3314**
  Fire Watch During Construction — 303

PART 8
Appendices
Appendices A through N — 305

- **G103.6**
  Watercourse Alteration — 306

- **Appendix N**
  Guidelines for Replicable Buildings — 307

PART 9
2018 *International Existing Building Code (IEBC)*
Chapters 1 through 16 and Appendices A through C — 313

- **IEBC 303.1**
  Live Loads — 315

- **IEBC 303.3.2, IEBC Appendix A5**
  Earthquake Hazard Reduction in Existing Concrete Buildings — 317

- **IEBC 305; Chapters 4, 5, 6, 13, 14**
  Reorganization — 319

- **IEBC 405.2.1.1**
  Snow Damage — 321

- **IEBC 502.4**
  Loading of Existing Structural Elements — 322

- **IEBC 502.7, 503.15, 804, 1105**
  Carbon Monoxide Detectors in Group I-1, I-2, I-4, and R Occupancies — 324

- **IEBC 502.8, 1106, 1301.2.3.1**
  Storm Shelters in Group E Additions — 326

- **IEBC 503.7**
  Anchorage for Concrete and Reinforced Masonry Walls — 328

- **IEBC 503.10**
  Anchorage of Unreinforced Masonry Partitions — 329

- **IEBC 505.4, 701.4**
  Emergency Escape Opening Operation — 330

- **IEBC 506.4**
  Structural Loads — 332

- **IEBC 507.4**
  Structural Loads in Historic Buildings — 335

- **IEBC 805.3.1.1**
  Single-Exit Buildings — 337

- **IEBC 904.1.4**
  Automatic Sprinkler System at Floor of Alteration — 340

- **IEBC 906.7**
  Anchorage of Unreinforced Masonry Partitions — 341

- **IEBC 1006**
  Seismic Loads and Access to Risk Category IV Structures — 342

- **IEBC 1103**
  Changes to Loads with an Addition — 344

Index — 347

Preface

The purpose of *Significant Changes to the International Building Code® 2018 Edition* is to familiarize building officials, fire officials, plans examiners, inspectors, design professionals, contractors, and others in the construction industry with many of the important changes in the 2018 *International Building Code®* (IBC®). This publication is designed to assist those code users in identifying the specific code changes that have occurred and, more important, understanding the reasons behind the changes. It is also a valuable resource for jurisdictions in their code-adoption process.

Only a portion of the total number of code changes to the IBC are discussed in this book. The changes selected were identified for a number of reasons, including their frequency of application, special significance, or change in application. However, the importance of those changes not included is not to be diminished. Further information on all code changes can be found in the Complete Revision History to the 2018 I-Codes, available from the International Code Council® (ICC®) online store at http://shop.iccsafe.org. The revision history provides the published documentation for each successful code change contained in the 2018 IBC since the 2015 edition.

The IBC discussion in this book is organized into eight general categories, each representing a distinct grouping of code topics. It is arranged to follow the general layout of the IBC, including code sections and section number format. The table of contents, in addition to providing guidance in use of this publication, allows for quick identification of those significant code changes that occur in the 2018 IBC.

This edition of *Significant Changes to the International Building Code* includes a ninth Part that addresses a limited number of selected code changes that occurred in the 2018 edition of the *International Existing Building Code* (IEBC). Applicable to all existing buildings, the IEBC is intended to provide flexibility to permit the use of alternative approaches to achieve compliance with minimum requirements to safeguard the public health, safety, and welfare. Both structural and nonstructural changes are addressed in this publication.

Throughout the book, each change is accompanied by a photograph, an application example, or an illustration to assist and enhance the reader's understanding of the specific change. A summary and a discussion of the significance of the changes are also provided. Each code change is identified by type, be it an addition, modification, clarification, or deletion.

The code change itself is presented in a format similar to the style utilized for code-change proposals. Deleted code language is shown with a strike-through, whereas new code text is indicated by underlining. As a result, the actual 2018 code language is provided, as well as a comparison with the 2015 language, so the user can easily determine changes to the specific code text.

As with any code-change text, *Significant Changes to the International Building Code 2018 Edition* is best used as a study companion to the 2018 IBC. Because only a limited discussion of each change is provided, the code itself should always be referenced in order to gain a more comprehensive understanding of the code change and its application.

The commentary and opinions set forth in this text are those of the authors and do not necessarily represent the official position of the ICC. In addition, they may not represent the views of any enforcing agency, as such agencies have the sole authority to render interpretations of the IBC. In many cases, the explanatory material is derived from the reasoning expressed by the code-change proponent.

Comments concerning this publication are encouraged and may be directed to the ICC at significantchanges@iccsafe.org.

About the *International Building Code*®

Building officials, design professionals, and others involved in the building construction industry recognize the need for a modern, up-to-date building code addressing the design and installation of building systems through requirements emphasizing performance. The *International Building Code* (IBC), in the 2018 edition, is intended to meet these needs through model code regulations that safeguard the public health and safety in all communities, large and small. The IBC is kept up to date through the open code-development process of the International Code Council (ICC). The provisions of the 2015 edition, along with those code changes approved through 2016, make up the 2018 edition.

The ICC, publisher of the IBC, was established in 1994 as a nonprofit organization dedicated to developing, maintaining, and supporting a single set of comprehensive and coordinated national model building construction codes. Its mission is to provide the highest-quality codes, standards, products, and services for all concerned with the safety and performance of the built environment.

The IBC is 1 of 15 International Codes® published by the ICC. This comprehensive building code establishes minimum regulations for building systems by means of prescriptive and performance-related provisions. It is founded on broad-based principles that make possible the use of new materials and new building designs. The IBC is available for adoption and use by jurisdictions internationally. Its use within a governmental jurisdiction is intended to be accomplished through adoption by reference, in accordance with proceedings establishing the jurisdiction's laws.

About the Authors

> Douglas W. Thornburg, AIA, CBO
> International Code Council
> Vice-President and Technical Director of Products and Services

Douglas W. Thornburg, AIA, CBO, is currently Vice-President and Technical Director of Products and Services for the International Code Council (ICC) where he provides administrative and technical leadership for the ICC product development activities. Prior to employment with ICC in 2004, he was in private practice as a code consultant and educator on building codes for nine years. Doug also spent ten years with the International Conference of Building Officials (ICBO) where he served as Vice-President/Education.

In his current role, Doug also continues to create and present building code seminars nationally and has developed numerous educational texts and resource materials. He was presented with ICC's inaugural Educator of the Year Award in 2008, recognizing his outstanding contributions in education and training.

A graduate of Kansas State University and a registered architect, Doug has over 37 years of experience in building code training and administration. He has authored a variety of code-related support publications, including the *IBC Illustrated Handbook* and the *Significant Changes to the International Building Code*.

> Sandra Hyde, P.E.
> International Code Council
> Senior Staff Engineer

Sandra Hyde is a Senior Staff Engineer with the ICC's Product Development Department. She develops technical resources in support of the structural provisions of the International Building, Existing Building and Residential Codes. Sandra reviews publications authored by ICC and engineering groups, while also developing publications and technical seminars on the structural provisions of the I-codes for building departments, design engineers, and special inspectors.

Prior to ICC, Sandra worked for Weyerhaeuser/Trus Joist in research and development of engineered lumber products. She has a Master's Degree in Structural Engineering from Portland State University and is a Registered Civil Engineer in Idaho and California. She has authored and reviewed support publications including *Significant Changes to the International Residential Code, Special Inspection Manual* and, in conjunction with APA, *Guide to the IRC Wall Bracing Provisions*.

About the Contributors

> Kevin H. Scott
> KH Scott and Associates
> President

Kevin H. Scott, President of KH Scott and Associates, LLC, has extensive experience in the development of fire safety, building safety, and hazardous materials regulations. With over 30 years in the development of fire code, building code and fire safety regulations at the local, state, national and international levels, Kevin develops and presents a variety of code-based seminars and is the author of ICC's publication *Significant Changes to the International Fire Code 2018 Edition*.

> Hamid Naderi, P.E, CBO
> International Code Council
> Senior Vice-President of Product Development

Hamid A. Naderi, P.E, C.B.O., is presently the Senior Vice President of Product Development with the International Code Council (ICC) where he is responsible for the research and development of technical resources, managing the development of multiple technical projects by expert authors, and coordinating the partnerships with outside technical organizations and publishers.

About the ICC

The International Code Council is a member-focused association dedicated to helping the building safety community and construction industry provide safe, sustainable, affordable, and resilient construction through the development of codes and standards used in the design, build, and compliance process. Most U.S. communities and many global markets choose the International Codes. ICC Evaluation Service (ICC-ES), a subsidiary of the International Code Council, has been the industry leader in performing technical evaluations for code compliance, fostering safe and sustainable design and construction.

Headquarters:
500 New Jersey Avenue, NW, 6th Floor
Washington, DC 20001-2070

Regional Offices:
Birmingham, AL; Chicago, IL; Los Angeles, CA

1-888-422-7233
www.iccsafe.org

PART 1
Administration

Chapters 1 and 2

- **Chapter 1**   Scope and Administration
  No changes addressed
- **Chapter 2**   Definitions

The provisions of Chapter 1 address the application, enforcement, and administration of subsequent requirements of the code. In addition to establishing the scope of the *International Building Code* (IBC), the chapter identifies which buildings and structures come under its purview. A building code, as with any other code, is intended to be adopted as a legally enforceable document to safeguard health, safety, property and public welfare. A building code cannot be effective without adequate provisions for its administration and enforcement. Chapter 2 provides definitions for terms used throughout the IBC. Codes, by their very nature, are technical documents, and as such, literally every word, term, and punctuation mark can add to or change the meaning of the intended result. ■

202
Definition of Greenhouse

202
Definition of Repair Garage

202
Definition of Sleeping Unit

CHAPTER 2
Removal of Definition References

202

Definition of Greenhouse

CHANGE TYPE: Addition

CHANGE SUMMARY: A definition of greenhouse has been added to the code in order to recognize that the primary characteristic of such a structure is its unique environment for growing plants, and not the structure itself or merely the presence of plants.

2018 CODE: **GREENHOUSE.** A structure or thermally isolated area of a building that maintains a specialized sunlit environment used for and essential to the cultivation, protection or maintenance of plants.

CHANGE SIGNIFICANCE: A greenhouse has historically been recognized as a structure intended for the growing of plants. The primary difference between a greenhouse and other structures is that the environment in a greenhouse is specific to this intended use. Although a variety of other activities can occur within a greenhouse, the key issue is the maintenance of a controlled environment within the facility necessary for proper plant growth. A definition of greenhouse has been added to the code in order to recognize that the primary characteristic of such a structure is its unique environment, not the structure itself or merely the presence of plants.

Buildings intended for human habitation are required to provide minimum features such as lighting, ventilation, heating, and cooling that are suitable for the health and welfare of the occupants, and oftentimes plants can coexist in such environments. In a greenhouse, the environment is such that it is maintained exclusively for, and essential to, the aggressive propagation of plants used by commercial growers for plant production. It is acceptable for other activities to be conducted in a greenhouse, such as retail sales, educational research, conservation, education, and assembly use. The occupancy classification of a greenhouse will be determined accordingly. However, in all cases, the environment of the greenhouse must be such that the plants will thrive.

Greenhouse

The extent of the definition is limited to those structures designed and used specifically for the growing, care, and maintenance of plants. As such, sunrooms, solariums, glass-enclosed walkways, atria, and other types of spaces that permit ample sunlight and prominently feature plants for aesthetic purposes are not considered as greenhouses under the new definition.

202

Definition of Repair Garage

CHANGE TYPE: Addition

CHANGE SUMMARY: The IFC definition of repair garage is now introduced to the IBC in order to provide clarity and consistency in the application of provisions related to the repair and maintenance of motor vehicles.

2018 CODE: REPAIR GARAGE. A building, structure or portion thereof used for servicing or repairing motor vehicles.

CHANGE SIGNIFICANCE: The term "repair garage" is used throughout the code, including in key provisions such as Section 311.2 (occupancy classification), Section 406.8 (special detailed requirements), and Section 903.2.9.1 (automatic sprinkler systems). However, since the 2000 edition, there has been no definition of a repair garage in the IBC. Without the definition in the 2003 IBC and future editions, guidance in applying code requirements has typically been provided through the use of the repair garage definition found in the *International Fire Code* (IFC). The IFC definition is now replicated in the IBC in order to provide clarity and consistency in the application of provisions related to the repair and maintenance of motor vehicles.

Originally in the 2000 IBC, the definition included any building or part thereof which is used for painting, body and fender work, engine overhauling or other major repair of motor vehicles. This definition was removed in the 2003 edition and the term had not been specifically defined in the IBC since then. The IFC had a differing definition which expanded the IBC's repair garage scope to include the servicing of motor vehicles. This includes maintenance activities such as brake work, oil changes, and similar activities. The IFC definition is now included in the IBC such that the definition is quite broad, including both repair and maintenance operations.

Repair garage

Significant Changes to the IBC 2018 Edition 202 ■ Definition of Sleeping Unit

202
Definition of Sleeping Unit

CHANGE TYPE: Clarification

CHANGE SUMMARY: The revised definition of dwelling unit clarifies that each individual bedroom within a residential suite is not to be considered as a sleeping unit, but rather the entire suite is to be deemed as one sleeping unit.

2018 CODE: SLEEPING UNIT. A ~~room or space in which people sleep, which can also include~~ <u>single unit that provides rooms or spaces for one or more persons, includes</u> permanent provisions for <u>sleeping and can include provisions for</u> living, eating, and either sanitation or kitchen facilities but not both. Such rooms and spaces that are also part of a dwelling unit are not sleeping units.

CHANGE SIGNIFICANCE: The single required characteristic of a sleeping unit is that it is used as the primary location for sleeping purposes. Guestrooms of Group R-1 hotels and motels are typically considered sleeping units. Sleeping units are also commonly found in congregate living facilities, such as dormitories, sorority houses, and fraternity houses, and are regulated as Group R-2 occupancies. Several of the varied Group I occupancies also contain resident or patient sleeping units. Of major importance, sleeping units are required by Section 420 to be separated from

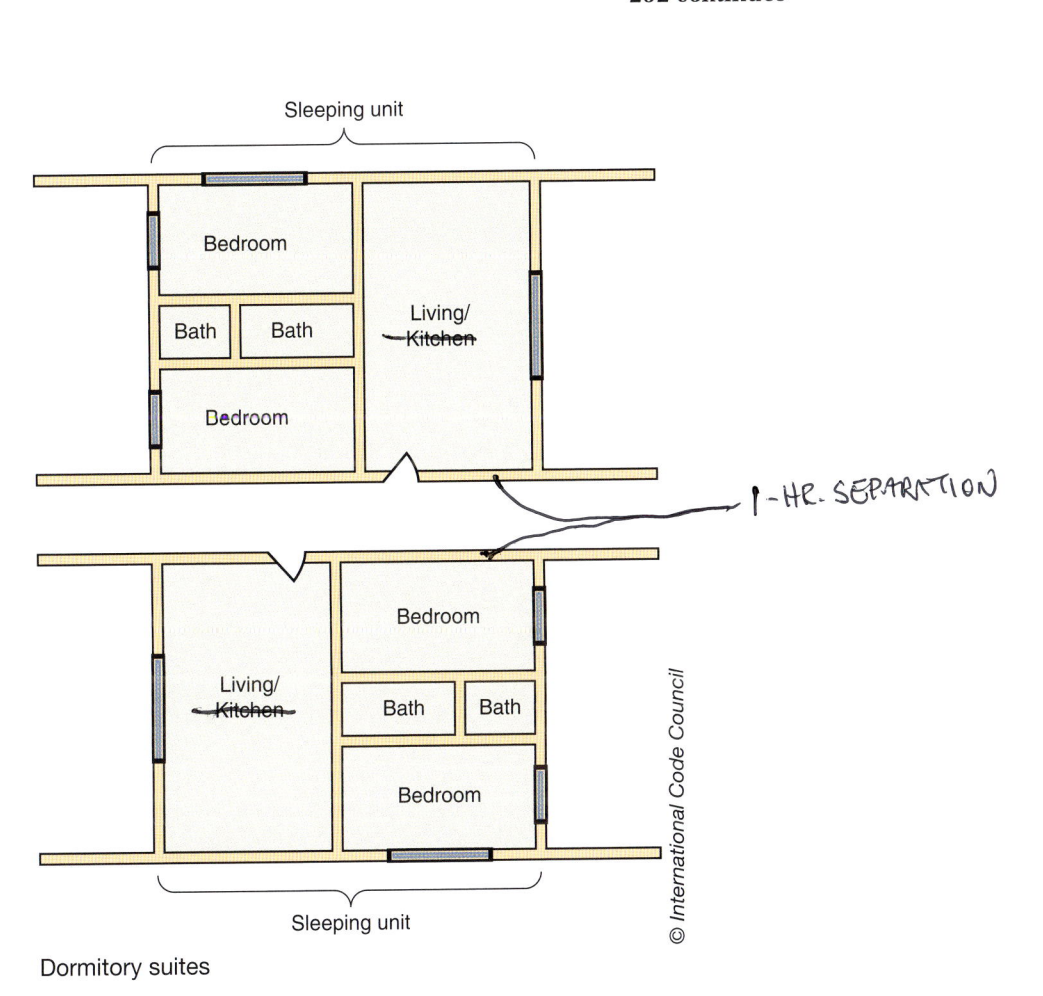

202 continues

Dormitory suites [with handwritten annotation: "1-HR. SEPARATION"]

202 continued

each other, as well as other contiguous occupancies, through the use of fire partitions, horizontal assemblies, or both. Therefore, it is critical that the extent of each sleeping unit be clearly identified in order to properly determine where the fire-resistance-rated separations are required. The definition of sleeping unit has been revised in order to clarify the varied configurations of rooms and spaces intended to be considered as sleeping units.

Traditionally, dormitories and similar congregate living facilities have consisted of multiple bedrooms, with common living, dining, cooking, and sanitation facilities. In such cases, the bedrooms are individually regulated as sleeping units. However, current residential environments often include "suites" where two or more bedrooms share their own living space and bathroom facilities. These rooms and spaces within the suite act as a group, similar to an apartment unit, and as such the entire suite should be evaluated as a single sleeping unit. The revised definition clarifies that each individual bedroom within a suite is not to be considered as a sleeping unit. Where the bedrooms are considered as a portion of the suite, the entire suite is to be deemed as one sleeping unit.

Chapter 2
Removal of Definition References

CHANGE TYPE: Deletion

CHANGE SUMMARY: All definition lists located throughout the code have now been removed based on the general recognition of the format of the IBC and the ongoing use of italics to identify terms that are defined in Chapter 2.

2018 CODE: ~~304.2 Definitions. The following terms are defined in Chapter 2.~~

~~AMBULATORY CARE FACILITY.~~

~~CLINIC, OUTPATIENT.~~

(The lists of terms in this and subsequent sections throughout the code have been deleted.)

CHANGE SIGNIFICANCE: Throughout the IBC, specific terms are used in a manner that differs from their ordinarily accepted meaning. Such terms are necessarily defined in order to clarify their meaning within the context of the code. In the first four editions of the IBC, general terms applying throughout the code were defined in Chapter 2, while those terms more specific to an individual subject area in the code, such as means of egress or fire protection systems, were defined at the beginning of the appropriate chapter. In some cases, the definitions were also scattered throughout the chapter. In the 2012 edition of the IBC, all definitions were relocated to Chapter 2 for consistency and usability purposes. However, the individual defined terms remained throughout the various chapters of the code. All definition lists scattered about the code have now been removed based on the general recognition of the format of the IBC and the ongoing use of italics to identify terms used throughout the code that are defined in Chapter 2.

There are more than 700 definitions in the IBC. Historically, approximately 10 percent of these terms were defined in Chapter 2, with the remaining 90 percent scattered in more than 40 locations throughout the

Chapter 2 continues

~~502.1 Definitions. The following terms are defined in Chapter 2:~~
~~AREA, BUILDING.~~
~~BASEMENT.~~
~~EQUIPMENT PLATFORM.~~
~~HEIGHT, BUILDING.~~
~~MEZZANINE.~~

© International Code Council

Definition references removed

Chapter 2 continued — other chapters in the code. This approach changed with the introduction of the 2012 IBC, where all definitions were relocated to Chapter 2. However, the specifically defined terms remained listed in their previous locations throughout the IBC to remind the code user that a definition of the term can be found in Chapter 2. This formatting approach was eliminated in the 2018 edition so that there are no direct references to Chapter 2, other than the continued identification of specifically defined terms as italicized text.

PART 2

Building Planning

Chapters 3 through 6

- Chapter 3 Occupancy Classification and Use
- Chapter 4 Special Detailed Requirements Based on Occupancy and Use
- Chapter 5 General Building Heights and Areas
- Chapter 6 Types of Construction

The application of the *International Building Code* to a structure is typically initiated through the provisions of Chapters 3, 5, and 6. Chapter 3 establishes one or more occupancy classifications based upon the anticipated uses of a building. The appropriate classifications are necessary to properly apply many of the code's non-structural provisions. The requirements of Chapter 6 deal with classification as to construction type, based on a building's materials of construction and the level of fire resistance provided by such materials. Limitations on a building's height and area, set forth in Chapter 5, are directly related to the occupancies it houses and its type of construction. Chapter 5 also provides the various methods available to address conditions in which multiple uses or occupancies occur within the same building. Chapter 4 contains special detailed requirements based on unique conditions or uses that are found in some buildings. ■

302.1
Classification of Outdoor Areas

303.4
Assembly Use of Greenhouses Classification

309.1
Mercantile Use of Greenhouses Classification

310.3, 310.4
Classification of Congregate Living Facilities

310.4.2
Owner-Occupied Lodging Houses

311.1.1
Classification of Accessory Storage Spaces

311.2
Classification of Self-Service Storage Facilities

312.1
Classification of Communication Equipment Structures

312.1.1
Classification of Agricultural Greenhouses

403.2.1.1
Type of Construction in High-Rise Buildings

404.6
Enclosure of Atriums

406.1
Motor-Vehicle-Related Occupancies

406.3
Regulation of Private Garages

406.6.2
Ventilation of Enclosed Parking Garages

407.5
Maximum Smoke Compartment Size

407.5.4
Required Egress from Smoke Compartments

420.7
Corridor Protection in Assisted Living Units

420.8
Group I-1 Cooking Facilities

420.10
Dormitory Cooking Facilities

422.6
Electrical Systems in Ambulatory Care Facilities

424.1
Children's Play Structures

427
Medical Gas Systems

428
Higher Education Laboratories

503.1, 706.1
Scope of Fire Wall Use

503.1.4
Allowable Height and Area of Occupied Roofs

505.2.1.1
Mezzanine and Equipment Platform Area Limitations

TABLE 506.2, NOTE i
Allowable Area of Type VB Greenhouses

507.4
Sprinklers in Unlimited Area Group A-4 Buildings

508.3.1
Group I-2, Condition 2 Nonseparated Occupancies

508.4.1, TABLE 508.4
Separated Occupancies vs. Fire Area Separations

TABLE 509
Incidental Uses

510.2
Horizontal Building Separation

TABLE 601, NOTE b
Fire Protection of Structural Roof Members

TABLE 602, NOTE i
Group R-3 Fire Separation Distance

602.3, 602.4.1
FRT Wood Sheathing in Exterior Wall Assemblies

302.1 Classification of Outdoor Areas

CHANGE TYPE: Clarification

CHANGE SUMMARY: It has been clarified that occupied roofs are to be assigned one or more occupancy classifications in a manner consistent with the classification of uses inside the building, based upon the fire and life safety hazards posed by the rooftop activities.

2018 CODE: 302.1 ~~General.~~ **Occupancy classification.** Occupancy classification is the formal designation of the primary purpose of the building, structure or portion thereof. Structures ~~or portions of structures~~ shall be classified ~~with respect to occupancy in~~ into one or more of the occupancy groups listed in this section based on the nature of the hazards and risks to building occupants generally associated with the intended purpose of the building or structure. ~~A~~ An area, room or space that is intended to be occupied at different times for different purposes shall comply with all ~~of the~~ applicable requirements ~~that are applicable to each of the purposes for which the room or space will be occupied~~ associated with such potential multipurpose. Structures ~~with multiple occupancies~~ or uses containing multiple occupancy groups shall comply with Section 508. Where a structure is proposed for a purpose that is not specifically ~~provided for in this code~~ listed in this section, such structure shall be classified in the ~~group that the~~ occupancy it most nearly resembles ~~according to~~ based on the fire safety and relative hazard ~~involved~~. Occupied roofs shall be classified in the group that the occupancy most nearly resembles, according to the fire safety and relative hazard, and shall comply with Section 503.1.4.

CHANGE SIGNIFICANCE: The initial step in analyzing a building for compliance with the IBC is to determine the appropriate occupancy classification for each area of the building based upon its use. Such determination can be done on a room-by-room basis or, in a more global sense,

302.1 continues

Rooftop restaurant

302.1 continued

by reviewing multiple spaces as a single use. However, the code has historically been silent in regard to the occupancy classification of occupied roofs. Chapter 10 regulating the means of egress, along with Chapter 11 addressing accessibility, contains references to the application of its provisions to occupied roofs. However, there has been no specific mention as to the proper means to classify such spaces based upon their use. It has been clarified that occupied roofs are to be assigned one or more occupancy classifications in a manner consistent with the classification of uses inside the building, based upon the fire and life safety hazards posed by the rooftop activities.

The IBC is considered as an "occupancy-based" code, where the primary difference in requirements between buildings is due to the varying uses that are anticipated. As such, it is critical that an occupancy classification be assigned to any occupied portion of a building in order that the appropriate fire and life safety criteria are applied. For example, where a rooftop contains a restaurant having dining seating for 50 or more persons, the occupied roof would be classified as a Group A-2 occupancy in order to address those hazards associated with such an assembly use. The determination of the occupancy classification, or classifications, of an occupied roof would be performed in a manner consistent with the classification of uses inside the building.

303.4 Assembly Use of Greenhouses Classification

CHANGE TYPE: Clarification

CHANGE SUMMARY: Where the use of a greenhouse is assembly in nature due to public access for the viewing of plants, classification as a Group A-3 occupancy is appropriate.

2018 CODE: 303.4 Assembly Group A-3. Group A-3 occupancy includes assembly uses intended for worship, recreation or amusement and other assembly uses not classified elsewhere in Group A including, but not limited to:

<u>Greenhouses for the conservation and exhibition of plants that provide public access</u> — BOTANICAL GARDEN

(No changes to other listed items.)

CHANGE SIGNIFICANCE: By definition, a greenhouse is now defined as "a structure or thermally isolated area of a building that maintains a specialized sunlit environment used for and essential to the cultivation, protection or maintenance of plants." Although a number of varied uses can occur within a greenhouse, the key issue is the maintenance of a controlled environment within the facility necessary for proper plant growth. The primary characteristic of such a structure is its unique environment, not the structure itself or merely the presence of plants. Therefore, when reviewing a greenhouse for occupancy classification purposes, the focus should be on the fire and life safety characteristics of activities expected to occur within the structure. Where the use of the space or building is assembly in nature, classification as a Group A occupancy is typically appropriate.

In a greenhouse, the environment is such that it is maintained exclusively for, and essential to, the aggressive propagation of plants used by commercial growers for plant production. However, this activity does not

303.4 continues

Crystal Palace arboretum

303.4 continued

preclude that a more occupant-intensive use be considered as the major function of the space. It is acceptable for other activities to be conducted in a greenhouse, including assembly uses such as botanical gardens, municipal parks, and public conservatories. The potential hazards in these types of structures are primarily occupant-based and consistent with other assembly uses. Therefore, the occupancy classification of such a greenhouse will be determined based on the expected occupant load, with a Group A-3 classification typically assigned where the established occupant load is 50 or more. Consistent with other assembly uses, a Group B classification is appropriate where the occupant load is less than 50.

309.1 Mercantile Use of Greenhouses Classification

CHANGE TYPE: Clarification

CHANGE SUMMARY: Where a greenhouse is provided with public access for the purpose of the display and sale of plants, a Group M occupancy shall be assigned.

2018 CODE: 309.1 Mercantile Group M. Mercantile Group M occupancy includes, among others, the use of a building or structure or a portion thereof for the display and sale of merchandise, and involves stocks of good, wares or merchandise incidental to such purposes and accessible to the public. Mercantile occupancies shall include, but not be limited to, the following:

<u>Greenhouses for display and sale of plants that provide public access</u>

(No changes to other listed items.)

CHANGE SIGNIFICANCE: Although a number of varied uses can occur within a greenhouse, a controlled environment within the facility necessary for proper plant growth is consistently present. The primary characteristic of such a structure is its unique environment, not the type of structure or merely the presence of plants. Therefore, when analyzing a greenhouse for occupancy classification purposes, the focus should be on the fire and life safety characteristics of activities expected to occur within the structure. Where the use of the space or building is retail sales, classification as a Group M occupancy is appropriate.

Although the environment in a greenhouse is such that it is maintained for the aggressive propagation of plants used by commercial growers for plant production, it does not preclude that a more

309.1 continues

Retail sales of plants

309.1 continued occupant-intensive use be considered as the major function of the space. It is acceptable for other activities to be conducted in a greenhouse, including mercantile uses such as retail stores and home improvement centers. The potential hazards in these types of structures are primarily occupant-based and consistent with other mercantile uses. Therefore, the occupancy classification of such a greenhouse is to be Group M.

Significant Changes to the IBC 2018 Edition 310.3, 310.4 ■ Classification of Congregate Living Facilities

310.3, 310.4
Classification of Congregate Living Facilities

CHANGE TYPE: Modification

CHANGE SUMMARY: Dormitories and similar nontransient uses now are to be considered as Group R-3 occupancies where the occupant load is 16 or less. In addition, transient lodging houses, such as bed-and-breakfast establishments, can only be considered as Group R-3 occupancies where their total occupant load is 10 or less. [handwritten: GUESTS ONLY?]

2018 CODE: ~~310.4~~ **310.3 Residential Group R-2.** Residential Group R-2 occupancies containing sleeping units or more than two dwelling units where the occupants are primarily permanent in nature, including:

Apartment houses
~~Boarding houses (nontransient) with more than 16 occupants~~
Congregate living facilities (nontransient) with more than 16 occupants
    <u>Boarding houses (nontransient)</u>
    Convents
    Dormitories
    Fraternities and sororities
    <u>Monasteries</u>
Hotels (nontransient)
Live/work units
~~Monasteries~~
Motels (nontransient)
Vacation timeshare properties

310.3, 310.4 continues

Group R-3 transient boarding house

310.3, 310.4 continued

~~310.5~~ **310.4 Residential Group R-3.** Residential Group R-3 occupancies where the occupants are primarily permanent in nature and not classified as Group R-1, R-2, R-4 or I, including:

> Buildings that do not contain more than two dwelling units
>
> ~~Boarding houses (nontransient) with 16 or fewer occupants~~
>
> ~~Boarding houses (transient) with 10 or fewer occupants~~
>
> Care facilities that provide accommodations for five or fewer persons receiving care
>
> Congregate living facilities (nontransient) with 16 or fewer occupants
> > <u>Boarding houses (nontransient)</u>
> >
> > <u>Convents</u>
> >
> > <u>Dormitories</u>
> >
> > <u>Fraternities and sororities</u>
> >
> > <u>Monasteries</u>
>
> Congregate living facilities (transient) with 10 or fewer occupants
> > <u>Boarding houses (transient)</u>
>
> Lodging houses with five or fewer guest rooms <u>and 10 or fewer occupants</u>

CHANGE SIGNIFICANCE: Although apartment houses are by far the most common Group R-2 occupancy, other types of nontransient residential uses also fall into this classification. Congregate living facilities having an occupant load of 17 or more, such as fraternity and sorority houses, dormitories, convents, and monasteries, are also considered as Group R-2 occupancies. Previously, such types of uses were listed separate from congregate living facilities. This created a condition where their classification could be viewed as Group R-2 regardless of occupant load. The provisions have been reformatted to clarify that dormitories and the other listed nontransient uses are simply a subset of congregate living facilities and all such uses are to be considered as Group R-3 occupancies where the occupant load is 16 or less.

Lodging houses, most commonly bed-and-breakfast establishments, were first regulated as Group R-3 occupancies in the 2015 IBC. Defined as "a one-family dwelling where one or more occupants are primarily permanent in nature and rent is paid for guest rooms," lodging houses were previously considered under the Group R-1 transient classification. In order to be considered as a Group R-3 lodging house, the 2015 provisions mandate there be a limit of five guest rooms. An additional condition now mandates that the lodging house also have a total occupant load of 10 or less. Where either the number of guest rooms in a lodging house exceeds five or the total number of occupants exceeds 10, a classification of Group R-1 is appropriate.

The limit of 10 occupants needed to achieve a Group R-3 classification is consistent with the limitation applied to other types of transient residential uses such as boarding houses and congregate living facilities. In addition, the insertion of the term "total" indicates that in the determination of the total occupant load both the occupants of the guest rooms as well as the occupant load assigned to the family quarters be considered.

310.4.2 Owner-Occupied Lodging Houses

CHANGE TYPE: Modification

CHANGE SUMMARY: The criteria permitting compliance with the IRC for the design and construction of owner-occupied lodging houses has been expanded by now also requiring that the total number of lodging house occupants be limited to 10.

2018 CODE: ~~310.5.2~~ 310.4.2 Lodging houses. Owner-occupied lodging houses with five or fewer guest rooms and 10 or fewer total occupants shall be permitted to be constructed in accordance with the *International Residential Code*.

CHANGE SIGNIFICANCE: A lodging house is defined as "a one-family dwelling where one or more occupants are primarily permanent in nature and rent is paid for guest rooms." The most common example of a lodging house is a bed-and-breakfast facility that serves as a single-family dwelling while also providing sleeping rooms for guests. A provision was added to the 2015 *International Building Code* allowing owner-occupied lodging houses to be designed and constructed in accordance with the *International Residential Code* provided the lodging house has no more than five guest rooms. The criteria permitting compliance with the IRC has been expanded by now also requiring that the total number of lodging house occupants be limited to 10.

The IBC has historically limited the number of occupants in transient-oriented congregate living facilities and boarding houses to 10 in order to be classified as a Group R-3 occupancy. Higher occupant loads will cause the occupancy to be classified as Group R-2. In order to be consistent with the occupant load limit for Group R-3 transient occupancies, the additional limit of 10 persons was placed on lodging houses permitted to be regulated under the IRC.

310.4.2 continues

Bed-and-breakfast lodging

310.4.2 continued

It is important to note that when determining the occupant load to be applied, the intent of the code change is that the total number of occupants assigned to the lodging house be applied, which would include the owner and other family members who reside there. This approach is consistent with most other evaluations of occupant load insomuch that the entire occupant load of the building or space be considered.

311.1.1 Classification of Accessory Storage Spaces

CHANGE TYPE: Modification

CHANGE SUMMARY: Regardless of size, storage rooms and storage spaces that are accessory to other uses are to be classified as part of the occupancy to which they are accessory.

2018 CODE: 311.1.1 Accessory storage spaces. A room or space used for storage purposes that is ~~less than 100 square feet (9.3 m²) in area and~~ accessory to another occupancy shall be classified as part of that occupancy. ~~The aggregate area of such rooms or spaces shall not exceed the allowable area limits of Section 508.2.~~

CHANGE SIGNIFICANCE: The proper occupancy classification of storage rooms has historically been one of the most elusive issues in the IBC. Although Group S occupancies are recognized in Section 311.1 as buildings, or portions thereof, used for storage purposes, there has always been some disagreement as to the classification of smaller storage areas, closets and similar spaces that are accessory to one or more other uses in the building. Assigning a Group S occupancy classification to a warehouse, or other significant storage area, has never been questioned. However, where the room or space poses little, if any, hazard above that created by the occupancy to which the storage use is accessory, there was some consensus that a unique Group S classification need not be applied. This approach to classifying a storage area is now formally addressed such that storage rooms or spaces that are accessory to other uses are to be classified as part of the occupancy to which they are accessory, regardless of the size of the storage area.

Storage rooms were specifically regulated in the initial edition of the IBC as incidental use areas where they exceeded 100 square feet. As such, they were not considered as distinct Group S occupancies. However, they

311.1.1 continues

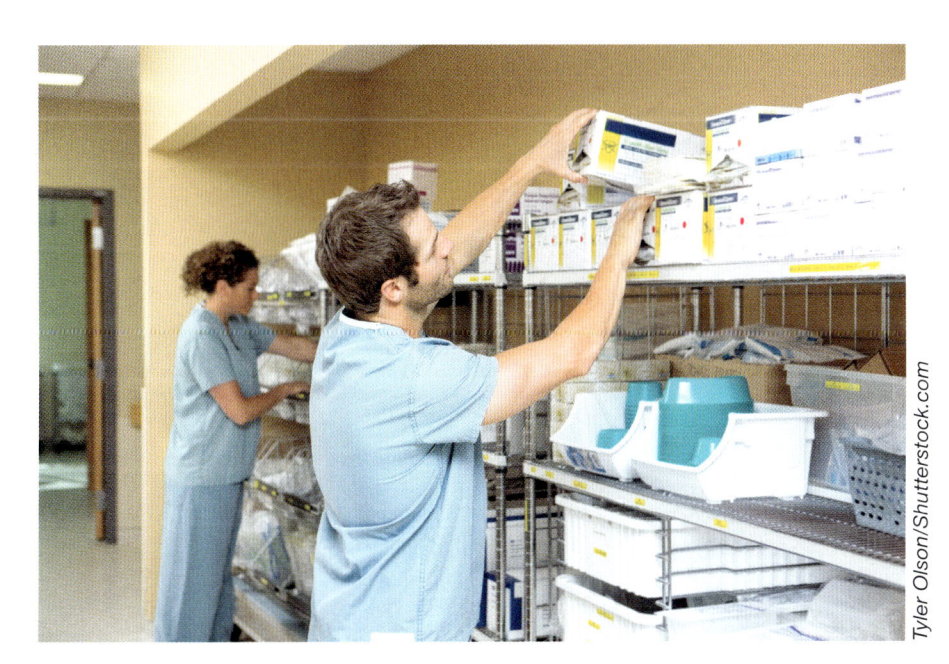

Hospital storage room

311.1.1 continued

were required to be separated from the remainder of the building by minimum 1-hour fire barriers. Due to the contradictions that occurred due to the potential for storage rooms to be classified as Group S occupancies eligible to be regulated under the nonseparated occupancy provisions, storage rooms were no longer considered as incidental use areas in the 2009 IBC. At that point, they were simply regulated under the general occupancy provisions of Chapter 3. The 2015 IBC introduced Section 311.1.1 recognizing that accessory storage spaces less than 100 square feet in area were to be classified as a part of the occupancy to which they are accessory. However, it has been typically viewed that the new provision implied that those storage spaces of 100 square feet or more should be classified as Group S. The 2015 provision has been revised to reflect that the Group S classification should not apply to accessory storage spaces.

The new approach to classifying storage spaces does not vary based upon the size of the storage space. There is no square footage or percentage threshold, such as 100 square feet or 10%, over which the Group S classification will be applied. Where the storage use is considered as accessory to the other uses in building, it shall be classified in accordance with those other uses. The key point is the hazard level that storage brings to the building. It is assumed that accessory storage uses pose little additional hazard above the occupancies which they serve. Where storage activities pose a significantly higher hazard than the other uses in the building, they would typically not be considered accessory and therefore classified as a Group S occupancy.

311.2 Classification of Self-Service Storage Facilities

CHANGE TYPE: Clarification

CHANGE SUMMARY: Due to the reasonable expectation that self-storage facilities will contain a considerable amount of combustible materials, such facilities are now specifically identified as Group S-1 occupancies.

2018 CODE: 311.2 Moderate-hazard storage, Group S-1. Storage Group S-1 occupancies are buildings occupied for storage uses that are not classified as Group S-2, including, but not limited to, storage of the following:

Aerosols products, Levels 2 and 3

Self-service storage facility (mini-storage)

(No changes to other listed items.)

CHANGE SIGNIFICANCE: Hazards created by storage uses are primarily contents-related as opposed to occupant-related. The general public is seldom exposed to the risks imposed by storage activities, as the occupants are typically employees who are familiar with their surroundings. However, the presence of significant fire loads and hazardous materials can make storage uses a significant concern. Where a storage use is not considered as a Group H high-hazard occupancy, it shall be classified as Group S, either a Group S-1 moderate-hazard occupancy or a Group S-2 low-hazard occupancy. Due to the reasonable expectation that self-service storage facilities will contain a considerable amount of combustible materials, such facilities are now specifically identified as Group S-1 occupancies.

Self-service storage facilities, sometimes referred to as mini-storage units, typically consist of multiple buildings housing numerous small garage-type storage spaces. In some cases, they can be multistory

311.2 continues

Self-storage units

Vizual Studio/Shutterstock.com

311.2 continued

buildings with hundreds of individual storage rooms. Regardless of their configuration, such facilities are used by individuals and businesses to store a wide variety of goods and materials. As would be expected, the fire load created by the stored items could be just as varied. Because the specific items being stored are typically unknown, it is necessary to make an educated guess at the potential hazard created within these facilities.

The classification as Group S-2 would be inappropriate due to the reasonable expectation that a considerable fire load is probable due to the items in storage. A Group S-2 occupancy anticipates the exclusive storage of noncombustible items. In contrast, classification as a Group H-3 storage facility is considered unreasonable due to the historical use of such facilities. Although it is certainly possible that some hazardous materials will be stored, classifying all self-storage facilities as high-hazard occupancies would seem to be an overreach. Therefore, the Group S-1 classification is deemed the most appropriate decision in order to address the anticipated hazards.

312.1
Classification of Communication Equipment Structures

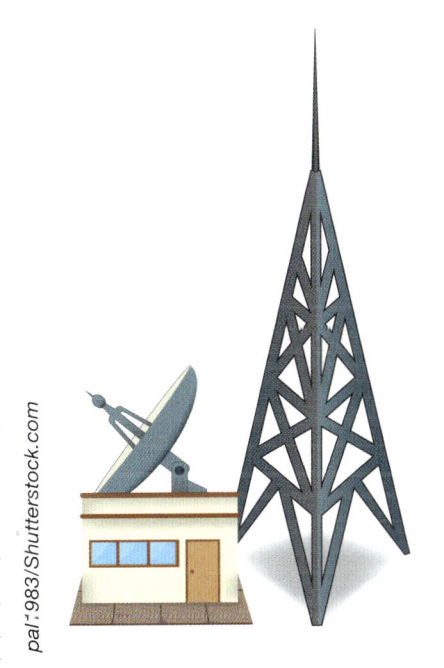

CHANGE TYPE: Modification

CHANGE SUMMARY: Classification as a Group U occupancy is now appropriate for those communication equipment structures that are less than 1,500 square feet in floor area.

2018 CODE: 312.1 General. Buildings and structures of an accessory character and miscellaneous structures not classified in any specific occupancy shall be constructed, equipped and maintained to conform to the requirements of this code commensurate with the fire and life hazard incidental to their occupancy. Group U shall include, but not be limited to, the following:

> Communication-equipment structures with a gross floor area of less than 1,500 square feet (139 m^2)

(No changes to other listed items.)

CHANGE SIGNIFICANCE: The classification of Group U is to be applied to those buildings and structures that pose a limited hazard. Such occupancies have no public occupancy and are generally regarded to be accessory or miscellaneous in nature. In those uses where occupancy does occur, the occupants are typically employees and their time spent in the building is limited. Examples of Group U occupancies include agricultural buildings, barns, livestock shelters, sheds, stables, fences, tanks, and towers. Communications equipment structures have traditionally met the conditions of a Group U occupancy, particularly those of limited size, but without clarification in the code they have often been classified as Group S or F as well. Classification as a Group U occupancy is now appropriate for those communication equipment structures that are less than 1,500 square feet in floor area.

The size limitation established for such equipment structures is unique to the Group U category as the other structures classified as Group U are only regulated for size based upon the provisions of Chapter 5 addressing allowable building area. The 1,500-square-foot limit should be inclusive of the typical equipment structures that are visited infrequently by only authorized and knowledgeable personnel.

Communications structures

312.1.1
Classification of Agricultural Greenhouses

CHANGE TYPE: Clarification

CHANGE SUMMARY: Because a Group U occupancy includes those low-hazard structures that do not conform to any other specific occupancy classification, it has been clarified that greenhouses are only to be considered as Group U where they are not more appropriately classified as one of the other occupancies established in the IBC.

2018 CODE: 312.1 General. Buildings and structures of an accessory character and miscellaneous structures not classified in any specific occupancy shall be constructed, equipped and maintained to conform to the requirements of this code commensurate with the fire and life hazard incidental to their occupancy. Group U shall include, but not be limited to, the following:

~~Greenhouses~~

(No changes to other listed items.)

312.1.1 Greenhouses. Greenhouses not classified as another occupancy shall be classified as Group U.

CHANGE SIGNIFICANCE: The classification of Group U is to be applied to those buildings and structures that pose a limited hazard. Such occupancies have no public occupancy and are generally regarded to be accessory or miscellaneous in nature. In those uses where occupancy does occur, the occupants are typically employees and their time spent in the building is limited. Examples of Group U occupancies include agricultural buildings, barns, livestock shelters, sheds, stables, fences, tanks, and towers. Of significant importance is the recognition that a Group U occupancy also includes those low-hazard structures that do not conform

Commercial greenhouse

to any other specific occupancy classification. As such, it has been clarified that greenhouses are only to be considered as Group U where they are not more appropriately classified as one of the other occupancies established in the IBC.

A greenhouse has historically been recognized as a structure intended for the growing of plants. The primary difference between a greenhouse and other structures is that the environment in a greenhouse is specific to this intended use. Although a variety of other activities can occur within a greenhouse, the key issue is the maintenance of a controlled environment within the facility necessary for proper plant growth. A definition of greenhouse has been added to the code in order to recognize that the primary characteristic of such a structure is its unique environment, not the structure itself or merely the presence of plants.

Buildings intended for human habitation are required to provide minimum features such as lighting, ventilation, heating, and cooling that are suitable for the health and welfare of the occupants, and oftentimes plants can co-exist in such environments. In a greenhouse, the environment is such that it is maintained exclusively for, and essential to, the aggressive propagation of plants used by commercial growers for plant production. It is acceptable for other activities to be conducted in a greenhouse, such as retail sales, educational research, conservation, education, and assembly use. The occupancy classification of a greenhouse will be determined accordingly. However, in all cases, the environment of the greenhouse must be such that the plants will thrive.

The extent of the definition is limited to those structures designed and used specifically for the growing, care and maintenance of plants. As such, sunrooms, solariums, glass-enclosed walkways, atria and other types of spaces that permit ample sunlight and prominently feature plants for aesthetic purposes are not considered as greenhouses under the new definition.

Where greenhouses are used for assembly, sales, or other activities that are more extensive in scope than addressed in the definition, their classification as Group U is not appropriate. In such cases, a Group A or M classification is warranted. Therefore, the term "greenhouses" was deleted from the list of Group U occupancies. Section 312.1.1 was added to recognize that greenhouses that do not conform to another occupancy classification will continue to retain the Group U listing.

403.2.1.1

Type of Construction in High-Rise Buildings

CHANGE TYPE: Modification

CHANGE SUMMARY: The reduction in the minimum required fire-resistance ratings for certain building elements of high-rise buildings is no longer applicable to Group H-2, H-3, and H-5 occupancies due to the high physical hazard level such uses pose.

2018 CODE: 403.2.1.1 Type of construction. The following reductions in the minimum fire-resistance rating of the building elements in Table 601 shall be permitted as follows:

2. In other than Group F-1, H-2, H-3, H-5, M and S-1 occupancies, the fire-resistance rating of the building elements in Type IB construction shall be permitted to be reduced to the fire-resistance ratings in Type IIA.

(No changes to other listed reductions.)

CHANGE SIGNIFICANCE: Primarily because a sprinklered high-rise building is provided with an increased level of fire protection supervision and control, the IBC permits a reduction in the minimum required fire-resistance ratings for various building elements. The requirement for supervisory initiating devices and water-flow initiating devices on each floor, in addition to the secondary on-site supply of water mandated in high-rise buildings subject to a moderate to high level of seismic risk, justifies the modification established in Item 2 of Section 403.2.1.1. The allowance for Type IB buildings has previously been applied to all occupancies except those with the potential for significant fire loads, Groups

R and D business park

F-1, M, and S-1. Consistent with these three occupancy classifications, the reduction is also no longer applicable to Groups H-2, H-3, and H-5 occupancies due to the high physical hazard level such uses pose.

Although Groups F-1, M, and S-1 occupancies are considered as moderate-hazard occupancies, they can pose a sizable fire hazard where large quantities of combustible materials are present. For this reason, high-rise buildings that house such occupancies have not been granted a reduction in fire-resistance ratings as established in Table 601. Because Group H occupancies are viewed as a higher hazard than these manufacturing, sales, and storage uses, it is appropriate that they also be excluded from applying the reduction. Group H-1 occupancies are not listed because they are exempted from the high-rise provisions by Section 403.1, Exception 5, and Group H-5 occupancies are not included because they pose health hazards rather than physical hazards.

404.6
Enclosure of Atriums

CHANGE TYPE: Modification

CHANGE SUMMARY: The requirement that those spaces not separated from an atrium be accounted for in the design of the smoke control system now applies only in those cases where the atrium is required to be provided with a smoke control system.

2018 CODE: 404.6 Enclosure of atriums. Atrium spaces shall be separated from adjacent spaces by a 1-hour fire barrier constructed in accordance with Section 707 or a horizontal assembly constructed in accordance with Section 711, or both.

Exceptions:

3. A fire barrier is not required between the atrium and the adjoining spaces of ~~any~~ <u>up to</u> three floors of the atrium provide such spaces are accounted for in the design of the smoke control system.

4. <u>A fire barrier is not required between the atrium and the adjoining spaces where the atrium is not required to be provided with a smoke control system.</u>

(No changes to other exceptions.)

CHANGE SIGNIFICANCE: As a general rule, an enclosure separation is required between an atrium and the remainder of the building. The basic requirement of a minimum 1-hour fire-resistance-rated fire barrier with protected openings provides protection somewhat equivalent to the otherwise mandated shaft protection. The separation between adjacent spaces and the atrium may be omitted on a maximum of up to three floor levels, provided the volume of the spaces open to the atrium is included in computations for the design of the smoke control system.

Atrium in airport terminal

It has been questioned as to the application of this requirement for those atriums where smoke control is not required. A fourth exception has been provided to clearly indicate that the requirement that those spaces not separated from the atrium be accounted for in the design of the smoke control system applies only in those cases where the atrium is required to be provided with smoke control.

In other than Group I-2 and Group I-1, Condition 2 occupancies, smoke control is not required in atriums that connect only two stories. Exception 3 of Section 404.6 allows for adjoining spaces to be open to the atrium provided their volume is taken into account in the design of a smoke control system. At question has been how to address an atrium that does not require smoke control where such atrium is not separated from the adjoining spaces. The new exception recognizes that under such conditions there is no mandate to provide complying separation or to install a smoke control system.

406.1
Motor Vehicle-Related Occupancies

CHANGE TYPE: Clarification

CHANGE SUMMARY: Provisions specific to motor-vehicle-related uses have been reformatted in a manner such that those requirements that apply to all such uses have been relocated in a single Section 406.1.

2018 CODE: 406.1 General. <u>All</u> motor-vehicle-related occupancies shall comply with ~~Sections 406.1 through 406.8~~ <u>Section 406.2. Private garages and carports shall also comply with Section 406.3. Open public parking garages shall also comply with Sections 406.4 and 406.5. Enclosed public parking garages shall also comply with Sections 406.4 and 406.6. Motor fuel-dispensing facilities shall also comply with Section 406.7. Repair garages shall also comply with Section 406.8.</u>

Format revisions to Sections 406.2.1 through 406.2.9.3, as well as Sections 406.3 through 406.8.3, are too extensive to be included here. Refer to Code Change G95-15 for the entire text of the code modifications.

CHANGE SIGNIFICANCE: Although uncommon, fire hazards related to motor vehicles are a concern. Therefore, specific requirements are established in Section 406 to regulate occupancies containing motor vehicles, whether they are parked, under repair or being fueled. Provisions specific to motor-vehicle-related uses have been reformatted in a manner such that those requirements that apply to all such uses have been relocated in a single Section 406.2.

Special regulations applicable to motor-vehicle-related uses are varied due to the differing activities that occur. Facilities addressed include repair garages, motor fuel-dispensing operations, and three types

Open parking garage

406.1 ■ **Motor Vehicle-Related Occupancies**

Service station canopy

of parking garages. The provisions have previously been inconsistently formatted throughout Section 406. They have now been reorganized in the following manner to allow for better consistency in application, particularly in those areas where a requirement applies to all types of motor-vehicle-related uses. Section 406.2, applicable to all motor-vehicle-related occupancies, now establishes requirements for the following conditions:

- Automatic garage door openers and vehicular gates
- Clear height of vehicle and pedestrian traffic areas
- Accessible parking spaces
- Floor surfaces
- Sleeping rooms
- Fuel dispensing
- Electric vehicle charging stations
- Mixed occupancies and separation
- Equipment and appliances
- Elevation of ignition sources

406.3
Regulation of Private Garages

CHANGE TYPE: Clarification

CHANGE SUMMARY: Parking structures that meet the definition of private garages are now permitted to comply with the provisions for public parking garages as an alternative approach.

2018 CODE: 406.3 Private garages and carports. Private garages and carports shall comply with Sections ~~406.3.1 through 406.3.6~~ 406.2 and 406.3, or shall comply with Sections 406.2 and 406.4.

CHANGE SIGNIFICANCE: A private garage has historically been limited by the size of the garage facility, with an absolute limit of 3,000 square feet. In the 2015 IBC, a new definition was introduced describing a private garage as a building or portion of a building where motor vehicles used by the tenants of the building or buildings on the premises are stored or kept, without any limitation on floor area. As a result, large parking structures used exclusively for use of the tenants would qualify as private parking garages and be subject to the fire separation and Group U classification criteria of Section 406.3. An allowance has now been provided such that private garages are permitted to comply with the provisions for public parking garages as an alternative approach.

Private garages, as regulated by Section 406.3, are required to be classified as Group U occupancies and are limited to 1,000 square feet of floor area. Multiple garages are permitted in the same building, but only where separated by minimum 1-hour fire barriers and/or horizontal assemblies. Often there are conditions under which a much larger Group S-2 open or enclosed parking structure is intended to be limited for use only by the building's tenants. The code now allows those parking structures that meet the definition of private garage to be designed and constructed under the provisions for public garages.

Private garages serving assisted living facility

406.6.2 Ventilation of Enclosed Parking Garages

CHANGE TYPE: Clarification

CHANGE SUMMARY: Chapters 4 and 5 of the IMC are now specifically referenced to ensure that all IMC ventilation and exhaust requirements for enclosed parking garages are applied.

2018 CODE: 406.6.2 Ventilation. A mechanical ventilation system <u>and exhaust system</u> shall be provided in accordance with <u>Chapters 4 and 5 of</u> the *International Mechanical Code*.

> <u>**Exception:** Mechanical ventilation shall not be required for enclosed parking garages that are accessory to one- and two-family dwellings.</u>

CHANGE SIGNIFICANCE: Chapter 4 of the *International Mechanical Code* (IMC) mandates minimum ventilation rates for enclosed parking garages. The ventilation system can be designed to operate either continuously or intermittently, providing the necessary fresh air into the structure. Exhaust, as regulated in IMC Chapter 5, must also be provided to remove air to the outdoor atmosphere. Both chapters of the IMC are now referenced to ensure that all IMC requirements for enclosed parking garages are applied. In addition, an exception has been introduced exempting enclosed parking garages accessory to one- and two-family dwellings from any ventilation and exhaust requirements. This exception will expand the scope of the exemption beyond Group U private garages. With the limited scope of one- and two-family dwellings, there is little chance multiple vehicles will be operating at the same time, regardless of the size of the garage.

Public garage ventilation system

407.5
Maximum Smoke Compartment Size

CHANGE TYPE: Modification

CHANGE SUMMARY: The allowance for larger smoke compartments in hospitals and other Group I-2, Condition 2 occupancies has now been modified to only include compartments containing single-patient sleeping rooms and suites, as well as those compartments without patient sleeping rooms.

2018 CODE: **407.5 Smoke barriers.** Smoke barriers shall be provided to subdivide every story used by persons receiving care, treatment or sleeping <u>into not fewer than two smoke compartments. Smoke barriers shall be provided to subdivide</u> ~~and to divide~~ other stories with an occupant load of 50 or more persons, into not fewer than two smoke compartments. The smoke barrier shall be in accordance with Section 709.

407.5.1 Smoke compartment size. ~~Such~~ Stories shall be divided into smoke compartments with an area of not more than 22,500 square feet (2092 m²) in Group I-2 <u>occupancies.</u> ~~Condition 1, and not more than 40,000 square feet (3716 m²) in Group I-2, Condition 2, and the~~

Exceptions:

1. <u>A smoke compartment in Group I-2, Condition 2 is permitted to have an area of not more than 40,000 square feet (3716 m²) provided that all patient sleeping rooms within that smoke compartment are configured for single patient occupancy and any suite within the smoke compartment complies with Section 407.4.4.</u>

2. <u>A smoke compartment in Group I-2, Condition 2 without patient sleeping rooms is permitted to have an area of not more than 40,000 square feet (3716 m²).</u>

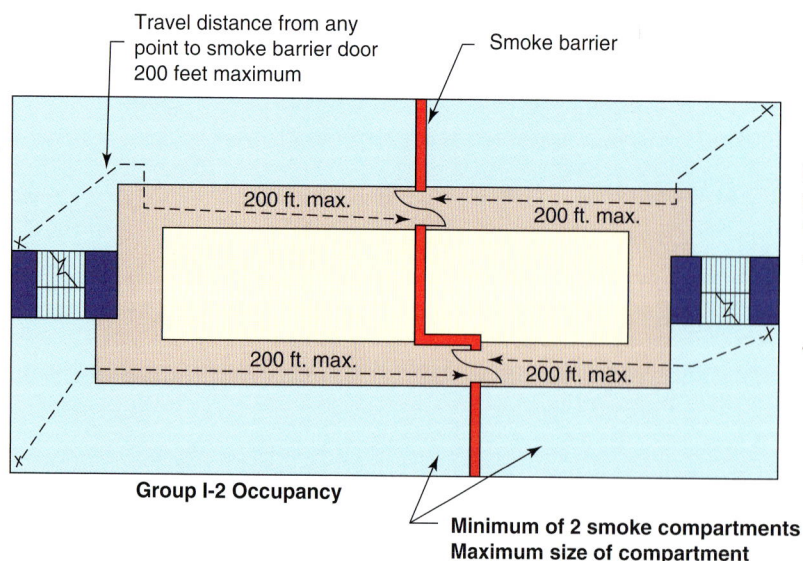

Smoke compartment floor area limits

407.5.2 Exit access travel distance. The distance of travel from any point in a smoke compartment to a smoke barrier door shall be not greater than 200 feet (60 960 mm).

CHANGE SIGNIFICANCE: Evacuation of a building such as a hospital or nursing home is a virtual impossibility in the event of a fire or other hazardous condition, particularly in multistory structures. Horizontal relocation, on the other hand, is possible with a properly trained staff. As a result, the code makes provisions for horizontal compartmentation, so that if necessary, care recipients can be moved from one compartment to another. The maximum size of each smoke compartment has historically been limited to 22,500 square feet. However, the 2015 edition of the IBC revised the size limitations for all hospitals and similar medical care facilities (Group I-2, Condition 2 occupancies) to 40,000 square feet. The scope of the 40,000-square-foot limitation has now been modified to only include compartments containing single-patient sleeping rooms and suites, as well as those compartments without patient sleeping rooms.

The increase in allowable smoke compartment size introduced in the 2015 IBC was based upon several basic issues, including the increased use of single-patient rooms and the operational considerations related to patient treatment. However, the revised provisions had no scoping specific to those conditions. The base requirement for maximum smoke compartment size for all Group I-2 occupancies, including both nursing homes and hospitals, has again been set at 22,500 square feet. However, two new exceptions specifically address those smoke compartments in hospitals and other Condition 2 occupancies where a larger compartment is acceptable.

Exception 1 allows for a maximum smoke compartment size of 40,000 square feet in all patient sleeping rooms and suites that are configured for single-patient occupancy. The decision as to single- vs. multiple-occupancy rooms is intended to be determined by design, rather than by an administrative decision. This exception recognizes the reduction in patient density that is accomplished through single-patient-occupancy rooms. Exception 2 recognizes that smoke compartments without sleeping rooms create a lesser degree of hazard to the occupants, as they are typically ambulatory and bed movement will not be required. A 40,000-square-foot limit is also considered acceptable under these conditions. It is important to note that the increased floor area allowance is only applicable to hospitals and other Group I-2, Condition 2 occupancies.

407.5.4

Required Egress from Smoke Compartments

CHANGE TYPE: Modification

CHANGE SUMMARY: In Group I-2 occupancies, any smoke compartment that does not have an exit from the compartment must now provide direct access to a minimum of two adjacent smoke compartments.

2018 CODE: ~~407.5.2~~ **407.5.4 Independent egress.** A means of egress shall be provided from each smoke compartment created by smoke barriers without having to return through the smoke compartment from which means of egress originated. <u>Smoke compartments that do not contain an exit shall be provided with direct access to not less than two adjacent smoke compartments.</u>

CHANGE SIGNIFICANCE: In protect-in-place uses, such as hospitals and nursing homes, it is necessary to provide multiple smoke compartments to allow for movement of patients during fires and other emergency events. Previously, the only limiting requirement for a means of egress system from a smoke compartment was that the egress path could not return through the compartment of origin. An additional condition now indicates that any smoke compartment not having an exit must provide direct access to a minimum of two adjacent compartments.

Where there is no exit, such as a horizontal exit, interior exit stairway or exterior door at grade level, directly from a smoke compartment, the resulting condition creates somewhat of a "dead-end smoke compartment." The IBC is now consistent with federal Medicare requirements in regard to the recognition of two alternative approaches to the design of the means of egress from a smoke compartment. Each compartment must be provided with a minimum of one direct exit, or direct access to at least two smoke compartments is required.

Egress from smoke compartments

420.7 Corridor Protection in Assisted Living Units

CHANGE TYPE: Modification

CHANGE SUMMARY: Shared living spaces, group meeting spaces, and multipurpose therapeutic spaces are now permitted to be open to fire-rated corridors in Group I-1 assisted living housing facilities provided specific conditions are met.

2018 CODE: 420.7 Group I-1 assisted living housing units. In Group I-1 occupancies, where a fire-resistance-rated corridor is provided in areas where assisted living residents are housed, shared living spaces, group meeting or multipurpose therapeutic spaces open to the corridor shall be in accordance with all of the following criteria:

1. The walls and ceiling of the space are constructed as required for corridors.
2. The spaces are not occupied as resident sleeping rooms, treatment rooms, incidental uses in accordance with Section 509, or hazardous uses.
3. The open space is protected by an automatic fire detection system installed in accordance with Section 907.
4. In Group I-1, Condition 1, the corridors into which the spaces open are protected by an automatic fire detection system installed in accordance with Section 907, or the spaces are equipped throughout with quick-response sprinklers in accordance with Section 903.3.2.
5. In Group I-1, Condition 2, the corridors into which the spaces open, in the same smoke compartment, are protected by an automatic fire detection system installed in accordance with

420.7 continues

Group I-1 living space open to corridor

420.7 continued

> Section 907, or the smoke compartment in which the spaces are located is equipped throughout with quick-response sprinklers in accordance with Section 903.3.2.
>
> **6.** The space is arranged so as not to obstruct access to the required exits.

CHANGE SIGNIFICANCE: Corridors in assisted living facilities classified as Group I-1 occupancies are intended to provide a direct egress path adequately separated from hazards in adjoining spaces by minimum 1-hour fire partitions and protected openings. In a fire event, such protection is intended to provide an environment relatively free of fire and smoke to allow for the evacuation and/or relocation of residents. However, in these assisted living facilities, necessary modifications have now been provided to facilitate the primary functions of these types of healthcare facilities. Consistent with the allowances introduced in the 2015 IBC for Group I-2 nursing homes, these modifications recognize the special needs of these facilities to provide the most efficient and effective care services. Shared living spaces, group meeting spaces, and multipurpose therapeutic spaces may be open to a fire-rated corridor provided specific conditions are met.

In Group I-1 assisted living facilities, residents are encouraged to spend time outside of their rooms. By providing a variety of shared living spaces open to the circulation/means-of-egress system, socialization and interaction are encouraged. Further, being able to preview activities that are occurring helps to encourage joining and allows reluctant participants to join at their own pace. Finally, a more open plan allows staff to more easily monitor residents throughout the day. For these reasons, the required physical and fire-resistive separation of shared resident spaces from corridors has been eliminated.

In order to address the concerns of having common resident spaces open to the corridor system, several conditions have been established. The walls and ceilings of the shared spaces must be constructed with minimum 1-hour fire partitions as required for corridors. The shared spaces are limited in use, as they cannot be occupied as sleeping rooms, treatment rooms, incidental uses, or hazardous uses.

From a fire protection standpoint, the open space must be protected by an automatic fire protection system. In addition, the corridors into which the spaces open must be protected by an automatic fire detection system. As an alternative, quick-response sprinklers may be provided. In Condition 1 facilities, such sprinklers need only be provided throughout the shared spaces. In Condition 2 facilities, the quick-response sprinklers must be provided throughout the entire smoke compartment. Through the application of these conditions, the openness desired for Group I-1 occupancies can be safely achieved.

Significant Changes to the IBC 2018 Edition   420.8 ■ Group I-1 Cooking Facilities   41

420.8
Group I-1 Cooking Facilities

CHANGE TYPE: Addition

CHANGE SUMMARY: A room or space containing a cooking facility with domestic cooking appliances is now permitted to be open to a corridor in Group I-1 occupancies provided nine specific conditions are met.

2018 CODE: <u>**420.8 Group I-1 cooking facilities.** In Group I-1 occupancies, rooms or spaces that contain cooking facilities with domestic cooking appliances shall be in accordance with all the following criteria:</u>

1. <u>In Group I-1, Condition 1 occupancies, the number of care recipients served by one cooking facility shall not be greater than 30.</u>
2. <u>In Group I-1, Condition 2 occupancies, the number of care recipients served by one cooking facility and within the same smoke compartment shall not be greater than 30.</u>
3. <u>The types of domestic cooking appliances permitted shall be limited to ovens, cooktops, ranges, warmers and microwaves.</u>
4. <u>The space containing the domestic cooking facilities shall be arranged so as not to obstruct access to the required exit.</u>
5. <u>Domestic cooking hoods installed and constructed in accordance with Section 505 of the *International Mechanical Code* shall be provided over cooktops or ranges.</u>

420.8 continues

- Appliances limited to ovens, cooktops, ranges, warmers and microwaves
- Fuel and electrical supply to cooking equipment be provided with shut-off accessible only to staff
- Timer to deactivate cooking appliances within 2 hours

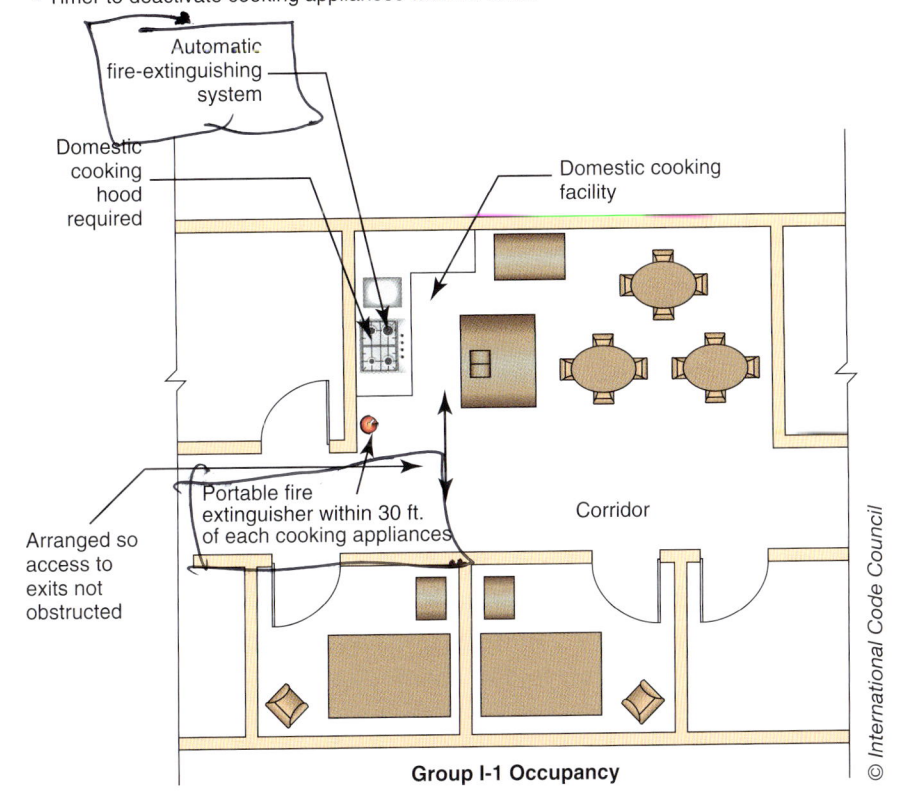

Group I-1 kitchen/dining spaces

420.8 continued

    6. Cooktops and ranges shall be protected in accordance with Section 904.13.

    7. A shut-off for the fuel and electrical supply to the cooking equipment shall be provided in a location that is accessible only to staff.

    8. A timer shall be provided that automatically deactivates the cooking appliances within a period of not more than 120 minutes.

    9. A portable fire extinguisher shall be provided. Installation shall be in accordance with Section 906 and the extinguisher shall be located within a 30-foot (9144 mm) distance of travel from each domestic cooking appliance.

420.8.1 Cooking facilities open to the corridor. Cooking facilities located in a room or space open to a corridor, aisle or common space shall comply with Section 420.8.

CHANGE SIGNIFICANCE: It is often desirable that kitchens in a typical Group I-1 memory care or assisted living facility be permitted open to contiguous spaces and rooms used for sleeping. Oftentimes, one or more kitchens are most efficient and effective when located directly off of a corridor system. Because corridors in Group I-1 occupancies are required to be fire-resistance-rated a minimum of 1 hour and provided with opening protectives, adjoining spaces must be separated through the use of fire partitions. New Section 420.8 now permits a room or space containing a cooking facility with domestic cooking appliances to be open to a corridor provided nine specific conditions are met.

    As assisted living and similar facilities transition from traditional models, the need for open and shared resident spaces is very important. A part of that group environment is a functioning kitchen that can also serve as the hearth of the home. Instead of a large, centralized, institutional kitchen where all meals are prepared and delivered to a central dining room or the resident's room, the "household model" uses decentralized kitchens and small dining areas to create the focus and feeling of home. Allowing kitchens that serve a small, defined group of residents to be open to common spaces and corridors is viewed as critically important to enhancing the feeling and memories of home for older adults.

    Because unattended cooking equipment is a leading cause of fires in residential facilities, it is important that necessary safeguards be put in place to address the hazards involved. In addition, limitations on occupancy are necessary to allow for efficient egress should an emergency condition occur. In Condition 1 occupancies, a single cooking facility can serve no more than 30 care recipients. In Condition 2 occupancies, the maximum occupant load of 30 is limited to the same smoke compartment. Ovens, cooktops, ranges, warmers, and microwaves are the only kitchen appliances permitted.

    Additional conditions deal primarily with the fire protection and mechanical systems related to the cooking activity. An IMC-compliant domestic hood must be installed over the cooktop or range, protected by an appropriate fire extinguishing system. Other safeguards required include the installation of an emergency shutoff for the fuel and electrical power supply, a timer that automatically deactivates the cooking appliances and the installation of a portable fire extinguisher.

420.10 Dormitory Cooking Facilities

CHANGE TYPE: Addition

CHANGE SUMMARY: The installation and use of domestic cooking appliances are now regulated in both common areas and sleeping rooms of Group R-2 college dormitories.

2018 CODE: 420.10 Group R-2 dormitory cooking facilities. Domestic cooking appliances for use by residents of Group R-2 college dormitories shall be in accordance with Sections 420.10.1 and 420.10.2.

420.10.1 Cooking appliances. Where located in Group R-2 college dormitories, domestic cooking appliances for use by residents shall be in compliance with all of the following:

1. The types of domestic cooking appliances shall be limited to ovens, cooktops, ranges, warmers, coffee makers and microwaves.
2. Domestic cooking appliances shall be limited to approved locations.
3. Cooktops and ranges shall be protected in accordance with Section 904.13.
4. Cooktops and ranges shall be provided with a domestic cooking hood installed and constructed in accordance with Section 505 of the *International Mechanical Code.*

420.10.2 Cooking appliances in sleeping rooms. Cooktops, ranges and ovens shall not be installed or used in sleeping rooms.

CHANGE SIGNIFICANCE: Studies have shown that cooking appliances are the leading cause of fires in residential settings. Electric ranges are by far the leading cause of home cooking appliance fires. Unattended cooking

420.10 continues

Shared kitchen in dormitory building

420.10 continued is a factor in the majority of home electric range fires. Physical conditions such as falling asleep or impairment by alcohol or drugs are other contributing factors. Distractions that pull the cook outside of the kitchen (doorbell, social interactions) are another. In spite of these concerns, there have never been requirements in the IBC that regulate such appliances in college residences. The installation and use of domestic cooking appliances are now regulated in both common areas and sleeping rooms of Group R-2 college dormitories.

The scope of the new provisions is limited to those appliances intended to be used by residents of Group R-2 college dormitories. It does not apply to residential dwelling units on college campuses that are not classified as dormitories. In addition to a limit on the types of appliances and their locations, cooktops and ranges are further regulated due to their increased hazard. They shall be protected through the installation of an approved automatic fire extinguishing system. A domestic cooking hood shall also be installed in accordance with Section 505 of the *International Mechanical Code*.

The types of appliances permitted to be installed and used include ovens, ranges, cooktops, warmers, coffee makers, and microwaves. These acceptable cooking appliances are consistent with those permitted in Group I-2, Condition 1 nursing home occupancies. Due to the heightened hazard posed by cooktops, ranges, and ovens, such appliances are not permitted within sleeping rooms.

422.6
Electrical Systems in Ambulatory Care Facilities

CHANGE TYPE: Addition

CHANGE SUMMARY: Reference is now made to IBC Chapter 27 addressing emergency and standby power systems, as well as NFPA 99, *Health Care Facilities Code*, regarding the design and construction requirements for essential electrical systems for electrical components, equipment, and systems in ambulatory care facilities.

2018 CODE: **422.6 Electrical systems.** In ambulatory care facilities, the essential electrical system for electrical components, equipment and systems shall be designed and constructed in accordance with the provisions of Chapter 27 and NFPA 99.

CHANGE SIGNIFICANCE: Ambulatory care facilities are minor surgery centers, dental surgery centers, and similar facilities where individuals are temporarily rendered incapable of self-preservation during medical, surgical, psychiatric, nursing, or similar care. The period of time under which the individual is under sedation, nerve blocks or anesthesia is typically quite short, and the procedures allow an individual to spend a limited amount of time within the facility. However, during such limited times where self-preservation is not impossible, it is still important that safeguards be in place to address the need for physical assistance in case of an emergency. Reference is now made to IBC Chapter 27 addressing emergency and standby power systems, as well as NFPA 99, *Health Care Facilities Code*, regarding the design and construction requirements for essential electrical systems for electrical components, equipment, and systems.

422.6 continues

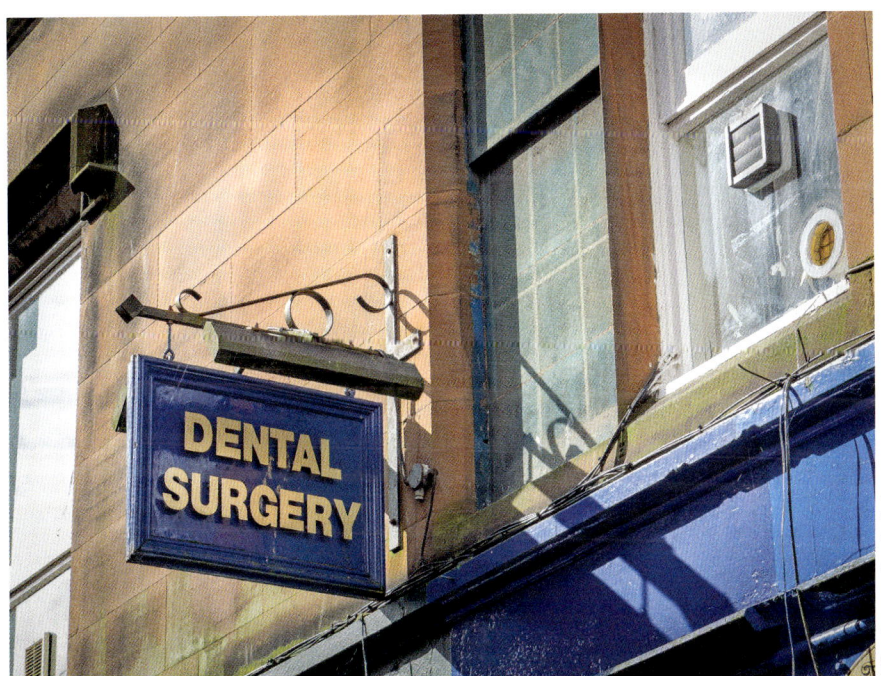

Ambulatory care facility

422.6 continued

The IBC has previously provided no guidance as to whether or not essential electrical systems, such as an emergency generator, are required in ambulatory care facilities. NFPA 99 has now been referenced as the document to be used in such an assessment. The *Health Care Facilities Code* provides a risk-based approach to determine the need for an essential electrical system, the class of system required, and the general design requirements for each type of system.

424.1
Children's Play Structures

CHANGE TYPE: Modification

CHANGE SUMMARY: The dimensional criteria under which children's play structures are scoped by the IBC have been revised, resulting in the potential for many more structures to be regulated for fire concerns.

2018 CODE: 424.1 ~~Children's play structures~~ General. Children's play structures installed inside all occupancies covered by this code that exceed 10 feet (3048 mm) in height ~~and~~ or 150 square feet (14m²) in area shall comply with Sections 424.2 through 424.5.

CHANGE SIGNIFICANCE: Play structures for children's activities, such as those structures occasionally found in fast-food restaurants, arcades, day-care facilities, and covered mall buildings, are regulated due to their combustibility. They must be constructed of noncombustible materials, or as an option if combustible, comply with alternate methods including the use of fire-retardant-treated wood, textiles complying with the designated flame propagation performance criteria, and plastic exhibiting an established maximum peak rate of heat release. The dimensional criteria under which children's play structures are scoped by the IBC have been revised, resulting in the potential for many more structures to be regulated for fire concerns.

Historically, requirements for children's play structures have only been applicable where the structure exceeds *both* the specified height and the specified floor area set forth in the code. Therefore, many such structures have not been regulated by the IBC because either their height or floor area fell slightly below the code threshold, even though they were

424.1 continues

Children's play structure

424.1 continued

extensive in size and fire hazard. By revising the compliance trigger in a manner such that only one aspect of their size, either height or floor area, need exceed the code's limits for the provisions to apply, a significant increase in the number of regulated structures is expected. Specifically, where the height of the children's play structure exceeds 10 feet, *or* where the floor area of the structure is greater than 150 square feet, the materials, fire protection, separation, and area limits of Section 424 must be met.

427 Medical Gas Systems

CHANGE TYPE: Addition

CHANGE SUMMARY: In order to provide a more comprehensive and efficient compilation of construction regulations, those IFC medical gas system requirements related directly to building construction have now been replicated in the IBC.

2018 CODE:

SECTION 427
MEDICAL GAS SYSTEMS

427.1 General. Medical gases at health care-related facilities intended for patient or veterinary care shall comply with Sections 427.2 through 427.2.3 in addition to requirements of Chapter 53 of the *International Fire Code*.

427.2 Interior supply location. Medical gases shall be located in areas dedicated to the storage of such gases without other storage or uses. Where containers of medical gases in quantities greater than the permitted amount are located inside the buildings, they shall be located in a 1-hour exterior room, 1-hour interior room or a gas cabinet in accordance with Sections 427.2.1, 427.2.2 or 427.2.3, respectively. Rooms or areas where medical gases are stored or used in quantities exceeding the maximum allowable quantity per control area as set forth in Tables 307.1(1) and 307.1(2) shall be in accordance with Group H occupancies.

427.2.1 One-hour exterior room. A 1-hour exterior room shall be a room or enclosure separated from the remainder of the building by fire barriers constructed in accordance with Section 707 or horizontal assemblies constructed in accordance with Section 711, or both, with a fire-resistance rating of not less than 1 hour. Openings between the room or enclosure and interior spaces shall be provided with self-closing smoke- and draft-control assemblies having a fire protection rating of not less than 1 hour. Rooms shall have not less than one exterior wall that is provided with not less than two vents. Each vent shall have a minimum free air opening of not less than 36 square inches (232 cm^2) for each 1,000 cubic feet (28 m^2) at normal temperature and pressure (NTP) of gas stored in the room and shall be not less than 72 square inches (465 cm^2) in aggregate free opening area. One vent shall be within 6 inches (152 mm) of the floor and one shall be within 6 inches (152 mm) of the ceiling. Rooms shall be provided with not fewer than one automatic sprinkler to provide container cooling in case of fire.

427.2.2 One-hour interior room. Where an exterior wall cannot be provided for the room, a 1-hour interior room or enclosure shall be provided and separated from the remainder of the building by fire barriers constructed in accordance with Section 707 or horizontal assemblies constructed in accordance with Section 711, or both, with a fire resistance rating of not less than 1 hour. Openings between the room

427 continues

427 continued

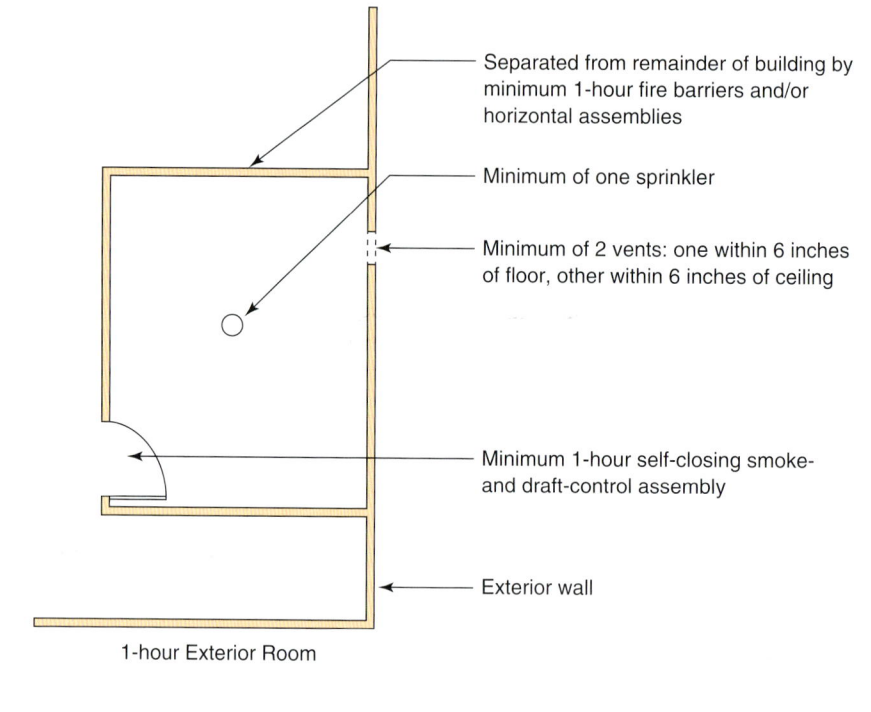

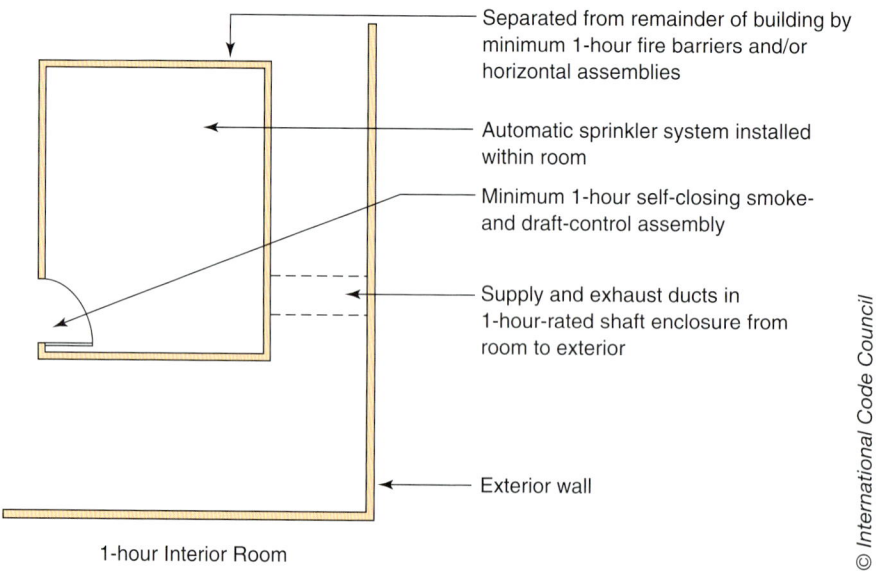

1-hour Exterior Room

- Separated from remainder of building by minimum 1-hour fire barriers and/or horizontal assemblies
- Minimum of one sprinkler
- Minimum of 2 vents: one within 6 inches of floor, other within 6 inches of ceiling
- Minimum 1-hour self-closing smoke- and draft-control assembly
- Exterior wall

1-hour Interior Room

- Separated from remainder of building by minimum 1-hour fire barriers and/or horizontal assemblies
- Automatic sprinkler system installed within room
- Minimum 1-hour self-closing smoke- and draft-control assembly
- Supply and exhaust ducts in 1-hour-rated shaft enclosure from room to exterior
- Exterior wall

Medical gas storage rooms

<u>or enclosure and interior spaces shall be provided with self-closing smoke- and draft-control assemblies having a fire protection rating of not less than 1 hour. Openings between the room or enclosure and interior spaces shall be provided with self-closing smoke- and draft-control assemblies having a fire protection rating of not less than 1 hour. An automatic sprinkler system shall be installed within the room. The room shall be exhausted through a duct to the exterior. Supply and exhaust ducts shall be enclosed in a 1-hour-rated shaft enclosure from the room to the exterior. Approved mechanical ventilation shall comply with the *International Mechanical Code* and be provided with a minimum rate of 1 cubic foot per minute per square foot (0.00508 m³/s/m²) of the area of the room.</u>

427.2.3 Gas cabinets. Gas cabinets shall be constructed in accordance with Section 5003.8.6 of the *International Fire Code* and shall comply with the following:

1. Cabinets shall be exhausted to the exterior through a dedicated exhaust duct system installed in accordance with Chapter 5 of the *International Mechanical Code.*

2. Supply and exhaust ducts shall be enclosed in a 1-hour-rated shaft enclosure from the cabinet to the exterior. The average velocity of ventilation at the face of access ports or windows shall be not less than 200 feet per minute (1.02 m/s) with a minimum of 150 feet per minute (0.076 m/s) at any point of the access port or window.

3. Cabinets shall be provided with an automatic sprinkler system internal to the cabinet.

CHANGE SIGNIFICANCE: Special construction provisions related to the storage of medical gases have historically been addressed in the *International Fire Code* (IFC). The scope of IFC Section 5306 includes the storage of compressed gases intended for inhalation or sedation, including analgesia systems for dentistry, podiatry, veterinary, and similar uses. Because most of the medical gas construction-related requirements in IFC reference the IBC, it was deemed logical that those requirements should be incorporated into the IBC itself. Therefore, those IFC medical gas system requirements related only to building construction have been replicated in the IBC.

Hospitals and most other healthcare facilities typically require the use of medical gases as a critical component of their functions. Oxygen, nitrous oxide and a variety of other compressed gases are piped into treatment rooms from medical gas storage rooms. These rooms have been regulated by the IFC for fire-resistive separation and ventilation purposes. Provisions address both exterior rooms, which must be located on an exterior wall, and interior rooms, where an exterior wall location cannot be provided. In addition, gas cabinet construction criteria are set forth. These requirements are now also located in Chapter 4 of the IBC in a manner consistent with other IFC provisions that have been replicated in the IBC in order to provide a more comprehensive and efficient set of construction regulations.

428
Higher Education Laboratories

CHANGE TYPE: Addition

CHANGE SUMMARY: Higher education laboratories using hazardous materials can now be considered Group B occupancies provided such laboratories comply with new Section 428 which provides an alternative approach to the existing control area provisions.

2018 CODE:

**SECTION 202
DEFINITIONS**

<u>**HIGHER EDUCATION LABORATORY.** Laboratories in Group B occupancies used for educational purposes above the 12th grade. Storage, use and handling of chemicals in such laboratories shall be limited to purposes related to testing, analysis, teaching, research or developmental activities on a nonproduction basis.</u>

**SECTION 428
HIGHER EDUCATION LABORATORIES**

<u>**428.1 Scope.** Higher education laboratories complying with the requirements of Sections 428.1 through 428.4 shall be permitted to exceed the maximum allowable quantities of hazardous materials in control areas set forth in Tables 307.1(1) and 307.1(2) without requiring classification as a Group H occupancy. Except as specified in Section 428, such laboratories shall comply with all applicable provisions of this code and the *International Fire Code.*</u>

<u>**428.2 Application.** The provisions of Section 428 shall be applied as exceptions or additions to applicable requirements of this code. Unless specifically modified by Section 428, the storage, use and handling of</u>

University laboratory

hazardous materials shall comply with all other provisions in Chapters 38 and 50 through 67 of the *International Fire Code* and this code for quantities not exceeding the maximum allowable quantity.

428.3 Laboratory suite construction. Where laboratory suites are provided, they shall be constructed in accordance with this section and Chapter 38 of the *International Fire Code*. The number of laboratory suites and percentage of maximum allowable quantities of hazardous materials in laboratory suites shall be in accordance with Table 428.3.

428.3.1 Separation from other nonlaboratory areas. Laboratory suites shall be separated from other portions of the building in accordance with the most restrictive of the following:

1. Fire barriers and horizontal assemblies as required in Table 428.3. Fire barriers shall be constructed in accordance with Section 707 and horizontal assemblies constructed in accordance with Section 711.

   Exception: Where an individual laboratory suite occupies more than one story, the fire resistance rating of intermediate floors contained within the laboratory suite shall comply with the requirements of this code.

2. Separations as required by Section 508.

428.3.2 Separation from other laboratory suites. Laboratory suites shall be separated from other laboratory suites in accordance with Table 428.3.

428.3.3 Floor assembly fire resistance. The floor assembly supporting laboratory suites and the construction supporting the floor of laboratory suites shall have a fire resistance rating of not less than 2 hours.

   Exception: The floor assembly of the laboratory suites and the construction supporting the floor of the laboratory suites are allowed to be 1-hour fire resistance rated in buildings of Types IIA, IIIA and VA construction, provided that the building is three or fewer stories.

428.3.4 Maximum number. The maximum number of laboratory suites shall be in accordance with Table 428.3. Where a building contains both laboratory suites and control areas, the total number of laboratory suites and control areas within a building shall not exceed the maximum number of laboratory suites in accordance with Table 428.3.

428.3.5 Means of egress. Means of egress shall be in accordance with Chapter 10.

428.3.6 Standby or emergency power. Standby or emergency power shall be provided in accordance with Section 414.5.2 where laboratory suites are located above the sixth story above grade plane or located in a story below grade plane.

428.3.7 Ventilation. Ventilation shall be in accordance with Chapter 7 of NFPA 45, and the *International Mechanical Code*.

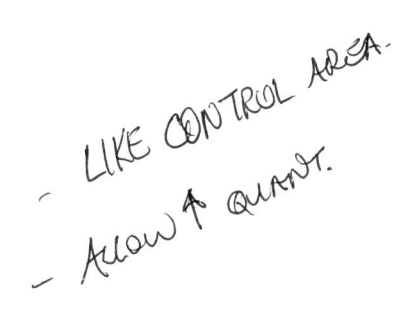

428 continues

428 continued

428.3.8 Liquid tight floor. Portions of laboratory suites where hazardous materials are present shall be provided with a liquid-tight floor.

428.3.9 Automatic fire-extinguishing systems. Buildings containing laboratory suites shall be equipped throughout with an approved automatic sprinkler system in accordance with Section 903.3.1.1.

428.4 Percentage of maximum allowable quantity in each laboratory suite. The percentage of maximum allowable quantities of hazardous materials in each laboratory suite shall be in accordance with Table 428.3

TABLE 428.3 Design and Number of Laboratory Suites Per Floor

Floor Level		Percentage of the Maximum Allowable Quantity Per Lab Suite[a]	Number of Lab Suites Per Floor	Fire-Resistance Rating for Fire Barriers in Hours[b]
Above Grade Plane	21+	Not allowed	Not Permitted	Not Permitted
	16-20	25	1	2[c]
	11-15	50	1	2[c]
	7-10	50	2	2[c]
	4-6	75	4	1
	3	100	4	1
	1-2	100	6	1
Below Grade Plane	1	75	4	1
	2	50	2	1
	Lower than 2	Not Allowed	Not Allowed	Not Allowed

a. Percentages shall be of the maximum allowable quantity per control area shown in Tables 307.1(1) and 307.1(2), with all increases allowed in the footnotes to those tables.
b. Fire barriers shall include walls, floors and ceilings necessary to provide separation from other portions of the building.
c. Vertical fire barriers separating laboratory suites from other spaces on the same floor shall be permitted to be 1-hour fire-resistance rated.

CHANGE SIGNIFICANCE: Colleges and universities often have chemistry, biology, medical, engineering, and other types of laboratories where significant amounts of hazardous materials are stored and used. The IBC and IFC have not historically addressed these teaching and research laboratories in a specific manner, instead requiring that they be regulated under the general hazardous materials provisions which often are not appropriate for specialized academic laboratory settings. Therefore, a collection of new provisions have been introduced, both in the IBC and the IFC, to address the unique circumstances encountered.

There are a number of conditions typically present in higher education laboratories that make them unique, thus requiring unique solutions:

- Lower hazardous materials density in individual laboratory spaces. In academic environments, there are typically a large number of laboratories, but each laboratory only contains a small amount of hazardous material. Individually the quantities in use and storage are relatively low, but the total quantity of hazardous materials on the story can be significant, even to the point of exceeding the maximum allowable quantities per control area. The lower density condition is considered as a means of mitigating the overall risk.

- Considerable staff oversight of activities utilizing hazardous materials. Most academic laboratories are well staffed with faculty members and support personnel who are well acquainted with hazardous material safety. They are an integral part of the preparation and review of laboratory safety documentations and safety audits.
- Mixed-occupancy buildings. Higher education laboratories are often found in campus buildings that also contain storage, business, and assembly uses. The traditional limits on the permissible amount of hazardous materials on upper floor levels are extremely restrictive, often requiring that lecture halls and classrooms be placed on upper floors so that the lower stories can be utilized for laboratories. This scenario places significant occupant loads on upper floors above those areas devoted to the use and storage of hazardous materials.

There are three primary considerations in regard to the new provisions for higher education laboratories. Specifically, new Section 428 addresses the following primary needs:

- Increased general laboratory safety. The introduction of an entirely new set of provisions establishes requirements and allowances specific to higher education laboratory uses. Along with IBC and IFC references to NFPA 45, *Standard on Fire Protection for Laboratories Using Chemicals*, the IFC now contains a new chapter addressing the unique issues of these types of laboratories in a manner even more comprehensive that the IBC.
- Control area limitations. In an increasingly number of cases, the buildings housing higher education laboratories are built taller and/or larger than in the past. In response to this reality, greater numbers of control areas and larger percentages of maximum allowable quantities are necessary. The new provisions provide for an alternate design approach for such scenarios where traditional control area limitations or construction as a Group H occupancy are not feasible. The new "laboratory suite" concept creates an option to allow increased flexibility in the storage and use of hazardous materials while continuing to maintain a Group B occupancy classification.
- Allowances for existing nonsprinklered buildings. Although not regulated due to the scope of the IBC, an approach to regulating existing academic laboratory buildings without fire sprinkler protection has been established in the IFC. Limited to very small quantities of hazardous materials, the allowance recognizes the many buildings built decades ago where retrofitting with sprinklers is not practical.

The IBC now recognizes that these higher education laboratories can be considered as a part of the Group B occupancy classification afforded to other portions of college and university buildings provided such laboratories comply with new Section 428. It is important that the IFC be consulted for any additional requirements related to the storage and use of hazardous materials. In addition, the activities are limited to the testing, analysis, teaching, research, and development on a nonproduction basis.

A key aspect of the academic laboratory provisions is the creation of laboratory suites. Similar in concept to control areas, such suites are fully enclosed by fire-resistance-rated construction in order to provide containment areas. The number of laboratory suites permitted on a story, as well as the amount of hazardous materials that can be used and/or stored within a laboratory suite, is significantly greater than allowed in control areas.

503.1, 706.1

Scope of Fire Wall Use

CHANGE TYPE: Modification

CHANGE SUMMARY: The use of fire walls is now strictly limited to only the determination of permissible types of construction, based upon allowable building area and height.

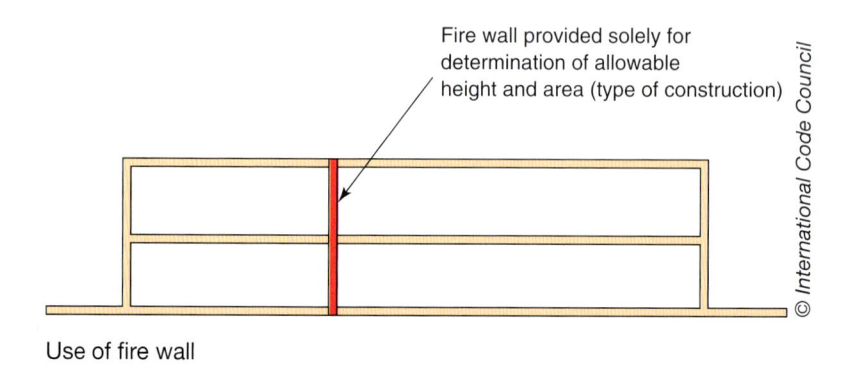

2018 CODE: 503.1 General. Unless otherwise specifically modified in Chapter 4 and this chapter, building height, number of stories and building area shall not exceed the limits specified in Sections 504 and 506 based on the type of construction as determined by Section 602 and the occupancies as determined by Section 302 except as modified hereafter. Building height, number of stories and building area provisions shall be applied independently. ~~Each~~ For the purposes of determining area limitations, height limitations and type of construction, each portion of a building separated by one or more fire walls complying with Section 706 shall be considered to be a separate building.

706.1 General. ~~Each portion of a building separated by one or more fire~~ Fire walls ~~that comply with the provisions of this section~~ shall be ~~considered a separate building~~ constructed in accordance with Sections 706.2 through 706.11. The extent and location of such fire walls shall provide a complete separation. Where a fire wall separates occupancies that are required to be separated by a fire barrier wall, the most restrictive requirements of each separation shall apply.

CHANGE SIGNIFICANCE: Fire walls are considered as the most protective of the various fire separation elements set forth in the IBC. The structural stability, materials, fire-resistance, and continuity requirements for fire walls provide for a substantial expectation that the fire separation created by a fire wall is at the highest level. There has always been some confusion as to the extent of a fire wall's use regarding the separation of a single structure into two or more smaller buildings. A fundamental concept of the code is that larger buildings typically have more restrictive requirements than smaller buildings. Therefore, using fire walls to create multiple smaller buildings under the same roof allows each small building to be regulated independently rather than as one large building. An issue was the extent of provisions in the IBC that can be applied to the smaller buildings created by one or more fire walls. The use of fire walls is now strictly limited to only the determination of permissible types of construction, based upon allowable building area and height.

Use of fire wall

Both Sections 503.1 and 706.1 previously indicated that the portions of a structure separated by one or more fire walls were required to be considered as separate buildings. Although it was possible to consider that the requirement located in Section 503.1 was limited in scope due to its inclusion in Chapter 5 addressing general building heights and areas, the statement in Section 706.1 was global in nature and implied that the smaller buildings created by fire walls were to be regulated as unique and individual buildings for all purposes of the code. In addition, there was an often-applied opinion that the various elements and systems on each side of a fire wall must be completely self-contained. The revised provisions now indicate that the use of a fire wall is solely predicated on the determination of the maximum allowable height and area calculations per Chapter 5. Using the provisions to control other building features or elements such as means of egress, fire protection systems, or building utilities is no longer appropriate.

[Handwritten notes:]
SELDOM USE OF F.W. ⇒ FIRE BARRIERS LIKELY.
F.W. UBC = NOT SEPARATING INTO 2 BUILDINGS.
PROPERTY LINES?

503.1.4

Allowable Height and Area of Occupied Roofs

CHANGE TYPE: Addition

CHANGE SUMMARY: New criteria are now provided establishing the appropriate methodology in the regulation of building height in stories above grade plane where one or more occupancies is located on the roof.

2018 CODE: <u>**503.1.4 Occupied roofs.** A roof level or portion thereof shall be permitted to be used as an occupied roof provided the occupancy of the roof is an occupancy that is permitted by Table 504.4 for the story immediately below the roof. The area of the occupied roofs shall not be included in the building area as regulated by Section 506.</u>

<u>**Exceptions:**</u>

<u>1. The occupancy located on an occupied roof shall not be limited to the occupancies allowed on the story immediately below the roof where the building is equipped throughout with an automatic sprinkler system in accordance with Section 903.3.1.1 or 903.3.1.2 and occupant notification in accordance with Section 907.5 is provided in the area of the occupied roof.</u>

<u>2. Assembly occupancies shall be permitted on roofs of open parking garages of Type I or Type II construction, in accordance with the exception to Section 903.2.1.6.</u>

<u>**503.1.4.1 Enclosures over occupied roof areas.** Elements or structures enclosing the occupied roof areas shall not extend more than 48 inches (1220 mm) above the surface of the occupied roof.</u>

<u>**Exception:** Penthouses constructed in accordance with Section 1510.2 and towers, domes, spires and cupolas constructed in accordance with Section 1510.5.</u>

CHANGE SIGNIFICANCE: The IBC regulates the size of buildings, both building area and building height, in order to limit to a reasonable level the magnitude of a fire that potentially may develop. A building's maximum allowable height in regard to number of stories above grade plane is

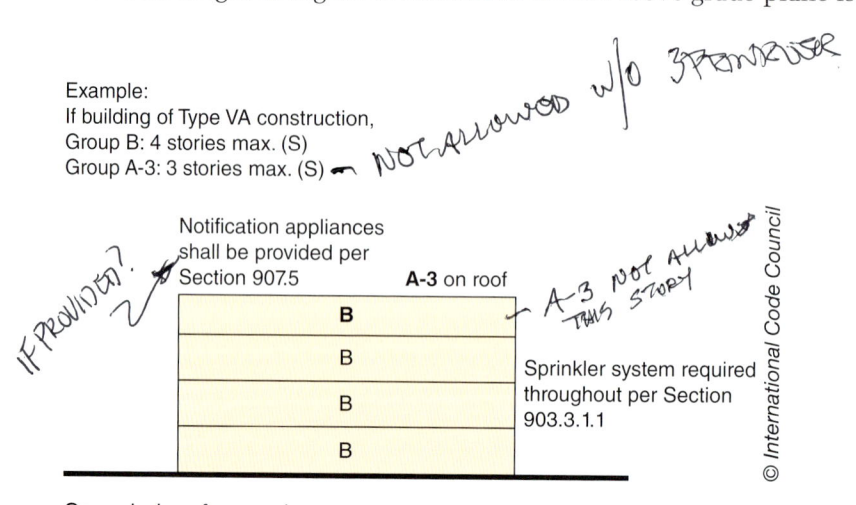

Example:
If building of Type VA construction,
Group B: 4 stories max. (S)
Group A-3: 3 stories max. (S)

Notification appliances shall be provided per Section 907.5   A-3 on roof

| B |
| B |
| B |
| B |

Sprinkler system required throughout per Section 903.3.1.1

Occupied roof example

determined based upon the building's type of construction and the occupancy classification of the uses involved. Where the roof of the building is occupiable, the code has previously been silent as to how this condition affects the allowable height determination. New criteria are now provided establishing the appropriate methodology in the regulation of building height in stories above grade plane where one or more occupancies is located on the roof.

Buildings are generally limited in the number of stories located above grade plane, based on the type of construction and occupancy, or occupancies, involved. The presence of one or more uses on the roof, often referred to as an "occupied roof," has caused differences in opinion as to how this would affect the building's allowable height in stories above grade plane. A story, by definition, is considered as that portion of a building between the upper surface of a floor and the upper surface of the floor or roof next above. Because a roof deck has no floor or roof above it, an occupied roof does not qualify as a story. However, the presence of occupants and fire loading on an occupied roof has always raised questions as to whether or not some degree of limitation should be provided. New provisions allow for an occupancy to be located on the roof provided the roof occupancy is permitted by Table 504.4 for the story directly below the roof.

The application of Exception 1 permits the placement of any occupancy classification, or classifications, on the roof provided two conditions are met:

1. The building is fully sprinklered in accordance with NFPA 13 or 13R as applicable, and
2. Under all conditions, occupant notification must be provided to the occupied portion of the roof.

Exception 2 addresses assembly occupancies located on the roofs of open parking garages. Where the garage is of noncombustible Type I or II construction, Group A assembly occupancies are permitted on the roof without applying the fire protection system conditions set forth in Exception 1.

In order to maintain the rooftop openness needed for the proper application of the new provisions, any elements that enclose the occupied roof area are limited to a maximum height of 4 feet above the roof's surface. Such limits are not applicable for those structures designed and constructed in compliance with the provisions of Section 1510 regarding penthouses and other rooftop structures.

In addition to the allowance that an occupied roof is not considered as a story for purposes of applying the IBC, it is also not considered as building area in the regulation of allowable floor area. Addressed in much the same manner as penthouses and other roof structures, the roof area is not a factor in determining the permissible size of the building.

505.2.1.1

Mezzanine and Equipment Platform Area Limitations

CHANGE TYPE: Clarification

CHANGE SUMMARY: Where both a mezzanine and an equipment platform are located in the same room, the general limitation for mezzanines cannot be exceeded when applying the two-thirds allowance.

2018 CODE: 505.2.1.1 Aggregate area of mezzanines and equipment platforms. Where a room contains both a mezzanine and an equipment platform, the aggregate area of the two raised floor levels shall be not greater than two-thirds of the floor area of that room or space in which they are located. The area of the mezzanine shall not exceed the area determined according to Section 505.2.1.

505.3.1 Area limitation. The aggregate area of all equipment platforms within a room shall not be greater than two-thirds of the area of the room in which they are located. Where an equipment platform is located in the same room as a mezzanine, the area of the mezzanine shall be determined by Section 505.2.1 and the combined aggregate area of the equipment platforms and mezzanines shall be not greater than two-thirds of the room in which they are located. The area of the mezzanine shall not exceed the area determined according to Section 505.2.1.

CHANGE SIGNIFICANCE: Where a floor level is relatively small compared to the floor level below, it may be possible to consider the upper floor level as a mezzanine rather than a story. A mezzanine is granted several significant allowances, including that it not be considered as contributing to allowable floor area or number of stories. As a general rule, the aggregate area of mezzanines cannot be larger than one-third the area of the room in which it is located. A greater allowance is available where the elevated areas are equipment platforms, up to two-thirds of the area of the room below. Provisions have been clarified where both a mezzanine and an equipment platform are located in the same room.

Historically, where a mezzanine and an equipment platform are located in the same room, their total floor area is permitted to be up to two-thirds the floor area of the room in which they are located. Where the

Example:
Assume both an equipment platform and a mezzanine are located in the same 24,000 sq. ft. room.

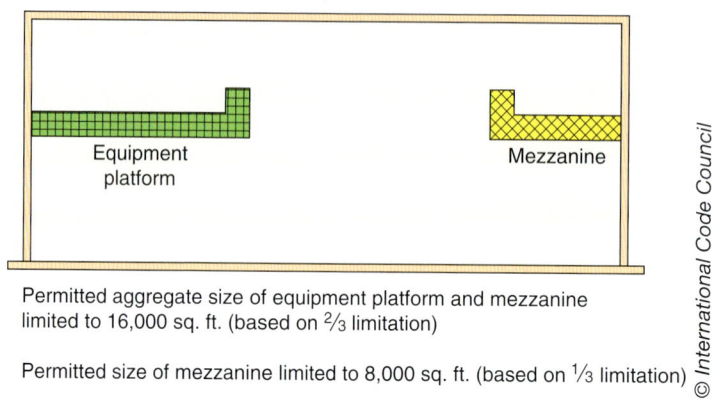

Permitted aggregate size of equipment platform and mezzanine limited to 16,000 sq. ft. (based on ⅔ limitation)

Permitted size of mezzanine limited to 8,000 sq. ft. (based on ⅓ limitation)

Mezzanine and equipment platform example

equipment platform is relatively small, the mezzanine could be much larger than permitted by the base requirement in the code (one-third the floor area) and still meet the two-thirds limitation. For example, the equipment platform could be 5% of the floor area of the room below, allowing the floor area of the mezzanine to be almost 62% of the area below. This potential result was not the intended application of the two-thirds allowance and the revised code text provides a clarification of the original intent. The reformatting and additional language now clearly indicates that the general limitation for mezzanines cannot be exceeded when applying the two-thirds allowance.

Table 506.2, Note i

Allowable Area of Type VB Greenhouses

CHANGE TYPE: Modification

CHANGE SUMMARY: The tabular allowable area for nonsprinklered single-story greenhouses classified as Group U occupancies has been substantially increased for Type VB buildings to be consistent with those greenhouses classified as Group B, M, F-2, and E.

2018 CODE:

TABLE 506.2 Allowable Area Factor

Occupancy Classification	See Footnotes	Type I		Type II		Type III		Type IV	Type V	
		A	B	A	B	A	B	HT	A	B
U	NS[i]	UL	35,500	19,000	8,500	14,000	8,500	18,000	9,000	5,500
	S1	UL	142,000	76,000	34,000	56,000	34,000	72,000	36,000	22,000
	SM	UL	106,500	57,000	25,500	42,000	25,500	54,000	27,000	16,500

Note:

i. The maximum allowable area for a single-story nonsprinklered Group U greenhouse is permitted to be 9,000 square feet, or the allowable area shall be permitted to comply with Table C102.1 of Appendix C.

(No changes to other portions of table and notes.)

CHANGE SIGNIFICANCE: Allowable building area has long been established as a fundamental code requirement in order to limit a building's size based upon its occupancy classification and type of construction. Table 506.2 recognizes the concept of equivalent risk, where the greater the fire hazard due to the building's occupancy results in a lesser permitted building area for a particular construction type. Group U occupancies, which include agricultural greenhouses, are significantly limited in allowable building area due to the lack of requirements for fire protection features. However, in some cases the area limitations are significantly more restrictive than similarly constructed greenhouses where the public is present. Therefore, the tabular allowable area for nonsprinklered single-story greenhouses has been substantially increased for Type VB buildings to be consistent with those greenhouses classified as Group B, M, F-2, and E.

The majority of commercial greenhouses are truly agricultural structures and classified as Group U. In most cases, the code requirements for human comfort, health, safety, and welfare are not applicable or necessary

Maximum allowable area increased to 9,000 sq. ft. (from 5,500 sq. ft.)

Type VB construction
Single-story
Nonsprinklered

Group U Greenhouse

Group U greenhouse allowable area example

for the construction or operation of such structures. Nearly all such greenhouses are built of Type VB nonsprinklered construction with a previous allowable area limit of 5,500 square feet prior to any frontage increase. The revised allowable area permitted for these types of structures has been increased to 9,000, providing consistency with other greenhouse occupancies. This increase reflects the equivalency of risk that is present in the various occupancy classifications that can be assigned to greenhouses. In addition, the new allowance provides for a very small increase in allowable area for greenhouses of construction types IIB and IIIB. The maximum allowable area for other types of structures classified as Group U remains unchanged.

Reference is further made to Table C102.1 in Appendix C which allows a maximum of 12,000 square feet in floor area for a single-story nonsprinklered Group U agricultural building. As a reminder, the provisions of any appendix chapter do not apply unless specifically adopted.

507.4 Sprinklers in Unlimited Area Group A-4 Buildings

CHANGE TYPE: Clarification

CHANGE SUMMARY: The sprinkler omission permitted for indoor participant sport areas of unlimited area Group A-4 buildings is now clearly not applicable to storage rooms, press boxes, concession areas, and other ancillary spaces.

2018 CODE: 507.4 Sprinklered, one-story buildings. The area of a Group A-4 building no more than one story above grade plane of other than Type V construction, or the area of a Group B, F, M or S building no more than one story above grade plane of any construction type, shall not be limited where the building is provided with an automatic sprinkler system throughout in accordance with Section 903.3.1.1 and is surrounded and adjoined by public ways or yards not less than 60 feet (18 288 m) in width.

Exceptions:

1. *(No change to first exception.)*

2. The automatic sprinkler system shall not be required in areas occupied for indoor participant sports, such as tennis, skating, swimming and equestrian activities in occupancies in Group A-4, provided that ~~both~~ all of the following criteria are met:

    2.1 Exit doors directly to the outside are provided for occupants of the participant sports areas.

    2.2 The building is equipped with a fire alarm system with manual fire alarm boxes installed in accordance with Section 907.

    2.3 <u>An automatic sprinkler system is provided in storage rooms, press boxes, concession booths or other spaces ancillary to the sport</u> activity space.

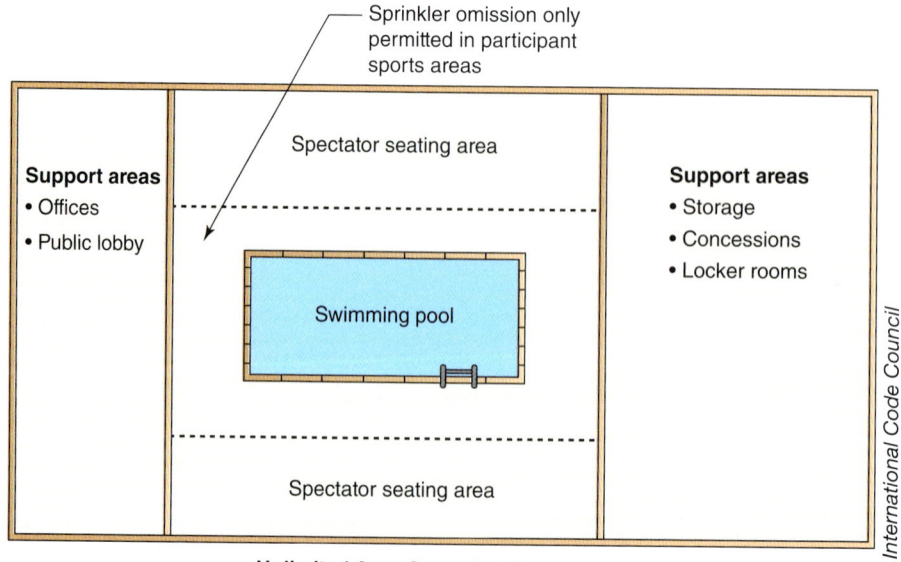

Sprinkler protection in unlimited area Group A-4 buildings example

CHANGE SIGNIFICANCE: Through the use of adequate safeguards such as sprinkler protection and significant building frontage, the IBC allows unlimited building areas for a variety of low- and moderate-hazard occupancies. One-story Group A-4 occupancies are among those buildings permitted to be unlimited in floor area provided adequate perimeter open space is provided, the type of construction is other than Type V and the building is provided with an automatic sprinkler system. The automatic sprinkler system need not be extended to participant sports areas, provided direct egress is available and the building has a fire alarm system. An additional provision now clarifies that the sprinkler omission is not applicable to storage rooms, press boxes, concession areas, and other ancillary spaces.

It is anticipated that sports areas such as tennis courts, skating rinks, and swimming pools will have little, if any, combustible loads if the uses are limited to those types described by the code. If there is a reasonable expectation that other types of uses could occur, it would be inappropriate to omit the sprinkler system in such areas. Many of these ancillary spaces are concealed and provide no awareness of a developing fire condition. Such spaces often have significant amounts of combustible contents. Therefore, sprinkler protection must be provided in these types of spaces, including concession stands and equipment storage rooms.

508.3.1.2

Group I-2, Condition 2 Nonseparated Occupancies

CHANGE TYPE: Modification

CHANGE SUMMARY: Additional limitations have now been established in mixed-occupancy buildings regulated under the nonseparated occupancy provisions where one of the occupancies involved is a Group I-2, Condition 2 hospital use.

2018 CODE: **508.3.1 Occupancy classification.** Nonseparated occupancies shall be individually classified in accordance with Section 302.1. The requirements of this code shall apply to each portion of the building based on the occupancy classification of that space. In addition, the most restrictive provisions of Chapter 9 that apply to the nonseparated occupancies shall apply to the total nonseparated occupancy area.

508.3.1.1 High-rise buildings. Where nonseparated occupancies occur in a high-rise building, the most restrictive requirements of Section 403 that apply to the nonseparated occupancies shall apply throughout the high-rise building.

508.3.1.2 Group I-2, Condition 2 occupancies. Where one of the nonseparated occupancies is Group I-2, Condition 2, the most restrictive requirements of Sections 407, 509 and 712 shall apply throughout the fire area containing the Group I-2 occupancy. The most restrictive requirements of Chapter 10 shall apply to the path of egress from the Group I-2, Condition 2 occupancy up to and including the exit discharge.

CHANGE SIGNIFICANCE: Where a building contains multiple occupancies, Section 508 requires that at least one of three established methodologies be applied to address the varied hazards. Nonseparated occupancies, one of the three available methods, allows for no physical or fire-resistance-rated separation between occupancies, provided the most restrictive fire protection and type of construction requirements apply to the total nonseparated occupancy area. Additional limitations have now been established where one of the nonseparated occupancies is a Group I-2, Condition 2 hospital use.

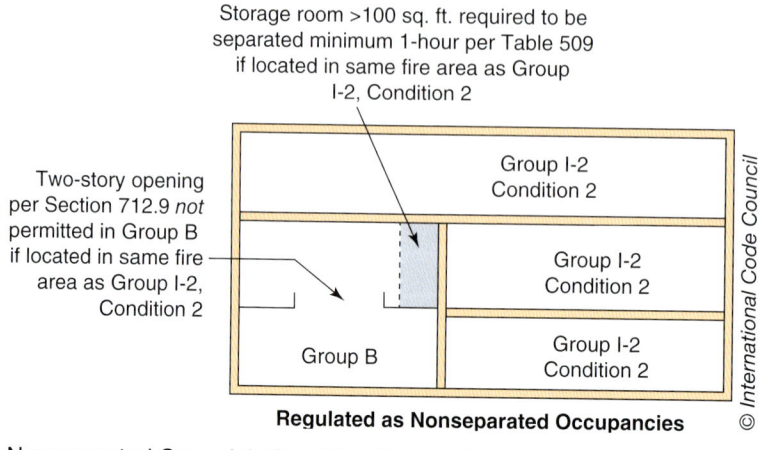

Nonseparated Group I-2, Condition 2 example

Where no fire separation is provided between a Group I-2, Condition 2 occupancy and other occupancies in the building due to the application of the nonseparated occupancy provisions, it is important that some critical fire protection features be extended beyond the healthcare portion of the building. Many of these restrictions directly support the defend-in-place concept that hospitals rely on. Within an individual fire area containing a Group I-2, Condition 2 occupancy, the more restrictive provisions of Sections 407, 509, and 712 now apply to all occupancies within the fire area. In addition, it has been clarified that the more restrictive means of egress provisions apply to the entire path of egress from the Group I-2, Condition 2 occupancy until arriving at the public way.

Section 407 contains provisions that are specific to Group I-2 occupancies that may not necessarily apply to the entire building. Within the fire area that contains the Group I-2, Condition 2 occupancy, provisions for corridor construction, smoke compartmentation, and hospital-specific egress must be maintained in order to support the defend-in-place concept. Separation and/or protection requirements for incidental uses that are specific to Group I-2 occupancies as established in Section 509 must also be provided where such uses occur within other occupancies in the same fire area as the hospital occupancy. The vertical opening limitations set forth in Section 712 for Group I-2 occupancies must also be applied to other occupancies in the same fire area, addressing the concern of unprotected vertical openings between adjacent stories. In all cases, the most restrictive applicable provisions of Sections 407, 509, and 712 are to be applied where one of the nonseparated occupancies is a Group I-2, Condition 2.

Means of egress concepts such as the minimum width appropriate for stretcher and bed traffic must be applied from the hospital area through the exit discharge. This mandate continues to be addressed in a general sense by Section 1004.4 indicating that "where two or more occupancies utilize portions of the same means of egress system, those egress components shall meet the more stringent requirements of all occupancies that are served."

It should be noted that while new in the IBC, these provisions are consistent with the requirements from the federal CMS for hospitals.

508.4.1, Table 508.4

Separated Occupancies vs. Fire Area Separations

CHANGE TYPE: Clarification

CHANGE SUMMARY: New provisions in Section 508.4.1 and Table 508.4 clarify that the fire separations used for mixed-occupancy purposes and those used for fire area purposes address different concerns, and as such the most restrictive fire-resistance-rated conditions shall apply.

2018 CODE: **508.4.1 Occupancy classification.** Separated occupancies shall be individually classified in accordance with Section 302.1. Each separated space shall comply with this code based on the occupancy classification of that portion of the building. <u>The most restrictive provisions of Chapter 9 that apply to the separate occupancies shall apply to the total nonfire-barrier-separated occupancy areas. Occupancy separations that serve to define fire area limits established in Chapter 9 for requiring a fire protection system shall also comply with Section 901.7.</u>

TABLE 508.4 Required Separation of Occupancies

Notes:

<u>f. Occupancy separations that serve to define fire area limits established in Chapter 9 for requiring fire protection systems shall also comply with Section 707.3.10 and Table 707.3.10 in accordance with Section 901.7.</u>

(No change to Table 508.4 and other notes to table.)

CHANGE SIGNIFICANCE: Where a building contains multiple occupancies, Section 508 requires that at least one of three established methodologies be applied to address the varied hazards. Separated occupancies, one of the three available methods, is based upon the similarities, or dissimilarities, of hazards posed by the occupancies being regulated. Where the hazards are deemed to be sufficiently dissimilar, some degree of fire-resistance-rated separation is required by Table 508.4. However, no fire-resistive separation is required by the table where the occupancies pose hazards that are somewhat similar.

Examples: Nonsprinklered mixed occupancy buildings regulated under separated occupancy provisions of Section 508.4

No sprinkler system

B	F-1
8,000 sq. ft.	10,000 sq. ft.

Minimum 3-hour fire barrier required

- Occupancy separation not required per separated occupancies and Table 508.4.
- Fire area separation of 3 hours required by Section 903.2.4 and Table 707.3.10.

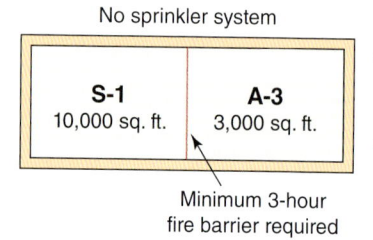

No sprinkler system

S-1	A-3
10,000 sq. ft.	3,000 sq. ft.

Minimum 3-hour fire barrier required

- Occupancy separation of 2 hours required per separated occupancies and Table 508.4.
- Fire area separation of 3 hours required by Sections 903.2.1.3 and 903.2.9 and Table 707.3.10.

Separated occupancies/fire area examples

Fire area separations, as regulated by Section 901.7, are selectively used to divide a building into limited-size compartments so as to not exceed the limits established in Section 903 for requiring an automatic sprinkler system. The fire area concept is based on a time-tested approach to limiting the spread of fire in a building. The degree of required fire separation, provided by fire barriers, horizontal assemblies, or both, is set forth in Section 707.3.10. New provisions in Section 508.4.1 and Table 508.4 clarify that the fire separations used for mixed-occupancy purposes and those used for fire area purposes address different concerns, and as such the most restrictive fire-resistance-rated conditions shall apply.

As an example, where using the separated occupancies method to address a mixed-occupancy building containing both Group F-1 and Group S-1 occupancies, no fire separation is mandated by Table 508.4 due to the similarity in hazards. However, if the fire area concept is applied to create conditions under which a sprinkler system is not required in the building, Table 707.3.10 would require a separation composed of minimum 3-hour fire barriers and/or horizontal assemblies. Therefore, the most restrictive condition, the minimum 3-hour separation, must be provided.

Table 509
Incidental Uses

CHANGE TYPE: Clarification

CHANGE SUMMARY: The current description in Table 509 regulating incidental uses in regard to rooms containing stationary storage battery systems has been revised to allow for ongoing consistency with the *International Fire Code* (IFC). In addition, a new entry dealing with rooms housing electrical installations and transformers references applicable provisions in *National Electrical Code* (NEC).

2018 CODE:

TABLE 509 Incidental Uses

Room or Area	Separation and/or Protection
Stationary storage battery systems having ~~a liquid electrolyte capacity of more than 50 gallons for flooded lead-acid, nickel cadmium or VRLA, or more than 1,000 pounds for lithium-ion and lithium metal polymer~~ <u>an energy capacity greater than the threshold quantity specified in Table 1206.2 of the *International Fire Code*</u>	1 hour in Group B, F, M, S and U occupancies; 2 hours in Group A, E, I and R occupancies
<u>Electrical installations and transformers</u>	<u>See Sections 110.26 through 110.34 and Sections 450.8 through 450.48 of NFPA 70 for protection and separation requirements</u>

(No changes to other portions of Table 509.)

CHANGE SIGNIFICANCE: Incidental uses, as defined and regulated by Table 509, are rooms and spaces that pose an increased hazard to other areas of the building, but that do not rise to the degree of a distinct and separate occupancy classification. The intent of the fire separation and/or fire protection requirements is to provide safeguards because of the increased hazard level presented by the incidental use. The current description in Table 509 regarding rooms containing stationary storage battery systems has been revised to allow for ongoing consistency with the *International Fire Code* (IFC). In addition, a new incidental use dealing with rooms housing electrical installations and transformers references applicable provisions in *National Electrical Code* (NEC).

Stationary battery systems used for facility standby, emergency or uninterruptable power, can pose significant hazards. During lead-acid battery electrolysis, oxygen and hydrogen gases can be released into the atmosphere forming flammable mixtures. Nickel cadmium and lithium ion batteries can have a thermal runaway resulting in high-temperature fires. Batteries can be a source of ignition or can explode. Batteries also may contain corrosive materials. Therefore, the IBC requires that rooms where such battery systems are located be regulated as incidental uses. The scoping provisions within the code have previously been specific in regard to the thresholds for applying the incidental use provisions. Table 509 now simply references IFC Table 1206.2 so that future revisions to the IFC dealing with the types of batteries and quantities of materials will be automatically addressed in the table.

A variety of construction requirements for electrical rooms are set forth in the *National Electrical Code* (NFPA 70). Their application could potentially be better known to architects and contractors if the provisions

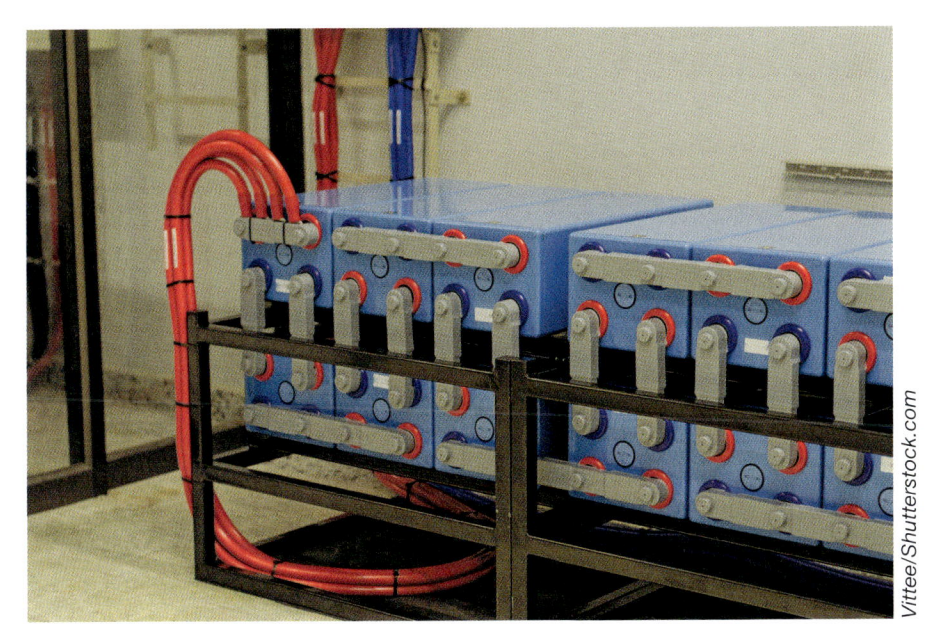

Storage battery system

were also located in the IBC. However, it is too cumbersome to try and address all of the combinations of protectives. In addition, modifications to the NEC requirements could result in a conflict with those found in the IBC. Therefore, Table 509 references specific sections in the NEC for the construction of electrical equipment rooms. This new reference accomplishes two major objectives: 1) make IBC users aware that important construction requirements can be found in another publication, and 2) make the requirements consistent with IBC language, format and references. Item 1 is accomplished by referencing the applicable sections of the NEC where construction requirements are located. Item 2 places the reference information in Table 509 addressing incidental uses where fire separations are required to be fire barriers with protected openings.

510.2
Horizontal Building Separation

CHANGE TYPE: Clarification

CHANGE SUMMARY: Vertical offsets are permitted in the horizontal fire-resistance-rated separation mandated for "podium buildings" provided the minimum required fire-resistance rating is maintained for the offsets and their supporting elements.

2018 CODE: **510.2 Horizontal building separation allowance.** A building shall be considered as separate and distinct buildings for the purpose of determining area limitations, continuity of fire walls, limitation on number of stories and type of construction where all of the following conditions are met:

1. The buildings are separated with a horizontal assembly have a fire-resistance rating of not less than 3 hours. Where vertical offsets are provided as part of a horizontal assembly, the vertical offset and the structure supporting the vertical offset shall have a fire-resistance rating of not less than 3 hours.
2. The building below, including the horizontal assembly, is of Type IA construction.

(No change to remaining listed conditions.)

CHANGE SIGNIFICANCE: The provisions of Section 510.2 are among the very few that recognize a horizontal fire-resistive separation as a means to create separate buildings when applying specified code requirements. The use of a minimum 3-hour fire-resistance-rated horizontal assembly is

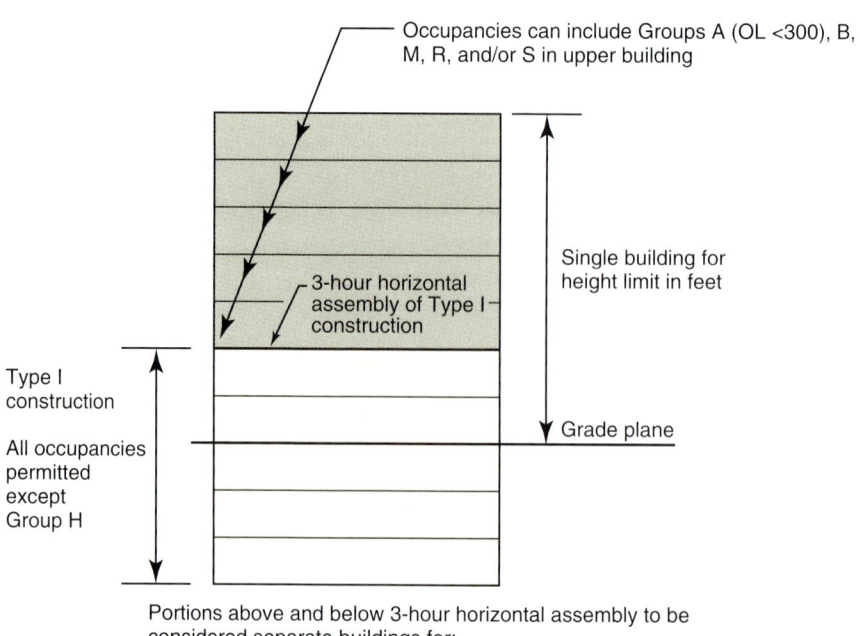

Horizontal building separation

viewed as an equivalent construction feature, in certain aspects, to a fire wall. This methodology is often referred to as "podium" or "pedestal" buildings. It has been clarified that vertical offsets are permitted in the horizontal separation, provided the minimum required fire-resistance rating is maintained for the offsets and their supporting elements.

Where the horizontal building separation allowance is applied for the purpose of determining area limitations, continuity of fire walls, limitation on number of stories and type of construction, six conditions must be met. The first condition requires that the separation be a minimum 3-hour horizontal assembly. It is not uncommon where the use of vertical offsets is beneficial due to the need to accommodate elevation changes for a particular site or different ceiling heights within a story. The IBC has not previously been specific as to how to deal with such offsets, resulting in multiple interpretations on what is intended. The additional code text now clearly allows vertical offsets, provided the fire-resistive integrity of the separation is maintained. By requiring the offsets and supporting construction to have a minimum fire-resistance rating equivalent to that required for the horizontal assembly, the necessary fire separation can be maintained.

Condition 2 addressing the minimum type of construction below the horizontal assembly has also been revised to clarify that the Type IA construction requirement is applicable to both the building below the horizontal assembly, as well as the horizontal assembly itself.

Table 601, Note b

Fire Protection of Structural Roof Members

CHANGE TYPE: Modification

CHANGE SUMMARY: All portions of the roof construction, including primary structural frame members such as girders and beams, are now selectively exempted from fire-resistance requirements based on Table 601 where every portion of the roof construction is at least 20 feet above any floor below.

2018 CODE:

TABLE 601 Fire-Resistance Rating Requirements for Building Elements

Building Element	Type I A	Type I B	Type II A	Type II B	Type III A	Type III B	Type IV HT	Type V A	Type V B
Primary structural frame[f]	3[a,b]	2[a,b]	1[b]	0	1[b]	0	HT	1[b]	0
Roof construction and associated secondary members	1½[b]	1[b,c]	1[b,c]	0[c]	1[b,c]	0	HT	1[b,c]	0

b. Except in Group F-1, H, M and S-1 occupancies, fire protection of structural members <u>in roof construction</u> shall not be required, including protection of <u>primary structural frame members,</u> roof framing and decking where every part of the roof construction is <u>20 feet</u> or more above any floor immediately below. Fire-retardant-treated wood members shall be allowed to be used for such unprotected members.

(No changes to other portions of Table 601 and notes.)

CHANGE SIGNIFICANCE: The provisions of Chapter 6 in regard to fire resistance are intended to address the structural integrity of the building elements under fire conditions. As a building increases in floor area, height and/or fire hazard, the fire-resistive protection of building elements is often required. The basic fire-resistance ratings for the various

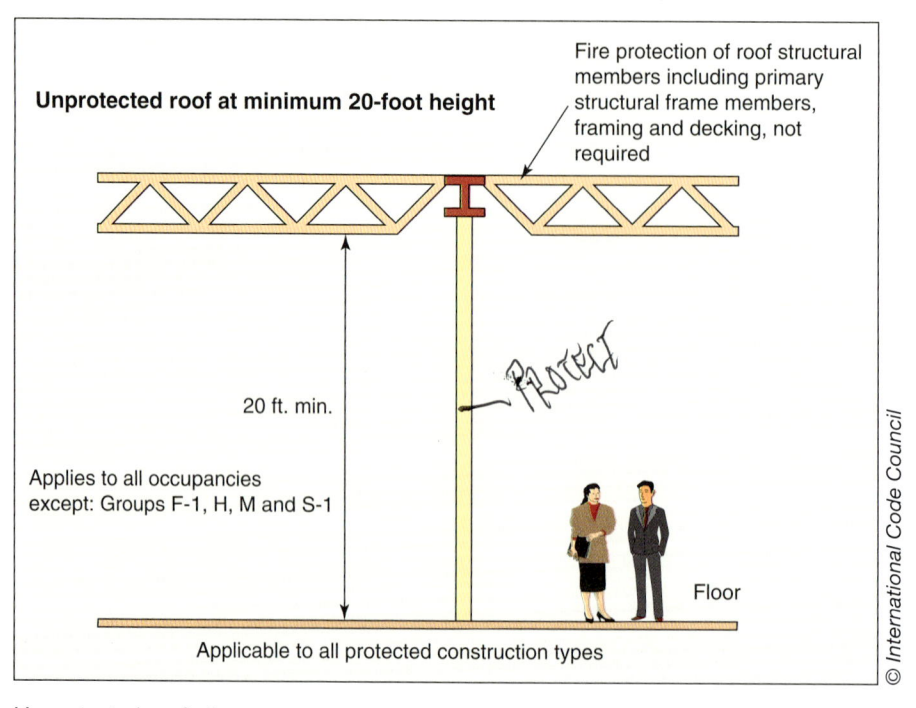

Unprotected roof allowance

types of construction are established in Table 601. Footnote b has historically modified the base requirements in the table, as they relate to the roof construction, by selectively eliminating the requirement for protecting roof structural members where the roof construction is at least 20 feet above the floor below. The reduction, applicable to all occupancies other than Groups F-1, H, M, and S-1, recognizes that the temperatures at this elevation during most fire incidents are quite low. Because footnote b was only applicable to the building element "roof construction and associated secondary members," and was not referenced in the requirements for "primary structural frame," its use was often not applied to roof girders, beams and similar primary structural members. By expanding the scope of the footnote to primary structural frame elements, as well as specifically mentioning in the footnote its application to primary structural frame members, it is very clear that all portions of the roof construction are exempt from fire-resistance requirements based on Table 601.

Table 602, Note i

Group R-3 Fire Separation Distance

CHANGE TYPE: Clarification

CHANGE SUMMARY: Where the building under consideration is of Type IIB or Type VB construction and houses a Group R-3 occupancy, it has been clarified that no fire-resistance rating is required for exterior walls due to location on the lot where the fire separation distance is a minimum of 5 feet.

2018 CODE:

TABLE 602 Fire-Resistance Rating Requirements for Exterior Walls Based on Fire Separation Distance

Fire Separation Distance	Type of Construction	Occupancy Group H	Occupancy Group F-1, M, S-1	Occupancy Group A, B, E, F-2, I, R^i, S-2, U
X < 5	All	3	2	1
5 ≤ X < 10	IA	3	2	1
	Others	2	1	1
10 ≤ X < 30	IA, IB	2	1	1
	IIB, VB	1	0	0
	Others	1	1	1
X ≥ 30	All	0	0	0

i. For a Group R-3 building of Type IIB or Type VB construction, the exterior wall shall not be required to have a fire-resistance rating where the fire separation distance is 5 feet (1523 mm) or greater.

(No changes to other portions of Table 602 and notes.)

CHANGE SIGNIFICANCE: Because an owner typically has no control over what occurs on an adjacent lot, the location of buildings on the owner's lot must be regulated relative to the lot line. The lot line concept provides a convenient means of protecting one building from another building due to radiant heat exposure. The regulations for exterior wall protection based on proximity to the lot line (fire separation distance) are contained in Table 602. In order to properly utilize Table 602, it is also necessary to identify the occupancies involved and the building's type of construction. As the fire separation distance increases, the fire-resistance rating requirements are reduced. Where the building under consideration is of Type IIB or Type VB construction and houses a Group R-3 occupancy, it has been clarified that no fire-resistance rating is required for exterior walls due to location on the lot where the fire separation distance is a minimum of 5 feet.

Although this provision has been similarly applied in past editions of the IBC, it was often overlooked due to the methodology involved in reaching the 5-foot fire separation distance allowance. The base requirement of Table 602 would require a minimum 10-foot fire separation distance for all Group R occupancies in Type IIB and VB buildings. However, footnote g indicates that no exterior wall fire-resistance rating is required where Table 705.8 permits nonbearing exterior walls with an

Table 602, Note i ■ Group R-3 Fire Separation Distance

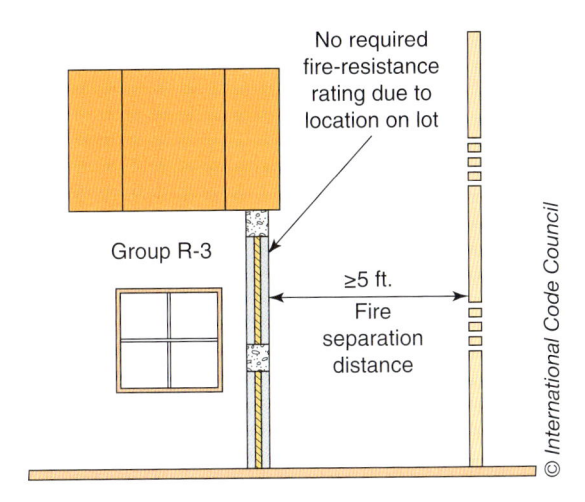

Group R-3 exterior wall condition

- No required fire-resistance rating due to location on lot
- Group R-3
- ≥5 ft. Fire separation distance

unlimited amount of unprotected openings. In referring to Table 705.8, footnote f states that the area of unprotected and protected openings is not limited for Group R-3 occupancies, provided the fire separation distance is 5 feet or greater. The new footnote provided in Table 602 now eliminates the difficulty in applying the provision by directly addressing the allowance.

602.3, 602.4.1

FRT Wood Sheathing in Exterior Wall Assemblies

CHANGE TYPE: Clarification

CHANGE SUMMARY: It has now been clarified that fire-retardant-treated wood sheathing, as well as wood framing, is permitted within exterior walls of Type III and IV buildings where the wall assembly does not exceed a 2-hour rating.

2018 CODE: 602.3 Type III. Type III construction is that type of construction in which the exterior walls are of noncombustible materials and the interior building elements are of any material permitted by this code. Fire-retardant-treated wood framing <u>and sheathing</u> complying with Section 2303.2 shall be permitted within exterior wall assemblies of a 2-hour rating or less.

602.4.1 Fire-retardant-treated wood in exterior walls. Fire-retardant-treated wood framing <u>and sheathing</u> complying with Section 2303.2 shall be permitted within exterior wall assemblies <u>not less than 6 inches (152 mm) in thickness</u> with a 2-hour rating or less.

CHANGE SIGNIFICANCE: The classification of buildings as Type III or IV construction was established at the turn of the 20th century to address conflagrations in heavily built-up areas where structures were erected side by side in congested downtown business districts. In order to control the spread of fire from building to building, the Type III and IV classifications require fire-resistance-rated noncombustible exterior walls. Although the fundamental limitation on exterior walls of Type III and IV buildings has been the requirement for noncombustible construction, the IBC has historically permitted the use of fire-retardant-treated wood framing where the wall assembly has a maximum fire-resistance rating

FRT wood sheathed building

of two hours. However, there has been some question as to whether or not fire-retardant-treated wood sheathing would also be permitted as a part of exterior wall construction. It has now been clarified that both framing and sheathing of fire-retardant-treated wood are permitted in exterior walls of Type III and IV buildings where the wall assembly does not exceed a 2-hour rating. In addition, one of several format changes to Section 602.4 addresses the minimum required wall assembly thickness applicable to the provision.

PART 3

Fire Protection

Chapters 7 through 9

- Chapter 7 Fire and Smoke Protection Features
- Chapter 8 Interior Finishes
- Chapter 9 Fire Protection and Life Safety Systems

The fire protection provisions of the *International Building Code* (IBC) are found primarily in Chapters 7 through 9. There are two general categories of fire protection: active and passive. The fire and smoke resistance of building elements and systems in compliance with Chapter 7 provides for passive protection. Chapter 9 contains requirements for various active systems often utilized in the creation of a safe building environment, including automatic sprinkler systems, standpipe systems and fire alarm systems. To further address the rapid spread of fire, the provisions of Chapter 8 are intended to regulate interior-finish materials, such as wall and floor coverings.

704.2, 704.4.1
Column Protection in Light-Frame Construction

TABLE 705.2
Extent of Projections

705.2.3, 705.2.3.1
Combustible Balconies, Projections, and Bay Windows

705.8.1
Measurement of Fire Separation Distance for Opening Protection

706.1.1
Party Walls Not Constructed as Fire Walls

706.2
Structural Continuity of Double Fire Walls

708.4
Continuity of Fire Partitions

708.4.2
Fireblocking and Draftstopping at Fire Partitions

713.8.1
Membrane Penetrations of Shaft Enclosures

716.2.6.5
Delayed-Action Self-Closing Doors

803.1.1, 803.1.2
Interior Wall and Ceiling Finish Testing

803.3
Interior Finish Requirements for Heavy Timber Construction

803.11, 803.12
Flame Spread Testing of Laminates and Veneers

901.6.2
Integrated Fire Protection System Testing

902.8
Fire Pump and Fire Sprinkler Riser Rooms

903.2.1
Sprinklers Required in Group A Occupancies

903.2.3
Sprinklers in Group E Occupancies

903.3.1.1.2
Omission of Sprinklers in Group R-4 Bathrooms

903.3.1.2.1
Sprinkler Protection at Balconies and Decks

903.3.1.2.3
Protection of Attics in Group R Occupancies

904.12
Commercial Cooking Operations

904.13
Domestic Cooking Protection in Institutional and Residential Occupancies

904.14
Aerosol Fire Extinguishing Systems

905.3.1
Class III Standpipes

905.4
Class I Standpipe Connection Locations

907.2.1
Fire Alarms in Group A Occupancies

907.2.10
Group R-4 Fire Alarm Systems

704.2, 704.4.1

Column Protection in Light-Frame Construction

CHANGE TYPE: Modification

CHANGE SUMMARY: In walls of light-frame construction where primary structural frame members require fire-resistive protection, columns extending only between the bottom and top plates do not need to be provided with individual encasement protection.

2018 CODE: 704.2 Column protection. Where columns are required to have protection to achieve a fire-resistance rating, the entire column shall be provided individual encasement protection by protecting it on all sides for the full column height, including connections to other structural members, with materials having the required fire-resistance rating. Where the column extends through a ceiling, the encasement protection shall be continuous from the top of the foundation or floor/ceiling assembly below through the ceiling space to the top of the column.

> **Exception:** Columns that meet the limitations of Section 704.4.1.

704.4.1 Light-frame construction. Studs, columns and boundary elements that are integral elements in load-bearing walls of light-frame construction and are located entirely between the top and bottom plates or tracks shall be permitted to have required fire-resistance ratings provided by the membrane protection provided for the load-bearing wall.

CHANGE SIGNIFICANCE: Primary structural frame members, including columns, require fire-resistive protection in buildings of Type I, IIA, IIIA, and VA construction. As a general requirement, Section 704.2 requires columns to be provided with individual encasement protection where such a fire-resistance rating is required. All sides must typically be protected for the full height of the column. In walls of light-frame construction where primary structural frame members require fire-resistive protection, it is now recognized that columns extending only between the bottom and top plates do not need to be provided with individual encasement protection.

Built-up element between the bottom plate and the top plates

© International Code Council

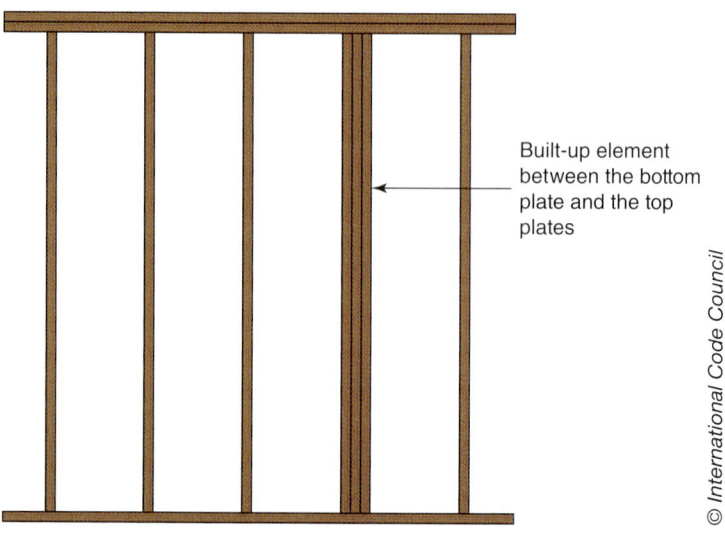

Built-up element of multiple studs

803.3
Interior Finish Requirements for Heavy Timber Construction

803.11, 803.12
Flame Spread Testing of Laminates and Veneers

901.6.2
Integrated Fire Protection System Testing

902.8
Fire Pump and Fire Sprinkler Riser Rooms

903.2.1
Sprinklers Required in Group A Occupancies

903.2.3
Sprinklers in Group E Occupancies

903.3.1.1.2
Omission of Sprinklers in Group R-4 Bathrooms

903.3.1.2.1
Sprinkler Protection at Balconies and Decks

903.3.1.2.3
Protection of Attics in Group R Occupancies

904.12
Commercial Cooking Operations

904.13
Domestic Cooking Protection in Institutional and Residential Occupancies

904.14
Aerosol Fire Extinguishing Systems

905.3.1
Class III Standpipes

905.4
Class I Standpipe Connection Locations

907.2.1
Fire Alarms in Group A Occupancies

907.2.10
Group R-4 Fire Alarm Systems

704.2, 704.4.1

Column Protection in Light-Frame Construction

CHANGE TYPE: Modification

CHANGE SUMMARY: In walls of light-frame construction where primary structural frame members require fire-resistive protection, columns extending only between the bottom and top plates do not need to be provided with individual encasement protection.

2018 CODE: 704.2 Column protection. Where columns are required to have protection to achieve a fire-resistance rating, the entire column shall be provided individual encasement protection by protecting it on all sides for the full column height, including connections to other structural members, with materials having the required fire-resistance rating. Where the column extends through a ceiling, the encasement protection shall be continuous from the top of the foundation or floor/ceiling assembly below through the ceiling space to the top of the column.

> **Exception:** Columns that meet the limitations of Section 704.4.1.

704.4.1 Light-frame construction. Studs, <u>columns</u> and boundary elements that are integral elements in ~~load-bearing~~ walls of light-frame construction <u>and are located entirely between the top and bottom plates or tracks</u> shall be permitted to have required fire-resistance ratings provided by the membrane protection provided for the ~~load-bearing~~ wall.

CHANGE SIGNIFICANCE: Primary structural frame members, including columns, require fire-resistive protection in buildings of Type I, IIA, IIIA, and VA construction. As a general requirement, Section 704.2 requires columns to be provided with individual encasement protection where such a fire-resistance rating is required. All sides must typically be protected for the full height of the column. In walls of light-frame construction where primary structural frame members require fire-resistive protection, it is now recognized that columns extending only between the bottom and top plates do not need to be provided with individual encasement protection.

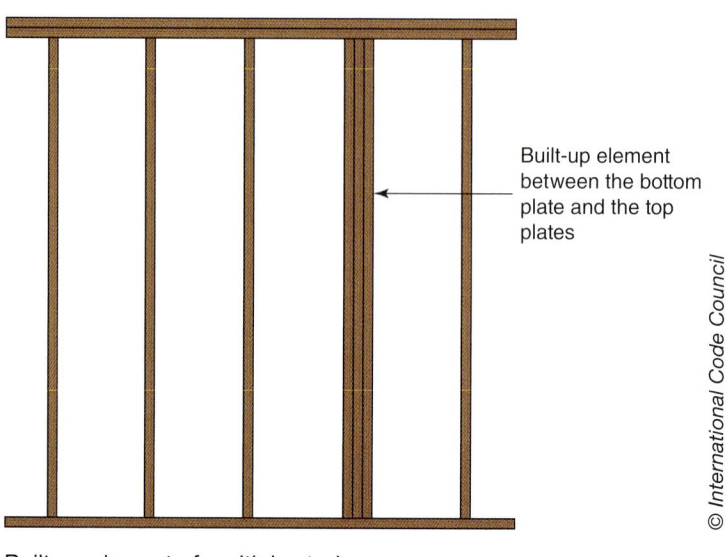

Built-up element of multiple studs

Reference is now made to Section 704.4.1 addressing required fire-resistance ratings in light-frame construction. In addition to studs and boundary elements, columns may now have the fire-resistance rating provided by membrane protection of the wall itself. It is mandated that the columns or studs must be entirely located between the top and bottom plates or tracks.

Many buildings are constructed of typical light-frame construction methods and the concentrated loads from trusses or beams must have a continuous load path to the foundation. It has often been interpreted that vertical construction elements such as built-up and solid structural elements are to be regulated as columns for fire-resistive purposes, and as such must be provided with individual encasement in order to provide the required fire protection.

It has been clarified that this allowance for membrane protection does not address continuous columns. However, built-up structural elements, such as multiple studs, located within fire-resistance-rated walls of light-frame construction can be a part of the fire-resistance-rated wall assembly without requiring individual encasement protection. The new reference to top and bottom tracks recognizes that the allowance also applies to light-frame steel framing systems.

Table 705.2
Extent of Projections

CHANGE TYPE: Modification

CHANGE SUMMARY: The minimum required clearance between the edge of a projection and the line used to determine the fire separation distance has been significantly decreased.

2018 CODE:

TABLE 705.2 Minimum Distance of Projection

Fire Separation Distance – FSD (FSD) (feet)	Minimum Distance from Line Used to Determine FSD
0 feet to less than 2 feet	Projections not permitted
Greater than 2 feet to less than 3 feet	24 inches
Greater than 3 feet to less than 30 5 feet	24 inches plus 8 inches for every foot of FSD beyond 3 feet or fraction thereof
30 feet 5 or greater	20 feet 40 inches

For SI: 1 foot = 304.8 mm; 1 inch = 25.4 mm.

CHANGE SIGNIFICANCE: Where architectural projections such as eave overhangs and balconies extend from walls in close proximity to a lot line, they create problems that are due to trapping the convected heat from a fire in an adjacent building. As this trapped heat increases the hazard for the building under consideration, the code mandates a minimum distance the leading edge of the projecting element must be separated from the line used to determine fire separation distance. The permitted extent of projections is established by Table 705.2 and based solely on the clear distance between the building's exterior wall and an interior lot line, centerline of a public way or assumed imaginary line between two buildings on the same lot. The minimum required clearance set forth in Table 705.2 between the edge of a projection and the line used to determine the fire separation distance has been greatly decreased from the clearance required by the 2015 IBC.

Projections are allowed to extend beyond the exterior wall, but only for a limited distance. The required clearance changes based on the fire separation distance measured from the exterior wall. The modification

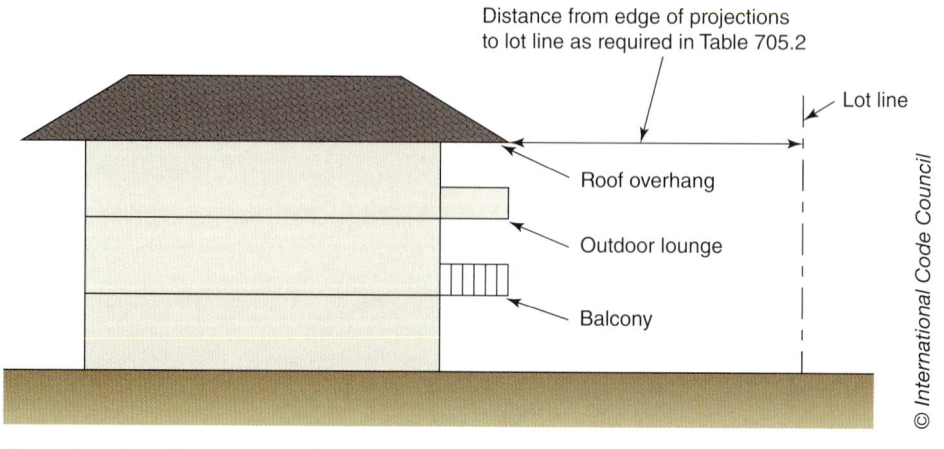

Fire separation distance at projections

occurs where the fire separation distance is between 5 feet and 30 feet, with the change becoming more significant as the distance approaches 30 feet. Where an exterior wall of a building has a fire separation distance of 30 feet, the 2015 IBC requires a minimum clearance fire separation distance of 20 feet measured from the edge of the projection. For that same condition, the 2018 IBC will only require a clearance of 40 inches between the projection's leading edge and the line used to determine the fire separation distance.

Provisions established in the 2015 edition of the IBC were intended to simplify the projection distance provisions by formatting the requirements in a table. The 2015 change attempted to address an identified anomaly within the table. However, that change created a much more restrictive requirement than what was in the 2012 IBC and earlier editions. It was determined that there was no technical justification for this more restrictive requirement. The maximum required separation of 40 inches has been reestablished and the table has been slightly reformatted in a manner that more consistently identifies the distance at which the provisions are to be applied.

705.2.3, 705.2.3.1

Combustible Balconies, Projections, and Bay Windows

CHANGE TYPE: Clarification

CHANGE SUMMARY: Construction requirements for balconies, porches, decks, bay windows, and oriel windows have been relocated from Section 1406 (Combustible Materials on the Exterior Side of Exterior Walls) to Section 705.2.3 (Combustible Projections).

2018 CODE: 705.2.3 Combustible projections. Combustible projections extending to within 5 feet (1524 mm) of the line used to determine the fire separation distance shall be of not less than 1-hour fire-resistance-rated construction, ~~Type IV~~ heavy timber construction complying with Section 2304.11, fire-retardant-treated wood or as ~~required~~ permitted by Section ~~1406.3~~ 705.2.3.1.

> **Exception:** Type VB construction shall be allowed for combustible projections in Group R-3 and U occupancies with a fire separation distance greater than or equal to 5 feet (1524 mm).

~~1406.3~~ **705.2.3.1 Balconies and similar projections.** Balconies and similar projections of combustible construction other than fire-retardant-treated wood shall be fire-resistance rated where required by Table 601 for floor construction or shall be of heavy timber construction in accordance with Section ~~602.4~~ 2304.11. The aggregate length of the projections shall not exceed 50 percent of the building's perimeter on each floor.

> **Exceptions:**
> 1. On buildings of Types I and II construction, three stories or less above grade plane, fire-retardant-treated wood shall be permitted for balconies, porches, decks and exterior stairways not used as required exits.
> 2. Untreated wood and plastic composites that comply with ASTM D 7032 and Section 2612 are permitted for pickets, rails and similar guard~~rail devices~~ components that are limited to 42 inches (1067 mm) in height.
> 3. Balconies and similar projections on buildings of Types III, IV and V construction shall be permitted to be of Type V construction, and shall not be required to have a fire-resistance rating where sprinkler protection is extended to these areas.
> 4. Where sprinkler protection is extended to the balcony areas, the aggregate length of the balcony on each floor shall not be limited.

~~1406.4~~ **705.2.4 Bay and oriel windows.** Bay and oriel windows constructed of combustible materials shall conform to the type of construction required for the building to which they are attached.

> **Exception:** Fire-retardant-treated wood shall be permitted on buildings three stories or less above grade plane of Type I, II, III or IV construction.

Combustible balcony construction

CHANGE SIGNIFICANCE: Exterior walls of buildings of Types I, II, III and IV are typically required to be of noncombustible construction. However, it is common for some limited combustible elements to be installed on the exterior side of such exterior walls. Section 1406 covers the use of combustible materials on the exterior side of exterior walls. More specifically, Section 1406.3 has historically recognized specific allowances addressing the use of combustible balconies and similar projections, while Section 1406.4 has dealt with bay and oriel windows. Because Section 705.2.3 addresses type of construction and fire-resistive rating issues as they relate to projections, the provisions from both Sections 1406.3 and 1406.4 have been relocated.

An additional change occurred in Exception 2 of Section 705.2.3.1 regarding those materials permitted for use as guard components. Plastic composites which comply with ASTM D 7032 *Standard Specification for Establishing Performance Ratings for Wood, Plastic Composite Deck Boards and Guardrail Systems (Guards or Rails)* and Section 2612.3 are permitted to be installed in those same locations where untreated wood could be used for balcony and projection construction. Their use is limited in application to pickets, rails, and guard components with a height of 42 inches or less.

As a note, the plastic composites, unless determined to be noncombustible, are required to be tested to ASTM E 84 or UL 723 and achieve a flame spread index of not more than 200.

705.8.1
Measurement of Fire Separation Distance for Opening Protection

CHANGE TYPE: Clarification

CHANGE SUMMARY: The allowable area of openings in fire-resistance-rated exterior walls is to be based on the fire separation distance for each story, determined individually, in the same manner as applied in the determination of the required wall rating.

2018 CODE: 705.8.1 Allowable area of openings. The maximum area of unprotected and protected openings permitted in an exterior wall in any story of a building shall not exceed the percentages specified in Table 705.8 based on the fire separation distance of each individual story.

Exceptions:

1. In other than Group H occupancies, unlimited unprotected openings are permitted in the first story above grade plane where the wall ~~either~~ faces one of the following:
   1.1. ~~Where the wall faces a~~ A street and has a fire separation distance of more than 15 feet (4572 mm); or
   1.2. ~~Where the wall faces an~~ An unoccupied space. The unoccupied space shall be on the same lot or dedicated for public use, shall be not less than 30 feet (9144 mm) in width and shall have access from a street by a posted fire lane in accordance with the *International Fire Code*.
2. Buildings whose exterior bearing walls, exterior nonbearing walls and exterior primary structural frame are not required to be fire-resistance rated shall be permitted to have unlimited unprotected openings.

Fire Separation Distance (feet)	Allowable Area (Percentage of the area of the exterior wall, per story)		
	Protected Openings	Unprotected Openings Sprinklered Building	Unprotected Openings Nonsprinklered Building
9	25%	25%	10%
14	45%	45%	15%
15	75%	75%	25%
19	75%	75%	25%
20	Unlimited	Unlimited	45%
25	Unlimited	Unlimited	70%

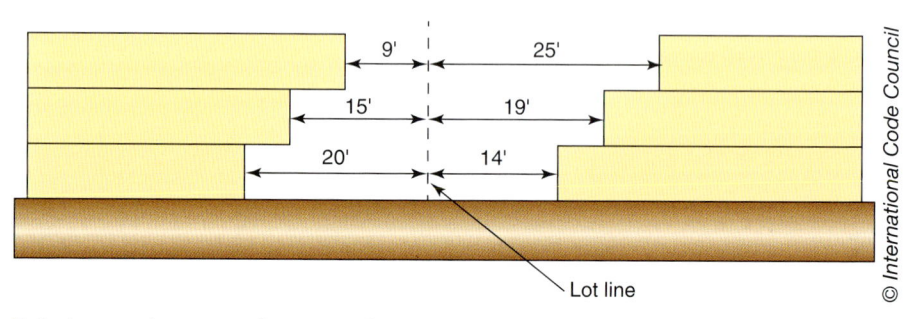

Exterior opening protection example

CHANGE SIGNIFICANCE: Openings in an exterior wall typically consist of windows and doors. The maximum permissible area of protected or unprotected openings permitted in each story of a fire-resistance-rated exterior wall is established in Table 705.8 based upon fire separation distance. It has been clarified that the limitations on openings are to be regulated on a story-by-story basis, in a manner similar to that applied for the exterior walls themselves.

This revision clarifies that the limitation of openings in exterior walls is based on the fire separation distance of each individual story. This approach is consistent with the method exterior walls have historically been evaluated for fire resistance based on fire separation distance. Footnote d of Table 602 has historically stated that "the fire-resistance rating of an exterior wall is determined based upon the fire separation distance of the exterior wall and the story in which the wall is located." Openings in exterior walls should be regulated in the same manner.

706.1.1

Party Walls Not Constructed as Fire Walls

CHANGE TYPE: Modification

CHANGE SUMMARY: Construction as a fire wall is no longer required for a party wall provided the aggregate height and area of the buildings on each side of the party wall are compliant with Chapter 5 and applicable easements and agreements are established addressing the maintenance of all fire and life safety systems of both buildings.

2018 CODE: 706.1.1 Party walls. Any wall located on a lot line between adjacent buildings, which is used or adapted for joint service between the two buildings, shall be constructed as a fire wall in accordance with Section 706. Party walls shall be constructed without openings and shall create separate buildings.

Exceptions:

1. Openings in a party wall separating an anchor building and a mall shall be in accordance with Section 402.4.2.2.1.

2. <u>Fire walls are not required on lot lines dividing a building for ownership purposes where the aggregate height and area of the portions of the building located on both sides of the lot line do not exceed the maximum height and area requirements of this code. For the building official's review and approval, he or she shall be provided with copies of dedicated access easements and contractual agreements that permit the owners of the portion of the building located on either side of the lot line access to the other side for purposes of maintaining fire and life safety systems necessary for the operation of the building.</u>

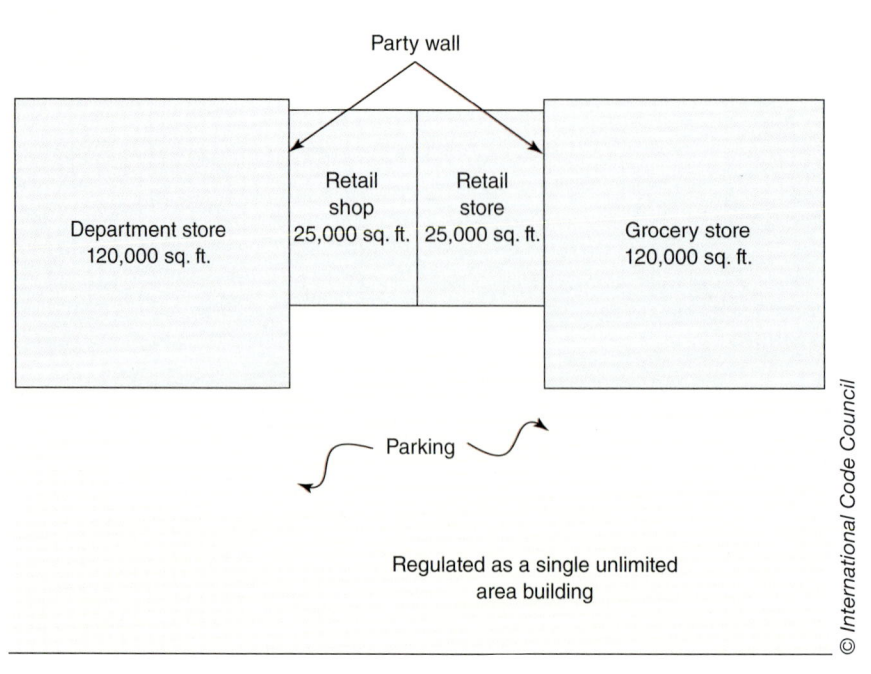

Example:

```
                            Party wall
                              ↙   ↘
  ┌──────────────┐     ┌──────┬──────┐     ┌──────────────┐
  │              │     │Retail│Retail│     │              │
  │ Department   │     │ shop │store │     │  Grocery store│
  │   store      │     │25,000│25,000│     │  120,000 sq. ft.│
  │ 120,000 sq.ft│     │sq.ft.│sq.ft.│     │              │
  │              │     │      │      │     │              │
  └──────────────┘     └──────┴──────┘     └──────────────┘

                        ↰  Parking  ↱

                  Regulated as a single unlimited
                          area building
```

Use of party walls

706.1.1 ■ Party Walls Not Constructed as Fire Walls

CHANGE SIGNIFICANCE: Where two separate structures are each built at a lot line, the opposing exterior walls of each structure are to be regulated individually based upon zero fire separation distance. This will result in each wall having a fire-resistance rating with no openings permitted. As an option, the code recognizes the presence of a joint-use party wall constructed without openings. The party wall is typically required to be regulated as a fire wall. Construction as a fire wall is no longer required for the common wall at the lot line provided the aggregate height and area of the opposing buildings are compliant with Chapter 5 and applicable easements and agreements are established addressing the maintenance of all fire and life safety systems of both buildings.

The new allowance recognizes the increasing frequency of property being subdivided with a lot line for ownership purposes. This concept has historically been acceptable for covered mall buildings, where anchor stores have lot lines specific to the anchor store established for financial purposes along the wall that separates the mall from the anchor store. However, this condition has not previously been addressed for other types of buildings and as a result, designers, building owners, and building officials have been left to deal with the issue on a case-by-case basis outside of the code.

The new exception specifies that where a party wall divides a building for ownership purposes, and the aggregate building height and area complies with the code, then the party wall does not need to be constructed as a fire wall. In other words, if the party wall did not exist, the entire building would comply with the allowable height in feet (Section 504.3), the allowable number of stories (Section 504.4), and the allowable area (Section 506.2).

In order to approve this approach, documents for easements and contracts between the various property owners shall be provided to the building official. These documents would be considered part of the construction documents. It is important that in dealing with fire and life safety issues the two buildings are considered as one, thus requiring access between buildings for the maintenance of all fire and life safety systems.

706.2 Structural Continuity of Double Fire Walls

CHANGE TYPE: Modification

CHANGE SUMMARY: In Seismic Design Categories D through F, floor and roof sheathing is permitted to continue through light-frame double fire wall assemblies where the sheathing does not exceed a thickness of ¾ inch.

2018 CODE: 706.2 Structural stability. Fire walls shall be designed and constructed to allow collapse of the structure on either side without collapse of the wall under fire conditions. Fire walls designed and constructed in accordance with NFPA 221 shall be deemed to comply with this section.

> **Exception:** In Seismic Design Categories D through F, where double fire walls are used in accordance with NFPA 221, floor and roof sheathing not exceeding ¾ inch (19.05 mm) thickness shall be permitted to be continuous through the wall assemblies of light frame construction.

CHANGE SIGNIFICANCE: A key element of a fire wall is its ability to allow, under fire conditions, the collapse of the structure on either side without collapse of the wall. Double fire walls complying with NFPA 221 are considered as being compliant with this requirement. The wall assemblies in a double fire wall design are intentionally separated. In fact, Section 6.6.5 of NFPA 221 limits connections between the two fire wall assemblies to only flashing at the top of the walls for weather protection purposes. This separation allows for significant movement or collapse of the roof or floor assembly on one side to cause failure of the associated fire wall while the other fire wall remains intact.

In light-frame construction, the sheathing which comprises floor or roof diaphragms is generally wood structural panels with a thickness between 7/16 inches and 23/32 inches. These panels are now permitted

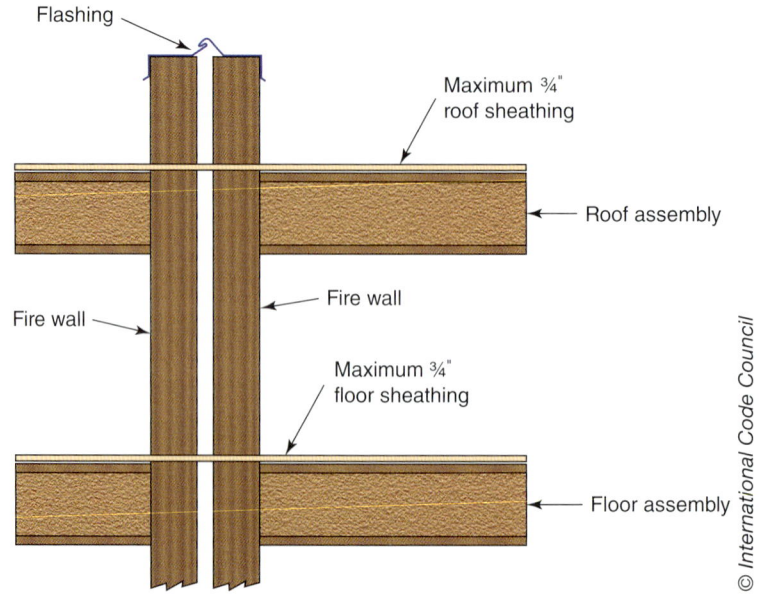

Sheathing continuity at double fire wall

to traverse the span between the fire wall assemblies to provide a continuous diaphragm for the floor assembly or the roof assembly. The ability to take advantage of a continuous diaphragm is a great benefit in the higher Seismic Design Categories of D, E, and F. The added benefit of performing the seismic function as a diaphragm is regarded as well worth the small risk of fire exposure from one side of a double fire wall.

Additionally, in these higher seismic design categories, the possibility exists for these two separate buildings to impact during a seismic event. Based on NFPA 221 Table A.5.7, the minimum separation between the double fire walls could be as small as 2½ inches. The continuous diaphragm can stabilize the two buildings and allow them to move as one unit, rather than two independent structures. This provides for greater safety during the seismic event in these more demanding seismic design categories. Note that in Seismic Design Categories A, B, and C, a complete separation between the two fire walls is still required.

708.4 Continuity of Fire Partitions

CHANGE TYPE: Clarification

CHANGE SUMMARY: The continuity requirements for fire partitions have been reformatted to provide for increased clarity of their construction requirements.

2018 CODE: ~~**708.4 Continuity.** Fire partitions shall extend from the top of the foundation or floor/ceiling assembly below to the underside of the floor or roof sheathing, slab or deck above or to the fire-resistance-rated floor/ceiling or roof/ceiling assembly above, and shall be securely attached thereto. In combustible construction where the fire partitions are not required to be continuous to the sheathing, deck or slab, the space between the ceiling and the sheathing, deck or slab above shall be fireblocked or draftstopped in accordance with Sections 718.2 and 718.3 at the partition line. The supporting construction shall be protected to afford the required fire-resistance rating of the wall supported, except for walls separating tenant spaces in covered and open mall buildings, walls separating dwelling units, walls separating sleeping units and corridor walls, in buildings of Type IIB, IIIB and VB construction.~~

Exceptions:

1. ~~The wall need not be extended into the crawl space below where the floor above the crawl space has a minimum 1-hour fire-resistance rating.~~

2. ~~Where the room-side fire-resistance-rated membrane of the corridor is carried through to the underside of the floor or roof sheathing, deck or slab of a fire-resistance-rated floor or roof above, the ceiling of the corridor shall be permitted to be protected by the use of ceiling materials as required for a 1-hour fire-resistance-rated floor or roof system.~~

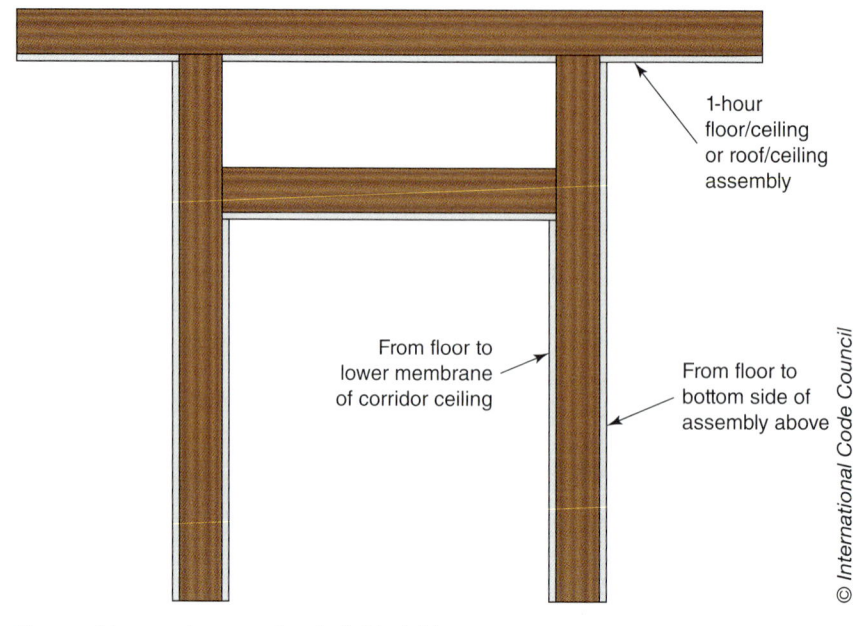

Room-side membrane extends full height

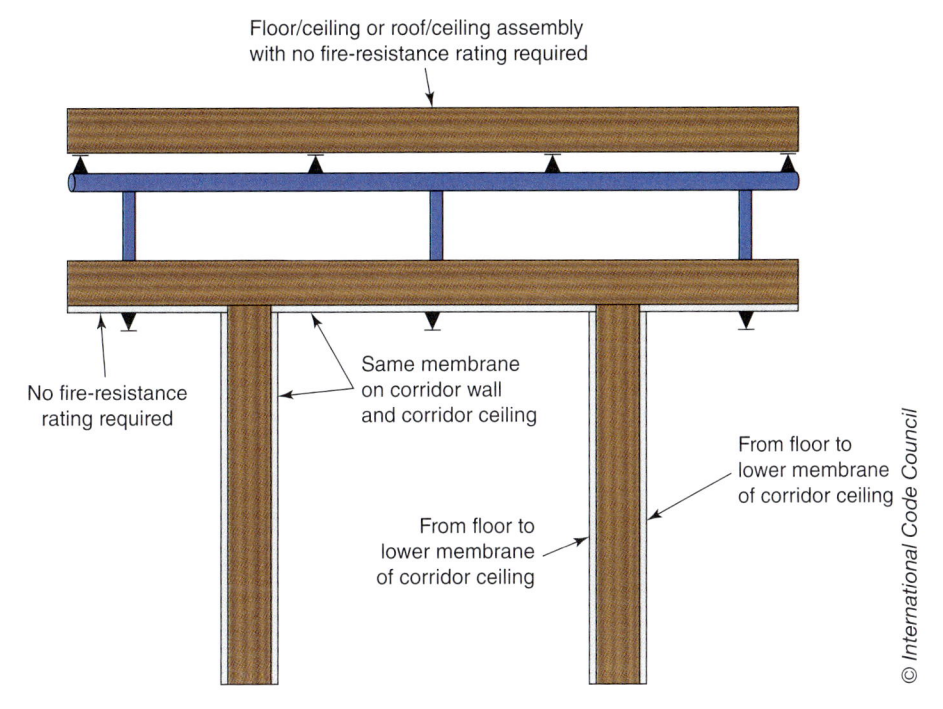

Sprinkler protection in above-ceiling space

> 3. ~~Where the corridor ceiling is constructed as required for the corridor walls, the walls shall be permitted to terminate at the upper membrane of such ceiling assembly.~~
> 4. ~~The fire partitions separating tenant spaces in a covered or open mall building, complying with Section 402.4.2.1, are not required to extend beyond the underside of a ceiling that is not part of a fire-resistance-rated assembly. A wall is not required in attic or ceiling spaces above tenant separation walls.~~
> 5. ~~Attic fireblocking or draftstopping is not required at the partition line in Group R-2 buildings that do not exceed four stories above grade plane, provided the attic space is subdivided by draftstopping into areas not exceeding 3,000 square feet (279 m²) or above every two dwelling units, whichever is smaller.~~
> 6. ~~Fireblocking or draftstopping is not required at the partition line in buildings equipped with an automatic sprinkler system installed throughout in accordance with Section 903.3.1.1 or 903.3.1.2, provided that automatic sprinklers are installed in combustible floor/ceiling and roof/ceiling spaces.~~

<u>**708.4 Continuity.** Fire partitions shall extend from the top of the foundation or floor/ceiling assembly below and be securely attached to one of the following:</u>

> <u>1. The underside of the floor or roof sheathing, deck or slab above.</u>
> <u>2. The underside of a floor/ceiling or roof/ceiling assembly having a fire-resistance rating that is not less than the fire-resistance rating of the fire partition.</u>

708.4 continues

708.4 continued

Exceptions:

1. Fire partitions shall not be required to extend into a crawl space below where the floor above the crawl space has a minimum 1-hour fire-resistance rating.

2. Fire partitions serving as a corridor wall shall not be required to extend above the lower membrane of a corridor ceiling provided that the corridor ceiling membrane is equivalent to corridor wall membrane, and either of the following conditions is met:

    2.1. The room-side membrane of the corridor wall extends to the underside of the floor or roof sheathing, deck or slab of a fire-resistance-rated floor or roof above.

    2.2. The building is equipped with an automatic sprinkler system installed throughout in accordance with Section 903.3.1.1 or 903.3.1.2, including automatic sprinklers installed in the space between the top of the fire partition and underside of the floor or roof sheathing, deck or slab above.

3. Fire partitions serving as a corridor wall shall be permitted to terminate at the upper membrane of the corridor ceiling assembly where the corridor ceiling is constructed as required for the corridor wall.

4. Fire partitions separating tenant spaces in a covered or open mall building complying with Section 402.4.2.1 shall not be required to extend above the underside of a ceiling. Such ceiling shall not be required to be part of a fire-resistance-rated assembly, and the attic or space above the ceiling at tenant separation walls shall not be required to be subdivided by fire partitions.

CHANGE SIGNIFICANCE: Fire partitions provide a limited degree of fire-resistive protection and are only mandated in very specific instances. Such wall assemblies are selectively required when separating dwelling units and sleeping units, separating tenant spaces in mall buildings, creating a fire-resistance-rated corridor, providing an elevator lobby separation, and providing egress balcony separation. The continuity requirements for fire partitions have been reformatted to provide for increased clarity of their construction requirements.

The requirements for fire partitions have been reformatted through the splitting of the section into three separate issues. Section 708.4 now only addresses the continuity of fire partitions in regard to their enclosure limits. Section 708.4.1 deals with the construction components supporting fire partitions, while Section 708.4.2 now addresses the fireblocking and draftstopping of fire partitions of combustible construction. All three of these issues were previously addressed in the single section.

The required extent of a fire partition begins at the top of the foundation or floor/ceiling assembly below. The upper terminus of the fire partition is now clearly stated as needing to terminate at either the underside of the floor or sheathing, deck or slab above, or the underside of the fire-resistance-rated floor/ceiling or roof/ceiling assembly, provided the rating is equivalent or greater than required for the fire partition.

As a part of the reformatting effort, the continuity exceptions were reworded to a limited degree. In addition, Exception 2 was expanded to allow another option where the fire partition need not extend above the lower membrane of a corridor ceiling. This vertical extent of the fire partition is now also not required where automatic sprinkler protection is extended to the concealed horizontal space above the top of the fire partition.

708.4.2

Fireblocking and Draftstopping at Fire Partitions

CHANGE TYPE: Clarification

CHANGE SUMMARY: Fireblocking and draftstopping requirements for fire partitions of combustible construction have been consolidated and modified.

2018 CODE: <u>**708.4.2 Fireblocks and draftstops in combustible construction.** In combustible construction where fire partitions do not extend to the underside of the floor or roof sheathing, deck or slab above, the space above and along the line of the fire partition shall be provided with one of the following:</u>

1. <u>Fireblocking up to the underside of the floor or roof sheathing, deck or slab above using materials complying with Section 718.2.1.</u>
2. <u>Draftstopping up to the underside of the floor or roof sheathing, deck or slab above using materials complying with Section 718.3.1 for floors or Section 718.4.1 for attics.</u>

Exceptions:

1. <u>Buildings equipped with an automatic sprinkler system installed throughout in accordance with Section 903.3.1.1, or in accordance with Section 903.3.1.2 provided that protection is provided in the space between the top of the fire partition and underside of the floor or roof sheathing, deck or slab above as required for systems complying with Section 903.3.1.1.</u>
2. <u>Where corridor walls provide a sleeping unit or dwelling unit separation, draftstopping shall only be required above one of the corridor walls.</u>

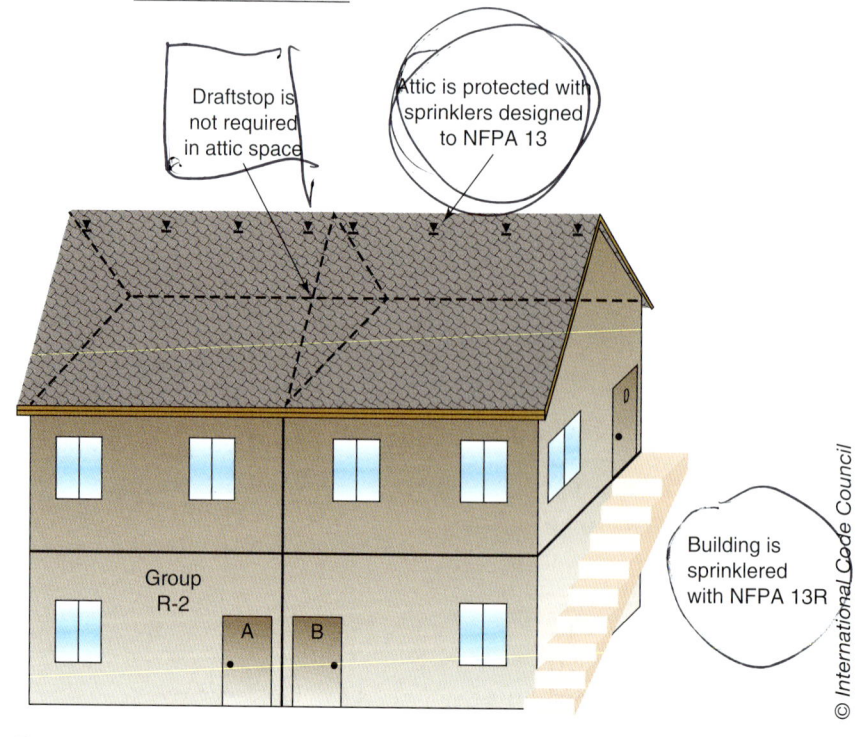

Residential attic protection

3. <u>In Group R-2 occupancies with fewer than four dwelling units, fireblocking and draftstopping shall not be required.</u>

4. <u>In Group R-2 occupancies up to and including four stories in height in buildings not exceeding 60 feet (18 288 mm) in height above grade plane, the attic space shall be subdivided by draftstops into areas not exceeding 3,000 square feet (279 m²) or above every two dwelling units, whichever is smaller.</u>

5. <u>In Group R-3 occupancies with fewer than three dwelling units, fireblocking and draftstopping shall not be required in floor assemblies.</u>

708.4 Continuity. ~~Fire partitions shall extend from the top of the foundation or floor/ceiling assembly below to the underside of the floor or roof sheathing, slab or deck above or to the fire-resistance-rated floor/ceiling or roof/ceiling assembly above, and shall be securely attached thereto. In combustible construction where the fire partitions are not required to be continuous to the sheathing, deck or slab, the space between the ceiling and the sheathing, deck or slab above shall be fireblocked or draftstopped in accordance with Sections 718.2 and 718.3 at the partition line. The supporting construction shall be protected to afford the required fire-resistance rating of the wall supported, except for walls separating tenant spaces in covered and open mall buildings, walls separating dwelling units, walls separating sleeping units and corridor walls, in buildings of Type IIB, IIIB and VB construction.~~

Exceptions:

1. ~~The wall need not be extended into the crawl space below where the floor above the crawl space has a minimum 1-hour fire-resistance rating.~~

2. ~~Where the room-side fire-resistance-rated membrane of the corridor is carried through to the underside of the floor or roof sheathing, deck or slab of a fire-resistance-rated floor or roof above, the ceiling of the corridor shall be permitted to be protected by the use of ceiling materials as required for a 1-hour fireresistance-rated floor or roof system.~~

3. ~~Where the corridor ceiling is constructed as required for the corridor walls, the walls shall be permitted to terminate at the upper membrane of such ceiling assembly.~~

4. ~~The fire partitions separating tenant spaces in a covered or open mall building, complying with Section 402.4.2.1, are not required to extend beyond the underside of a ceiling that is not part of a fire-resistance-rated assembly. A wall is not required in attic or ceiling spaces above tenant separation walls.~~

5. ~~Attic fireblocking or draftstopping is not required at the partition line in Group R-2 buildings that do not exceed four stories abovegrade plane, provided the attic space is subdivided by draftstopping into areas not exceeding 3,000 square feet (279 m²) or above every two dwellingunits, whichever is smaller.~~

708.4.2 continues

708.4.2 continued

6. ~~Fireblocking or draftstopping is not required at the partition line in buildings equipped with an automatic sprinkler system installed throughout in accordance with Section 903.3.1.1 or 903.3.1.2, provided that automatic sprinklers are installed in combustible floor/ceiling and roof/ceiling spaces.~~

718.3.2 Groups R-1, R-2, R-3 and R-4. ~~Draftstopping shall be provided in floor/ceiling spaces in Group R-1 buildings, in Group R-2 buildings with three or more dwelling units, in Group R-3 buildings with two dwelling units and in Group R-4 buildings. Draftstopping shall be located above and in line with the dwelling unit and sleeping unit separations.~~

Exceptions:

1. ~~Draftstopping is not required in buildings equipped throughout with an automatic sprinkler system in accordance with Section 903.3.1.1.~~

2. ~~Draftstopping is not required in buildings equipped throughout with an automatic sprinkler system in accordance with Section 903.3.1.2, provided that automatic sprinklers are installed in the combustible concealed spaces where the draftstopping is being omitted.~~

718.4.2 Groups R-1 and R-2. ~~Draftstopping shall be provided in attics, mansards, overhangs or other concealed roof spaces of Group R-2 buildings with three or more dwelling units and in all Group R-1 buildings. Draftstopping shall be installed above, and in line with, sleeping unit and dwelling unit separation walls that do not extend to the underside of the roof sheathing above.~~

Exceptions:

1. ~~Where corridor walls provide a sleeping unit or dwelling unit separation, draftstopping shall only be required above one of the corridor walls.~~

2. ~~Draftstopping is not required in buildings equipped throughout with an automatic sprinkler system in accordance with Section 903.3.1.1.~~

3. ~~In occupancies in Group R-2 that do not exceed four stories above grade plane, the attic space shall be subdivided by draftstops into areas not exceeding 3,000 square feet (279 m^2) or above every two dwelling units, whichever is smaller.~~

4. ~~Draftstopping is not required in buildings equipped throughout with an automatic sprinkler system in accordance with Section 903.3.1.2, provided that automatic sprinklers are installed in the combustible concealed space where the draftstopping is being omitted.~~

CHANGE SIGNIFICANCE: Fireblocking and draftstopping are required in combustible construction to cut off concealed draft openings (both vertical and horizontal). Experience has shown that some of the greatest damage occurs to conventional wood-framed buildings during a fire where the fire travels unimpeded through such concealed areas. Both fireblocks and draftstops are selectively required in combustible construction, including where fire partitions are provided, in order to limit the spread of fire, smoke and hot gases. The firestopping provisions previously located in Section 708.4 and the draftstopping provisions previously found in Sections 718.3.2 and 718.4.2 have been relocated to Section 708.4.2 as a part of the reformat of Section 708.4. In addition, a number of technical provisions were revised or added.

Section 708.4.2 is a new section which combines and relocates requirements from other sections of the code addressing fireblocking and draftstopping.

The new Exception 1 in Section 708.4.2 is a combination of the previous Exception 6 in Section 708.4, Exceptions 1 and 2 in Section 718.3.2 and Exceptions 2 and 4 in Section 718.4.2. This exception has been revised to specify that where the automatic sprinkler system is designed to NFPA 13R, and sprinklers are provided within the attic space, the sprinkler design in the attic space must comply with NFPA 13 *Standard for the Installation of Sprinkler Systems*. NFPA 13R does not contain criteria for installing sprinklers in the entire attic space, so the designer must go to NFPA 13 for that design.

Note that when Exception 1 is applied, the installation of sprinklers in the attic space when the building is protected with an automatic sprinkler system designed to NFPA 13R *Standard for the Installation of Sprinkler Systems in Low Rise Residential Occupancies* also complies with the new Section 903.3.1.2.3 regarding attic protection in Group R occupancies over 55 feet in height (Section 903.3.1.2.3, Item 3) and all Group R-4, Condition 2 occupancies (Section 903.3.1.2.3, Item 4).

The new Exception 2 in Section 708.4.2 was Exception 1 in Section 718.4.2.

The new Exception 3 in Section 708.4.2 comes from the charging language in previous Sections 718.3.2 and 718.4.2. This exception has also been revised. Previously, draftstopping or fireblocking was required in Group R-2 occupancies with three or more units. The new exception removes the draftstopping and fireblocking requirement from Group R-2 occupancies with three units or less.

The new Exception 4 in Section 708.4.2 was previously located in Section 708.4, Exception 5 and Section 718.4.2, Exception 3. This new exception has also been revised to include the limitation of 60 feet in height. This revision correlates with the scope of NFPA 13R. The standard is limited to application in buildings not exceeding four stories or 60 feet in height. As previously written, Section 718.4.2, Exception 3 could be applied to buildings that are taller than 60 feet, which was not intended.

The new Exception 5 in Section 708.4.2 comes from the charging language in previous Sections 718.3.2 and 718.4.2. This exception has also been revised. Previously, floor/ceiling assemblies were required to be protected in Group R-3 occupancies with two or more units. The new exception requires draftstopping and fireblocking for floor/ceiling assemblies in Group R-3 occupancies with three or more units.

713.8.1
Membrane Penetrations of Shaft Enclosures

CHANGE TYPE: Modification

CHANGE SUMMARY: Membrane penetrations not related to the purpose of a shaft enclosure are no longer prohibited from penetrating the outside of the enclosure.

2018 CODE: 713.8.1 Prohibited penetrations. Penetrations other than those necessary for the purpose of the shaft shall not be permitted in shaft enclosures.

> **Exception:** Membrane penetrations shall be permitted on the outside of shaft enclosures. Such penetrations shall be protected in accordance with Section 714.4.2.

CHANGE SIGNIFICANCE: Unless specifically permitted, penetrations have historically been prohibited at the fire-resistance-rated enclosure around a shaft. The strict limitations have been deemed necessary to ensure that the fire-resistive integrity of the shaft enclosure was not compromised by penetrations of the protective enclosure. Penetrations of the exterior membrane of the fire-resistance-rated assembly are now permitted provided they are in compliance with the membrane penetration provisions of Section 714.4.2.

Virtually all penetrations have been prohibited in the past, regardless of purpose with very limited exceptions. The prohibition applied to any penetration that was not necessary for the purpose of the shaft, such as electrical boxes. The new exception will no longer limit the type of or purpose for the penetration but will simply limit the location to the exterior membrane and require the proper protection.

The provisions are now consistent for both shaft enclosures and interior exit stairways. A similar allowance was added in the 2012 edition of the IBC that provides for membrane penetrations on the exterior side of enclosures for interior exit stairways.

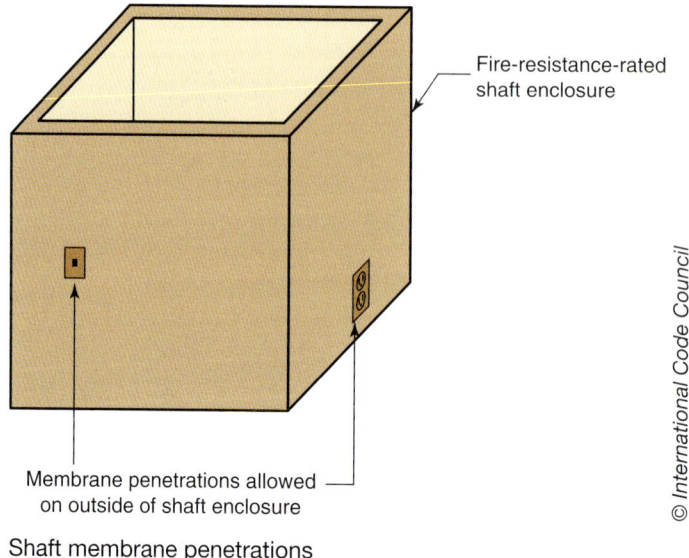

Shaft membrane penetrations

716.2.6.5 Delayed-Action Self-Closing Doors

CHANGE TYPE: Addition

CHANGE SUMMARY: Self-closing doors that are not also required to be automatic closing are now permitted to be equipped with delayed-action closers.

2018 CODE: <u>**716.2.6.5 Delayed-action closers.** Doors required to be self-closing and not required to be automatic closing shall be permitted to be equipped with delayed-action closers.</u>

202 DEFINITIONS

<u>**DELAYED-ACTION CLOSER.** A self-closing device that incorporates a delay prior to the initiation of closing. Delayed-action closers are mechanical devices with an adjustable delay.</u>

CHANGE SIGNIFICANCE: Doors required to be fire-protection-rated are also required to close and latch without a specific action by the door user. This is accomplished through the use of either self-closing doors or automatic-closing doors. There is an expectation that the door will move into the closed position and latch in order to maintain a complete fire separation for which the wall and door assemblies are intended. Self-closing doors not required to be automatic closing are now specifically allowed to have a delayed-action condition that will create a limited delay in the closing operation.

In addition, the definition of a "delayed-action closer" has been provided in Chapter 2. As an example, consider a door where an individual is leading a group of people. A door with a delayed-action closer allows the leader and the group to pass through the door before it closes, helping to keep the group intact.

Unlike automatic-closing doors which are commonly held in an open position, self-closing doors are normally in a closed position unless being used. Thus, in a fire situation, the doors with delayed-action closers would also be closed except when being used and for a relatively brief delay caused by the delayed-action closer. Note that automatic-closing

716.2.6.5 continues

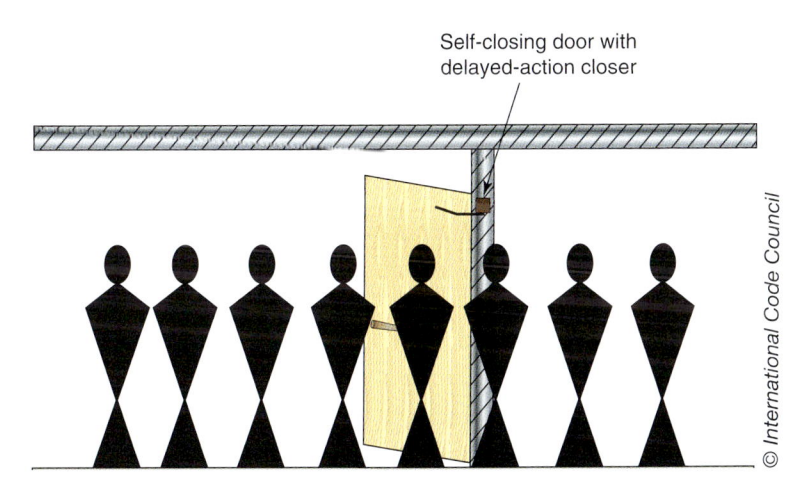

Travel through delayed-closing door

716.2.6.5 continued doors are not allowed to be equipped with a delayed-action closer. The new allowance only applies to those self-closing doors not required to be automatic closing.

The time delay of delayed-action closers is adjustable and the new provisions do not address the maximum allowable time delay. This will be a condition the building official needs to consider when approving these devices.

803.1.1, 803.1.2
Interior Wall and Ceiling Finish Testing

CHANGE TYPE: Clarification

CHANGE SUMMARY: Interior wall and ceiling finish testing criteria have been reorganized to enhance their application and enforcement.

2018 CODE: 803.1.1 Interior wall and ceiling finish materials tested in accordance with NFPA 286. Interior wall and ceiling finish materials shall be classified in accordance with ~~ASTM E-84 or UL 723~~ NFPA 286 and comply with Section 803.1.1.1. Materials complying with Section 803.1.1.1 shall be considered to also comply with the requirements of Class A. ~~Such interior finish materials shall be grouped in the following classes in accordance with their flame spread and smoke-developed indexes.~~

~~Class A: Flame spread index 0-25; smoke developed index 0-450.~~
~~Class B: Flame spread index 26-75; smoke developed index 0-450.~~
~~Class C: Flame spread index 76-200; smoke developed index 0-450.~~

~~**Exception:** Materials tested in accordance with Section 803.1.2.~~

[handwritten: B, C — IF PASSES CAN USE IT IN ALL OCCUP. WALL/CEILINGS]

~~803.1.2.1~~ **803.1.1.1 Acceptance criteria for NFPA 286.** The interior finish shall comply with the following:

1. During the 40 kW exposure, flames shall not spread to the ceiling.
2. The flame shall not spread to the outer extremity of the sample on any wall or ceiling.

803.1.1, 803.1.2 continues

ASTM E 84 test furnace

803.1.1, 803.1.2 continued

3. Flashover, as defined in NFPA 286, shall not occur.
4. The peak heat release rate throughout the test shall not exceed 800 kW.
5. The total smoke released throughout the test shall not exceed 1,000 m².

803.1.2 ~~Room corner test for interior~~ <u>Interior</u> **wall or ceiling finish materials** <u>**tested in accordance with ASTM E 84 and UL 723**</u>. Interior wall ~~or~~ <u>and</u> ceiling finish materials shall be ~~permitted to tested in accordance with NFPA 286. Interior wall or ceiling finish materials tested in accordance with NFPA 286 shall comply with Section 803.1.2.1.~~ <u>classified in accordance with ASTM E 84 or UL 723. Such interior finish materials shall be grouped in the following classes in accordance with their flame spread and smoke-developed indexes.</u>

<u>Class A: Flame spread index 0-25; smoke-developed index 0-450.</u>
<u>Class B: Flame spread index 26-75; smoke-developed index 0-450.</u>
<u>Class C: Flame spread index 76-200; smoke-developed index 0-450.</u>

Exception: <u>Materials tested in accordance with Section 803.1.1 and as indicated in Sections 803.1.3 through 803.13.</u>

<u>**803.1.3 Interior wall and ceiling finish materials with different requirements** The materials indicated in Sections 803.2 through 803.13 shall be tested as indicated in the corresponding sections.</u>

803.5 Textile wall coverings. Where used as interior wall finish materials, textile wall coverings, including materials having woven or nonwoven, napped, tufted, looped or similar surface and carpet and similar textile materials, shall be tested in the manner intended for use, using the product mounting system, including adhesive, and shall comply with the requirements of one of the following: Section ~~803.1.2, 803.1.3 or 803.1.4~~ <u>803.1.1, 803.5.1 or 803.5.2.</u>

~~803.1.3~~ <u>803.5.1</u> Room corner test for textile wall coverings and expanded vinyl wall coverings. Textile wall coverings and expanded vinyl wall coverings shall meet the criteria of Section ~~803.1.3.1~~ <u>803.5.1.1</u> when tested in the manner intended for use in accordance with the Method B protocol of NFPA, 265 using the product mounting system, including adhesive.

~~803.1.3.1~~ <u>803.5.1.1</u> Acceptance criteria for NFPA 265. The interior finish shall comply with the following:

1. During the 40 kW exposure, flames shall not spread to the ceiling.
2. The flame shall not spread to the outer extremities of the samples on the 8-foot by 12-foot (203 by 305 mm) walls.
3. Flashover, as defined in NFPA 265, shall not occur.
4. The total smoke released throughout the test shall not exceed 1,000 m².

~~803.1.4~~ 803.5.2 Acceptance criteria for textile and expanded vinyl wall or ceiling coverings tested to ASTM E 84 or UL 723. Textile wall and ceiling coverings and expanded vinyl wall and ceiling coverings shall have a Class A flame spread index in accordance with ASTM E 84 or UL 723 and be protected by an automatic sprinkler system installed in accordance with Section 903.3.1.1 or 903.3.1.2. Test specimen preparation and mounting shall be in accordance with ASTM E 2404.

CHANGE SIGNIFICANCE: The dangers of unregulated interior finish materials are twofold: 1) the rapid spread of fire and/or the production of large quantities of dense, black smoke, and 2) the contribution of additional fuel to the fire. As such, the materials used on walls and ceilings, as well as coverings applied to the floor, are strictly limited to those coverings that are in compliance with the appropriate test standard(s). Interior wall and ceiling finish criteria have been reorganized to enhance their application and enforcement.

All interior wall and ceiling finish materials are permitted to be tested to NFPA 286 *Standard Methods of Fire Tests for Evaluating Contribution of Wall and Ceiling Interior Finish to Room Fire Growth*. Therefore, the criteria referencing the application of NFPA 286 now are presented first as Section 803.1.1. This section has also been clarified to indicate that any material which passes NFPA 286 testing is considered as having a Class A flame spread rating. Since NFPA 286 does not classify materials into flame spread categories, the acceptance criteria for a successful test under NFPA 286 follows as Section 803.1.1.1.

Section 803.1.2 now addresses testing under ASTM E 84 and UL 723. The long-held flame spread classification ratings are included in this section.

Section 803.1.3 then addresses a variety of materials and conditions with other characteristics which cannot simply be tested to NFPA 286, NFPA 84, or UL 723; therefore, additional testing criteria is needed. These are items are now listed in Sections 803.2 through 803.13, where additional testing criteria are provided.

Sections 803.5 through 803.7 address textile wall coverings and expanded vinyl wall coverings. NFPA 265 *Standard Methods of Fire Tests for Evaluating Room Fire Growth Contribution of Textile or Expanded Vinyl Wall Covering on Full Height Panels and Walls* is the appropriate test standard for these materials and is, therefore, relocated to Section 803.5.1 along with the acceptance criteria in Section 803.5.1.1.

803.3 Interior Finish Requirements for Heavy Timber Construction

CHANGE TYPE: Modification

CHANGE SUMMARY: Materials considered heavy timber construction must now comply with interior finish requirements where exposed in interior exit stairways and exit passageways.

2018 CODE: 803.3 Heavy timber exemption. Exposed portions of building elements complying with the requirements for buildings of ~~Type IV~~ heavy timber construction in Section 602.4 or Section 2304.11 shall not be subject to interior finish requirements except in interior exit stairways, interior exit ramps, and exit passageways.

CHANGE SIGNIFICANCE: Heavy timber members exposed within a building have traditionally been unregulated for interior finish purposes. Any concerns over the contribution of heavy timber building elements to flame spread in a structure are dealt with in the general requirements of the IBC that allow the use of heavy timber construction.

Historically, the use of heavy timber members has been primarily limited to individual structural elements. However, heavy timber construction, including cross laminated timber, has become more commonly used to form the entire interior surfaces of buildings, including egress elements. As such, the provisions now specify that the heavy timber construction forming interior exit stairways, interior exit ramps, and exit passageways must comply with the interior finish requirements.

These high-level egress components form a protected egress path and this requirement provides for protection of the occupants as they exit. The interior finish requirements apply to all other materials and now also apply to heavy timber construction in limited applications. The flame spread criteria in Table 803.13 for heavy timber interior wall and ceiling finish are now applicable to these highly regulated means of egress elements.

Exposed heavy-timber construction in stair enclosure

Photo courtesy of American Wood Council

803.11, 803.12
Flame Spread Testing of Laminates and Veneers

CHANGE TYPE: Addition

CHANGE SUMMARY: Specific flame spread testing provisions have been added to the IBC to address the use of factory-produced laminated products with a wood substrate as well as facings and wood veneers applied over a wood substrate on site.

2018 CODE: 803.11 Laminated products factory produced with a wood substrate. Laminated products factory-produced with a wood substrate shall comply with one of the following:

1. The laminated product shall meet the criteria of Section 803.1.1.1 when tested in accordance with NFPA 286 using the product-mounting system, including adhesive, as described in Section 5.8 of NFPA 286.

2. The laminated product shall have a Class A, B or C flame spread index and smoke-developed index, based on the requirements of Table 803.13, in accordance with ASTM E84 or UL 723. Test specimen preparation and mounting shall be in accordance with ASTM E 2579.

803.12 Facings or wood veneers intended to be applied on site over a wood substrate. Facings or veneers intended to be applied on site over a wood substrate shall comply with one of the following:

1. The facing or veneer shall meet the criteria of Section 803.1.1.1 when tested in accordance with NFPA 286 using the product-mounting system, including adhesive, as described in Section 5.9 of NFPA 286.

2. The facing or veneer shall have a Class A, B or C flame spread index and smoke-developed index, based on the requirements of Table 803.13, in accordance with ASTM E84 or UL 723. Test specimen preparation and mounting shall be in accordance with ASTM E 2404.

Sections 803.11, 803.12 continues

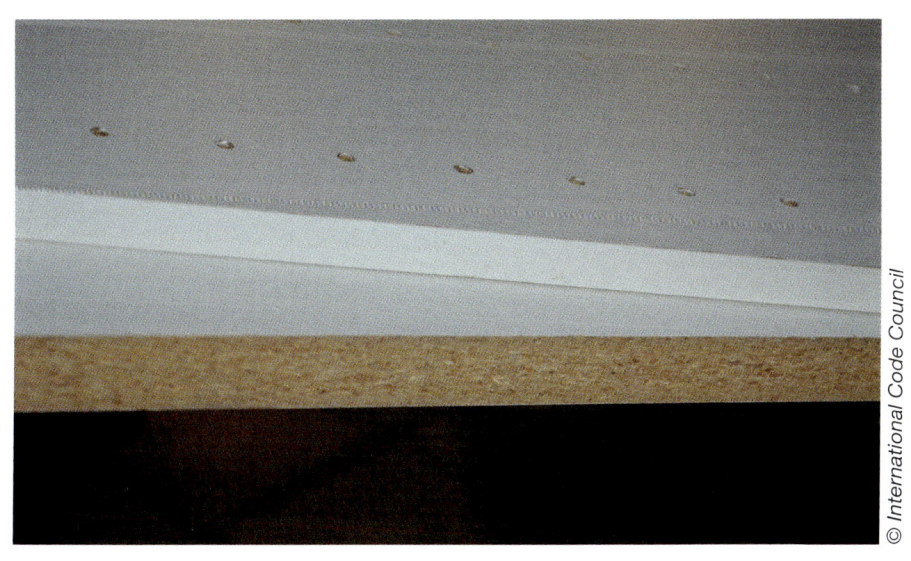

Factory-produced laminate product with wood substrate

Sections 803.11, 803.12 continued

CHANGE SIGNIFICANCE: ASTM has developed mounting methods for both "facings or wood veneer intended to be applied on site over a wood substrate" and "laminated products that are factory-produced and have a wood substrate." New Section 803.11 deals with factory-produced laminated products with a wood substrate. Facings produced as part of a commercial (factory-produced) panel are finished products and the manufacturer is responsible to ensure that the product itself is safe and there is no need to address the substrate. It has been shown that when veneers are applied over a wood substrate the resulting flame spread is much higher than when applied over gypsum board or over a noncombustible substrate. ASTM E 2579, *Standard Practice for Specimen Preparation and Mounting of Wood Products to Assess Surface Burning*, is referenced in Item 2 for preparation of the test specimen. ASTM E 2579 requires that the testing be done with the full product, including substrate; therefore, there is no need to retest for different substrates. The NFPA 286 *Standard Methods of Fire Test for Evaluating Contribution of Wall and Ceiling Interior Finish to Room Fire Growth* test is indicated in Item 1. NFPA 286 contains a section that addresses the testing of wall covering materials, including laminated products produced in the factory. The product must be tested in accordance with either Item 1 or 2.

Section 803.12, also new to the IBC, addresses products applied on site over a wood substrate. The concept is that these facings applied on site are basically the same as wall coverings and the manufacturer should be responsible for the facing only and needs to ensure that the material is safe and the test should occur over the appropriate substrate. It has been shown that, when veneers are applied over a wood substrate, the resulting flame spread is much higher than when applied over gypsum board or over a noncombustible substrate. ASTM E 2404, *Standard Practice for Specimen Preparation and Mounting of Textile, Paper or Polymeric (Including Vinyl) Wall or Ceiling Coverings, and of Facings and Wood Veneers Intended to be Applied on Site Over a Wood Substrate*, is referenced in Item 2. ASTM E 2404 requires the testing be done over a standard wood substrate and, thus, there will be no need to retest for different types of wood. The NFPA 286 *Standard Methods of Fire Test for Evaluating Contribution of Wall and Ceiling Interior Finish to Room Fire Growth* test is indicated in Item 1. NFPA 286 contains a section that addresses testing of facings applied on site. The product must be tested in accordance with either Item 1 or 2.

901.6.2 Integrated Fire Protection System Testing

CHANGE TYPE: Addition

CHANGE SUMMARY: Test criteria have been added to the code with a reference to new NFPA 4, *Standard for Integrated Fire Protection and Life Safety System Testing*, to ensure that where multiple fire protection systems or life safety systems are integrated, the acceptance process and subsequent testing must evaluate all of the integrated systems as a whole.

2018 CODE: 901.6.2 Integrated testing. Where two or more fire protection or life safety systems are interconnected, the intended response of subordinate fire protection and life safety systems shall be verified when required testing of the initiating system is conducted. In addition, integrated testing shall be performed in accordance with Sections 901.6.2.1 and 901.6.2.2.

901.6.2.1 High-rise buildings. For high-rise buildings, integrated testing shall comply with NFPA 4, with an integrated test performed prior to issuance of the certificate of occupancy and at intervals not exceeding 10 years, unless otherwise specified by an integrated system test plan prepared in accordance with NFPA 4. If an equipment failure is detected during integrated testing, a repeat of the integrated test shall not be required, except as necessary to verify operation of fire protection or life safety functions that are initiated by equipment that was repaired or replaced.

901.6.2.2 Smoke control systems. Where a fire alarm system is integrated with a smoke control system as outlined in Section 909, integrated testing shall comply with NFPA 4, with an integrated test performed prior to issuance of the certificate of occupancy and at intervals not exceeding 10 years, unless otherwise specified by an integrated system test plan prepared in accordance with NFPA 4. If an equipment failure is detected during integrated testing, a repeat of the integrated test shall not be required, except as necessary to verify operation of fire protection or life safety functions that are initiated by equipment that was repaired or replaced.

901.6.2 continues

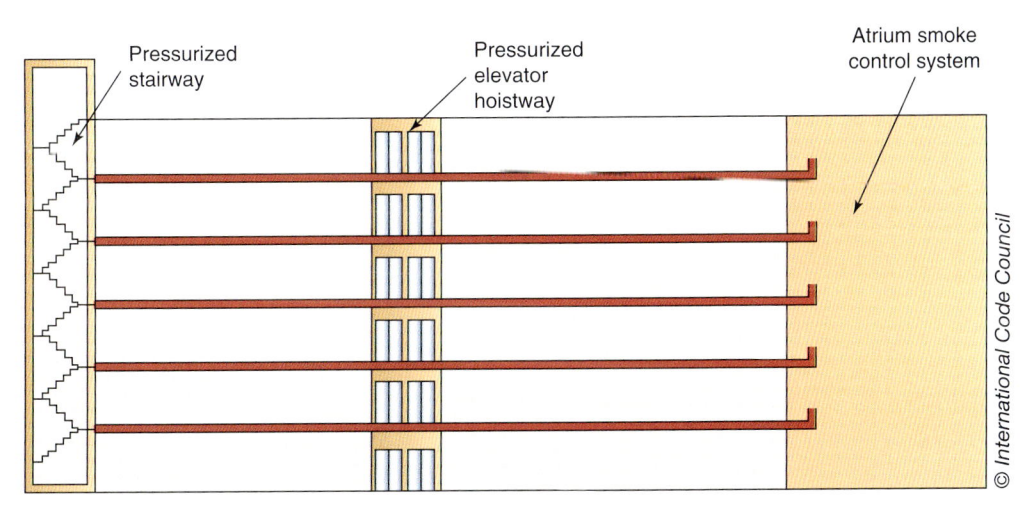

Multiple systems to be tested to work independently and simultaneously

901.6.2 continued

CHANGE SIGNIFICANCE: The code clearly intends to require proper operation of all fire protection and life safety systems within a building. In many cases, such as monitoring the fire sprinkler system with the fire alarm system and notifying a monitoring service when a fire sprinkler operates, the integration is relatively simple. However, in some cases, such as a fire alarm system initiating a complex combination of doors, dampers, elevators, and fans in a high-rise building, the integration can be highly complex and in most cases it involves the cooperation of many different trades, controls, and systems. New provisions ensure that the required testing of integrated features is scaled in a manner that is reasonable for a wide range of applications.

NFPA 4 *Standard for Integrated Fire Protection and Life Safety System Testing* is a newly published standard that deals with the testing of integrated systems. NFPA 4 is now referenced in Sections 901.6.2.1 and 901.6.2.2. These two sections require compliance with NFPA 4 for the more complex systems found in high-rise buildings and smoke control systems. Section 901.6.2 specifies that where there are multiple systems, they must be integrated, but it does not require compliance with NFPA 4 for less complex designs.

This creates a two-tiered approach to integrated testing which is designed to address both the highly complex scenarios and those that are not so complex where it is a simpler task to verify the functionality of integrated fire protection and life safety systems.

It should also be noted that the requirement established by Section 901.6.2 intends to convey that required testing of integrated features should follow a top-down approach, where testing of an initiating device or control warrants verification of subordinate systems or functions, but not necessarily the reverse. For example, where a smoke detector is intended to trigger an automatic damper (the damper being a "subordinate" device to the smoke detector), the test protocol should verify that the damper system responds as intended when the smoke detector activates. However, if a damper control system is tested simply to exercise dampers and to verify that dampers are operating, such testing should not require integrated testing of the initiating smoke detector or fire alarm system.

Lastly, for high-rise buildings and buildings with smoke control systems, the new sections specify that the initial acceptance testing must be completed prior to issuance of the Certificate of Occupancy.

902 Fire Pump and Fire Sprinkler Riser Rooms

CHANGE TYPE: Addition

CHANGE SUMMARY: A number of prescriptive requirements have been added regulating the design and construction of automatic sprinkler system riser rooms and fire pump rooms.

2018 CODE:

SECTION 902
FIRE PUMP AND RISER ROOM SIZE

901.8 902.1 Pump and riser room size. *(No change to relocated text.)*

902.1.1 Access. Automatic sprinkler system risers, fire pumps and controllers shall be provided with ready access. Where located in a fire pump room or automatic sprinkler system riser room, the door shall be permitted to be locked provided the key is available at all times.

902.1.2 Marking on access doors. Access doors for automatic sprinkler system riser rooms and fire pump rooms shall be labeled with an approved sign. The lettering shall be in a contrasting color to the background. Letters shall have a minimum height of 2 inches (51 mm) with a minimum stroke of 3/8 inch (10 mm).

902.1.3 Environment. Automatic sprinkler system riser rooms and fire pump rooms shall be maintained at a temperature of not less than 40°F (4°C). Heating units shall be permanently installed.

902.1.4 Lighting. Permanently installed artificial illumination shall be provided in the automatic sprinkler system riser rooms and fire pump rooms.

Fire sprinkler riser room

CHANGE SIGNIFICANCE: Instead of prescribing arbitrary dimensions for the design and construction of fire pump and riser rooms, the general provisions base the room area on clearances specified by the equipment manufacturers to ensure adequate space is available for its installation or removal. The design must provide enough area so that walls, finish materials, and doors are not required to be removed during maintenance activities. The provisions also prescribe that the size of the door serving a riser or pump room is of a size to accommodate the removal of the largest piece of equipment. The new requirements expand upon the performance objectives and establish some fundamental criteria relating to room access, identification, environment, and lighting.

Additional criteria regarding the design and construction of fire pump rooms and fire sprinkler riser rooms address some fundamental issues. Although the IBC and IFC do not require such rooms to be provided, in those cases where they are provided the rooms must meet the requirements in Section 902.8 of the IBC and Section 901.4.6 of the IFC.

Fire pump and fire sprinkler riser rooms must be readily accessible at all times, but access may be locked provided a key is always available. Heating equipment may be required since the minimum temperature in the room must be maintained at 40° Fahrenheit or higher to protect the piping and components from freezing. Permanent lighting must also be provided in fire pump rooms and fire sprinkler riser rooms.

903.2.1
Sprinklers Required in Group A Occupancies

CHANGE TYPE: Clarification

CHANGE SUMMARY: The extent to which automatic sprinkler systems are required in multistory Group A occupancies has been clarified.

2018 CODE: 903.2.1 Group A. An automatic sprinkler system shall be provided throughout buildings and portions thereof used as Group A occupancies as provided in this section. ~~For Group A-1, A-2, A-3 and A-4 occupancies, the automatic sprinkler system shall be provided throughout the story where the fire area containing the Group A-1, A-2, A-3 or A-4 occupancy is located, and throughout all stories from the Group A occupancy to, and including, the levels of exit discharge serving the Group A occupancy. For Group A-5 occupancies, the automatic sprinkler system shall be provided in the spaces indicated in Section 903.2.1.5.~~

903.2.1.1 Group A-1. An automatic sprinkler system shall be provided ~~for fire areas~~ <u>throughout stories</u> containing Group A-1 occupancies and ~~intervening floors~~ <u>throughout all stories from the Group A-1 occupancy to and including the levels</u> of ~~the building~~ exit discharge serving that occupancy where one of the following conditions exists:

1. The fire area exceeds 12,000 square feet (1115 m²).
2. The fire area has an occupant load of 300 or more.
3. The fire area is located on a floor other than a level of exit discharge serving such occupancies.
4. The fire area contains a multitheater complex.

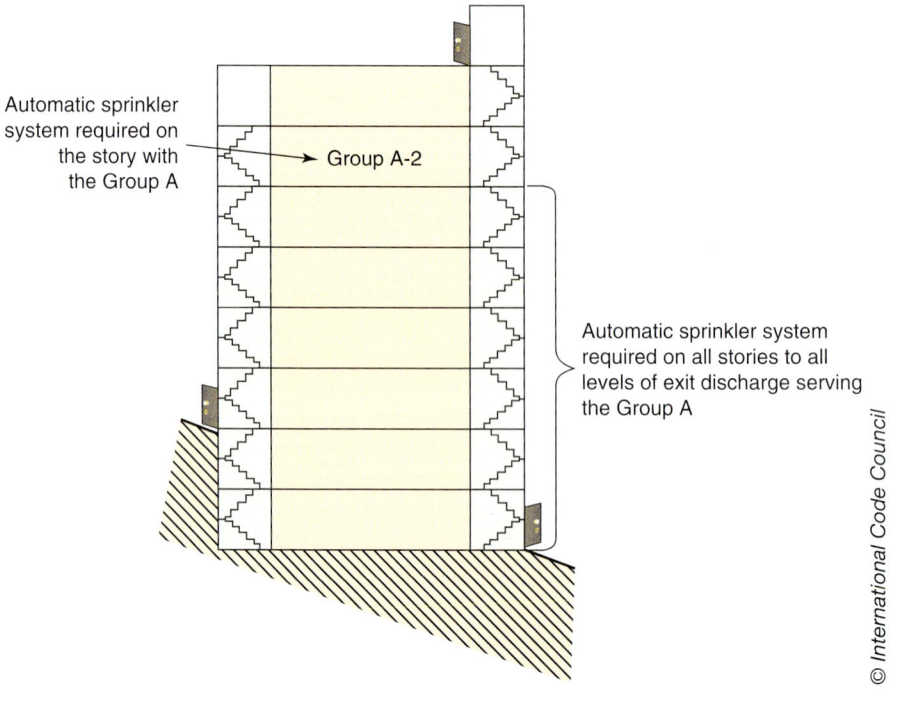

Extent of Group A sprinkler protection

903.2.1.2 Group A-2. An automatic sprinkler system shall be provided ~~for fire areas~~ throughout stories containing Group A-2 occupancies and ~~intervening floors~~ throughout all stories from the Group A-2 occupancy to and including the levels of ~~the building~~ exit discharge serving that occupancy where one of the following conditions exists:

1. The fire area exceeds 5,000 square feet (464 m²).
2. The fire area has an occupant load of 100 or more.
3. The fire area is located on a floor other than a level of exit discharge serving such occupancies.

903.2.1.3 Group A-3. An automatic sprinkler system shall be provided ~~for fire areas~~ throughout stories containing Group A-3 occupancies and ~~intervening floors~~ throughout all stories from the Group A-3 occupancy to and including the levels of ~~the building~~ exit discharge serving that occupancy where one of the following conditions exists:

1. The fire area exceeds 12,000 square feet (1115 m²).
2. The fire area has an occupant load of 300 or more.
3. The fire area is located on a floor other than a level of exit discharge serving such occupancies.

903.2.1.4 Group A-4. An automatic sprinkler system shall be provided ~~for fire areas~~ throughout stories containing Group A-4 occupancies and ~~intervening floors~~ throughout all stories from the Group A-4 occupancy to and including the levels of ~~the building~~ exit discharge serving that occupancy where one of the following conditions exists:

1. The fire area exceeds 12,000 square feet (1115 m²).
2. The fire area has an occupant load of 300 or more.
3. The fire area is located on a floor other than a level of exit discharge serving such occupancies.

903.2.1.5 Group A-5. An automatic sprinkler system shall be provided for all enclosed Group A-5 ~~occupancies in the following areas: concession stands, retail areas, press boxes and other~~ accessory use areas in excess of 1,000 square feet (93 m²).

903.2.1.5.1 Spaces under grandstands or bleachers. Enclosed spaces under grandstands or bleachers shall be equipped with an automatic sprinkler system in accordance with Section 903.3.1.1 where either of the following exist:

1. The enclosed area is 1,000 square feet (93 m²) or less and is not constructed in accordance with Section 1029.1.1.1.
2. The enclosed area exceeds 1,000 square feet (93 m²).

CHANGE SIGNIFICANCE: Assembly occupancies with sizable occupant loads or floor areas, as well as those located above or below the discharge level, require sprinkler protection due to the additional time needed for occupant egress. In addition, conditions such as low light levels,

903.2.1 continues

903.2.1 continued

overcrowding and multiple instances of potential obstructions can lead a hazardous environment that can be effectively addressed through the presence of an automatic sprinkler system. In order to provide further clarity as to the extent of such sprinkler protection, the conditions under which automatic sprinkler systems are required in Group A occupancies have been clarified. Furthermore, revised language details requirements for the fire sprinkler system used to protect spaces on the level of exit discharge.

Previous code language created an inconsistency among Sections 903.2.1 through 903.2.1.4. Section 903.2.1 stated sprinklers were required on the story with the Group A occupancy and on all stories to, and including, the level of exit discharge serving the Group A occupancy. Sections 903.2.1.1 through 903.2.1.4 use the term "intervening" floors when referring to the same requirement. However, the definition of "intervene" is "to occur or be between two things." Stories "intervening" or "between" the Group A occupancy and the level of exit discharge did not include the level of exit discharge. The conflicting provisions were revised to correct that inconsistency and relocate this key code provision into each section where it applies.

Section 903.2.1.5.1 was also added to clarify that the general sprinkler provisions of Section 903.2.1.5 include enclosed accessory spaces under grandstands and bleachers requirements. In addition, reference is made to Section 1029.1.1.1 regarding the protection methods available where the enclosed space beneath the grandstand or bleacher does not exceed 1,000 square feet. In such cases, all spaces under bleachers or grandstands, except toilet rooms and ticket booths less than 100 square feet, must be separated from the assembly seating areas by minimum 1-hour fire barriers and/or horizontal assemblies.

Section 903.2.1.5.1 Item 1 provides an alternative of fire sprinkler protection in lieu of the required fire-resistance-rated construction if the enclosed area does not exceed 1,000 square feet. In concert with Section 903.2.1.5, Item 2 of Section 903.2.1.5.1 requires enclosed spaces more than 1,000 square feet to be equipped with automatic sprinklers.

As a companion change, Section 1029.1.1.1 was revised to read as follows:

1029.1.1.1 Spaces under grandstands and bleachers. Spaces under grandstands or bleachers shall be separated by fire barriers complying with Section 707 and horizontal assemblies complying with Section 711 with not less than 1-hour fire-resistance-rated construction.

Exceptions:

1. Ticket booths less than 100 square feet (9.29 m²) in area.
2. Toilet rooms.
3. Other accessory use areas 1,000 square feet (92.9 m²) or less in area and equipped with an automatic sprinkler system in accordance with Section 903.3.1.1.

903.2.3 Sprinklers in Group E Occupancies

CHANGE TYPE: Modification

CHANGE SUMMARY: Criteria for occupant load threshold and location within the building have been added as conditions that could require sprinkler protection in a Group E educational occupancy.

2018 CODE: 903.2.3 Group E. An automatic sprinkler system shall be provided for Group E occupancies as follows:

1. Throughout all Group E fire areas greater than 12,000 square feet (1115 m²) in area.

2. ~~Throughout every portion of educational buildings below the lowest level of exit discharge serving that portion of the building.~~

   Exception: ~~An automatic sprinkler system is not required in any area below the lowest level of exit discharge serving that area where every classroom throughout the building has not fewer than one exterior exit door at ground level.~~

2. The Group E fire area is located on a floor other than a level of exit discharge serving such occupancies.

   Exception: In buildings where every classroom has not fewer than one exterior exit door at ground level, an automatic sprinkler system is not required in any area below the lowest level of exit discharge serving that area. ~~Where every classroom throughout the building has not fewer than one exterior exit~~ door ~~at ground level~~.

3. The Group E fire area has an occupant load of 300 or more.

[Handwritten note: MUCH LIKE ASSEMBLIES]

CHANGE SIGNIFICANCE: Sprinkler requirements for educational occupancies have traditionally been applied based solely on the size of the Group E fire area. Where the fire area containing the Group E occupancy

903.2.3 continues

Group E occupancy

903.2.3 continued

did not exceed 12,000 square feet in floor area, the installation of an automatic sprinkler system was not required by Section 903.2.3. Additional criteria triggering sprinkler protection based on occupant load and occupancy location have been introduced to better align the provisions with those for Group A occupancies.

Many school buildings serve multiple purposes, including parent/teacher conference night, open houses for prospective parents and students who may have never before been to the school, political elections, social and club meetings, religious gatherings, and adult education classes. If these uses are to be allowed, then the actual use of the school becomes more similar to a Group A occupancy rather than Group E.

Fire sprinklers are now required in any Group E fire area having an occupant load of 300 persons or more, regardless of fire area size. The new occupant load threshold of 300 for Group E fire areas is now consistent with those thresholds that have historically only been applied to Group A-1, A-3, and A-4 occupancies.

New Item 3 also now requires fire sprinklers where a Group E occupancy fire area is located on a floor other than the level of exit discharge. This condition is also consistent with historic sprinkler thresholds established for Group A-1, A-2, A-3, and A-4 occupancies. Where one or more of the three conditions exist, an automatic sprinkler system is now required.

903.3.1.1.2
Omission of Sprinklers in Group R-4 Bathrooms

CHANGE TYPE: Modification

CHANGE SUMMARY: The fire sprinkler requirements that previously extended to small bathrooms in Group R-4 occupancies have been deleted.

2018 CODE: 903.3.1.1.2 Bathrooms. In Group R occupancies ~~other than Group R-4 occupancies~~, sprinklers shall not be required in bathrooms that do not exceed 55 square feet (5 m²) in area and are located within individual dwelling units or sleeping units, provided that walls and ceilings, including the walls and ceilings behind a shower enclosure or tub, are of noncombustible or limited- combustible materials with a 15-minute thermal barrier rating.

CHANGE SIGNIFICANCE: Historically, in residential occupancies the required sprinkler protection has typically not been mandated to extend to small bathrooms within such occupancies. However, the allowance has not previously applied to Group R-4 occupancies. These occupancies consist of dwelling space for 6 to 16 persons on a 24-hour basis and provide some type of custodial care, such as alcohol and drug centers, assisted living facilities, congregate care facilities, group homes, halfway houses, residential board and care facilities, and social rehabilitation facilities. Sprinklers are no longer required in small bathrooms in these occupancies in order to be consistent with the other Group R occupancies.

Small bathroom in Group R-4 occupancy

903.3.1.2.1

Sprinkler Protection at Balconies and Decks

CHANGE TYPE: Clarification

CHANGE SUMMARY: Where nonrated balconies and similar combustible projections of dwelling and sleeping units are permitted in Type IIIA and VA buildings, it has been clarified that the sprinkler protection is to be extended to the area of the projections.

2018 CODE: 903.3.1.2.1 Balconies and decks. Sprinkler protection shall be provided for exterior balconies, decks and ground floor patios of dwelling units and sleeping units where either of the following conditions exist:

1. The building is of Type V construction, provided there is a roof or deck above.

2. Exterior balconies, decks and ground floor patios of dwelling units and sleeping units are constructed in accordance with Section 705.2.3.1, Exception 3.

Sidewall sprinklers that are used to protect such areas shall be permitted to be located such that their deflectors are within 1 inch (25 mm) to 6 inches (152 mm) below the structural members and a maximum distance of 14 inches (356 mm) below the deck of the exterior balconies and decks that are constructed of open wood joist construction.

CHANGE SIGNIFICANCE: As a general rule, wood balconies and similar combustible projections of other than fire-retardant-treated wood or heavy timber construction must be provided with fire-resistance-rated floor construction in buildings of Type IIIA and VA construction. However, Exception 3 of Section 705.2.3.1 permits such balconies and similar projections to be of Type V construction without any required fire-resistance

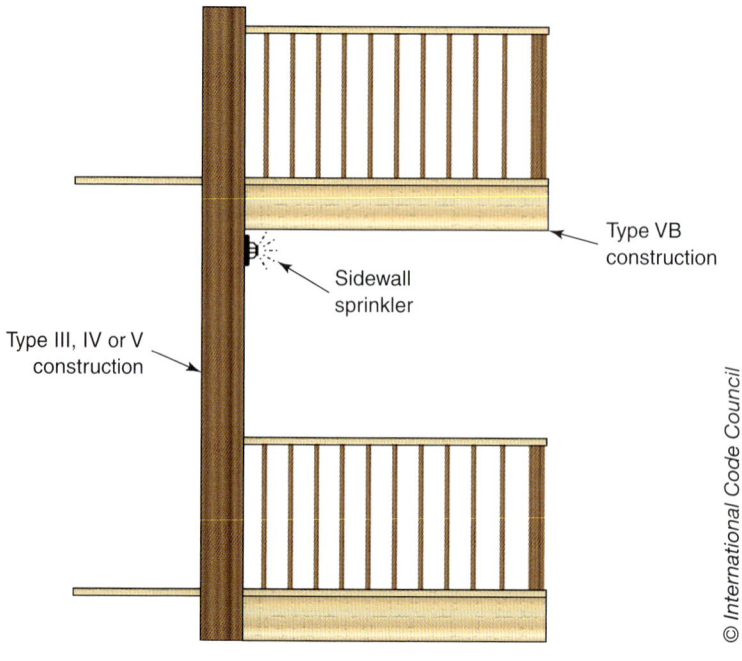

Sprinkler protection at exterior balcony

rating where sprinkler protection is extended to the area of the projections. New Item 2 of Section 903.3.1.2.1 now recognizes this requirement and specifies that sprinkler protection is required over the balcony, deck, or patio when this construction provision is utilized.

Fire sprinklers for exterior balconies, decks, and ground-floor patios with a roof above are intended to prevent exterior or exposure fires from spreading vertically or horizontally along the outside wall. Numerous significant losses have occurred in Group R multi-family occupancies where fires have spread up combustible exterior siding and entered unprotected openings, roof vents, or roof structures. Therefore, the sprinkler protection of balconies, decks, and ground-floor patios of dwelling units and sleeping units continues to be mandated in buildings of Type V construction.

903.3.1.2.3
Protection of Attics in Group R Occupancies

CHANGE TYPE: Addition

CHANGE SUMMARY: Sprinkler protection or acceptable alternative methods for the protection of attics are now addressed for mid-rise buildings housing multi-family occupancies and equipped with an NFPA 13R sprinkler system.

2018 CODE: **903.3.1.2.3 Attics.** Attic protection shall be provided as follows:

1. Attics that are used or intended for living purposes or storage shall be protected by an automatic sprinkler system.
2. Where fuel-fired equipment is installed in an unsprinklered attic, not fewer than one quick-response intermediate temperature sprinkler shall be installed above the equipment.
3. Where located in a building of Type III, Type IV or Type V construction designed in accordance with Section 510.2 or Section 510.4, attics not required by Item 1 to have sprinklers shall comply with one of the following if the roof assembly is located more than 55 feet (16 764 mm) above the lowest level of required fire department vehicle access:
   3.1. Provide automatic sprinkler system protection.
   3.2. Construct the attic using noncombustible materials.
   3.3. Construct the attic using fire-retardant-treated wood complying with Section 2303.2.
   3.4. Fill the attic with noncombustible insulation.

   The height of the roof assembly shall be determined by measuring the distance from the lowest required fire vehicle access road surface adjacent to the building to the eave of the highest

Construction of mid-rise residential building

pitched roof, the intersection of the highest roof to the exterior wall, or the top of the highest parapet, whichever yields the greatest distance. For the purpose of this measurement, required fire vehicle access roads shall include only those roads that are necessary for compliance with Section 503 of the *International Fire Code.*

4. Group R-4, Condition 2 occupancy attics not required by Item 1 to have sprinklers shall comply with one of the following:
   4.1. Provide automatic sprinkler system protection.
   4.2. Provide a heat detection system throughout the attic that is arranged to activate the building fire alarm system.
   4.3. Construct the attic using noncombustible materials.
   4.4. Construct the attic using fire-retardant-treated wood complying with Section 2303.2.
   4.5. Fill the attic with noncombustible insulation.

CHANGE SIGNIFICANCE: From 2015 to 2017, major fires in combustible, multistory, multi-family buildings have resulted in concerns about the effectiveness of automatic fire sprinklers in such structures. These buildings often were protected by fire sprinkler systems designed and installed in accordance with NFPA 13R *Standard for the Installation of Sprinkler Systems in Low Rise Residential Buildings* to protect occupant lives, but were not intended for property protection. Many of the fires started on the building exterior, spread vertically into unprotected combustible attics and then literally burned from the top down. New provisions address concerns about fire sprinkler control in taller residential buildings and "pedestal" buildings consisting of a noncombustible lower building topped by combustible living units.

Attic sprinkler protection is now required when the roof assembly is located more than 55 feet (16,764 mm) above the lowest level of required fire department vehicle access. Therefore, pedestal buildings that exceed a height of 55 feet (16,764 mm) above grade plane, including the pedestal, will be affected by the change. Some buildings with fewer stories that are located on sloped lots with fire department vehicle access roads required along the lower-elevation portion of the perimeter also may be affected by the new requirement.

The permissible attic protection options generally are modeled after existing Section 903.2.8.3 that was adopted in the 2015 edition of the IBC for Group R-4, Condition 2 buildings. Allowances for noncombustible construction materials, fire-retardant-treated wood, and filling with noncombustible insulation have for many years been permitted by NFPA 13 *Standard for the Installation of Sprinkler Systems* as an acceptable alternative to installing sprinklers in concealed spaces in otherwise fully sprinklered buildings.

Several roof conditions are addressed in order to establish the upper extent of the 55-foot measurement. The intent of identifying "required" fire apparatus access is to make it clear that, simply because access is available on an adjacent road or parking lot, the road need not be considered in the height measurement unless it is required as part of satisfying the code requirement for fire department vehicle access to the building.

903.3.1.2.3 continues

903.3.1.2.3 continued

In addition, the existing requirements in Section 903.2.8.3 for enhanced attic protection in Group R-4, Condition 2 occupancies have been relocated to the new Section 903.3.1.2.3 so that all supplemental and alternative protection requirements for NFPA 13R sprinkler systems are consolidated in one location.

The existing requirements for Group R-4, Condition 2 occupancies were also revised to clarify that, in an otherwise nonsprinklered attic, the entire attic would not require sprinklers based on the presence of fuel-fired equipment. Instead, the new scoping provision is consistent with Section 6.6.6.1 of the 2016 NFPA 13R in requiring that a sprinkler be installed only over the equipment.

904.12
Commercial Cooking Operations

CHANGE TYPE: Modification

CHANGE SUMMARY: The installation of fire-extinguishing systems as protection for commercial cooking operations must now also comply with NFPA 96. In addition, commercial cooking systems are now permitted to be protected with a water mist fire-extinguishing system complying with NFPA 750.

2018 CODE: 904.12 Commercial cooking systems. The automatic fire-extinguishing system for commercial cooking systems shall be of a type recognized for protection of commercial cooking equipment and exhaust systems of the type and arrangement protected. Preengineered automatic dry and wet chemical extinguishing systems shall be tested in accordance with UL 300 and listed and labeled for the intended application. Other types of automatic fire extinguishing systems shall be listed and labeled for specific use as protection for commercial cooking operations. The system shall be installed in accordance with this code, <u>NFPA 96,</u> its listing and the manufacturer's installation instructions. Automatic fire extinguishing systems of the following types shall be installed in accordance with the referenced standard indicated, as follows:

1. Carbon dioxide extinguishing systems, NFPA 12.
2. Automatic sprinkler systems, NFPA 13.
3. <u>Automatic water mist systems, NFPA 750</u>
~~3.~~<u>4.</u> Foam-water sprinkler system or foam-water spray systems, NFPA 16.
~~4.~~<u>5.</u> Dry-chemical extinguishing systems, NFPA 17.
~~5.~~<u>6.</u> Wet-chemical extinguishing systems, NFPA 17A.

904.12 continues

Automatic water mist system protecting commercial cooking operation

Photo courtesy of CaptiveAire

904.12 continued

Exception: Factory-built commercial cooking recirculating systems that are tested in accordance with UL 710B and listed, labeled and installed in accordance with Section 304.1 of the *International Mechanical Code.*

CHANGE SIGNIFICANCE: The protection of commercial cooking equipment and exhaust systems is perhaps the most common use of an alternative automatic fire-extinguishing system. The installation of the fire-extinguishing system is regulated through reference to a variety of criteria. An additional reference is now provided in the IBC regarding commercial cooking operations, NFPA 96 *Standard for Ventilation Control and Fire Protection of Commercial Cooking Operations.*

Having a direct reference from the *International Building Code* to NFPA 96, similar to the current references to other NFPA standards for specific system types, provides for a more efficient reference to applicable standards instead of sending the designer or code official through the *International Mechanical Code.* NFPA 96 has also been added to IBC Chapter 35 as a referenced standard. The purpose of NFPA 96 is to reduce the potential fire hazard of cooking operations and is relevant to the safety and operation of commercial cooking systems.

In addition, water mist systems have been added to the list of fire-extinguishing systems applicable for protection of commercial cooking systems. Several systems have been tested and approved for the protection of industrial cooking systems. NFPA 96 includes water mist extinguishing systems as an acceptable fire-extinguishing system for commercial cooking. Water mist systems consist of devices that emit a specific spray, or mist, that absorbs heat, displaces oxygen, and blocks heat transfer to control, suppress, or extinguish fires. NFPA 750 *Water Mist Fire Protection Systems* is referenced in regard to this new allowance.

FM Global has approved water mist systems from multiple companies for use in industrial oil cookers. The industrial oil cookers are commonly used in the food industry, containing large amounts of oil, up to 5,000 gallons, and operating at high temperatures. The FM Global approvals also apply to "downscaled" applications similar to what might be found in a commercial restaurant.

904.13 Domestic Cooking Protection in Institutional and Residential Occupancies

CHANGE TYPE: Modification

CHANGE SUMMARY: Where domestic-type cooking operations are present in Group I-1 occupancies and college dormitories classified as Group R-2, an automatic fire-extinguishing system is now mandated in conjunction with the required hood over any cooktop or range.

2018 CODE: 904.13 Domestic cooking systems. ~~in Group I-2 Condition 1. In Group I-2 Condition 1 occupancies where cooking facilities are installed in accordance with Section 407.2.6 of this code, the domestic cooking hood provided over the cooktop or range shall be equipped with an automatic fire-extinguishing system of a type recognized for protection of domestic cooking equipment. Preengineered automatic extinguishing systems shall be tested in accordance with UL 300A and listed and labeled for the intended application. The system shall be installed in accordance with this code, its listing and the manufacturer's instructions.~~ <u>Cooktops and ranges installed in the following occupancies shall be protected in accordance with Sections 904.13.1:</u>

1. <u>In Group I-1 occupancies where domestic cooking facilities are installed in accordance with Section 420.8.</u>
2. <u>In Group I-2, Condition 1 occupancies where domestic cooking facilities are installed in accordance with Section 407.2.6.</u>
3. <u>In Group R-2 college dormitories where domestic cooking facilities are installed in accordance with Section 420.10.</u>

~~**904.13.1 Manual system operation and interconnection.** Manual actuation and system interconnection for the hood suppression system shall be in accordance with Sections 904.12.1 and 904.12.2, respectively.~~

<u>**904.13.1 Protection from fire.** Cooktops and ranges shall be protected in accordance with Section 904.13.1.1 or 904.13.1.2.</u>

<u>**904.13.1.1 Automatic fire-extinguishing system.** The domestic recirculating or exterior vented cooking hood provided over the cooktop or range shall be equipped with an approved automatic fire-extinguishing system complying with the following:</u>

1. <u>The automatic fire-extinguishing system shall be of a type recognized for protection of domestic cooking equipment. Preengineered automatic fire-extinguishing systems shall be listed and labeled in accordance with UL 300A and installed in accordance with the manufacturer's instructions.</u>
2. <u>Manual actuation of the fire-extinguishing system shall be provided in accordance with Section 904.12.1.</u>
3. <u>Interconnection of the fuel and electric power supply shall be in accordance with Section 904.12.2.</u>

<u>**904.13.1.2 Ignition prevention.** Cooktops and ranges shall include burners that have been tested and listed to prevent ignition of cooking oil with burners turned on to their maximum heat settings and allowed to operate for 30 minutes.</u>

Domestic cooking equipment protected with UL 300A fire-extinguishing system

CHANGE SIGNIFICANCE: The new provisions in Section 904.13 relate directly to other code changes in Sections 420.8 and 420.10 related to the installation and use of domestic cooking equipment in assisted living facilities, group homes and college dormitories. They are generally consistent with the changes that occurred in the 2015 IBC addressing the presence of domestic cooking facilities in nursing homes and similar Group I-2, Condition 1 occupancies. The regulations have been extended to address similar hazards that occur where such cooking operations are present in Group I-1 occupancies and college dormitories classified as Group R-2. An automatic fire-extinguishing system is now mandated in such occupancies in conjunction with the required hood over any cooktop or range.

Occupants of these facilities are often capable of preparing small meals in cooking spaces less sophisticated than a full-scale commercial cooking operation. Even though present at a smaller scale, hazards still exist and occupants still need to be protected. New Section 904.13.1 provides an option, either through the installation of an automatic fire-extinguishing system recognized for protection of domestic cooking equipment, or through the installation of a cooking appliance meeting specific heating limitations.

Section 904.13.1.1 provides the criteria for an automatic fire-extinguishing system. The system must be designed and tested to UL 300A *Outline of Investigation for Extinguishing System Units for Residential Range Top Cooking Surfaces*. This standard is not equivalent to the UL 300 standard for commercial cooking facilities, but rather addresses domestic use and equipment.

Option 2 is established in Section 904.13.1.2 and requires the cooktops and ranges to meet specific heating limitations. These cooking appliances shall have listed ignition resistant burners that do not allow cooking oils to ignite during testing. Recent work by the Fire Protection Research Foundation confirms that burners meeting these specifications are highly unlikely to ignite cooking materials. Their report can be found at: http://www.nfpa.org/news-and-research/fire-statistics-and-reports/research-reports/other-research-topics/misc-reports/analytical-modeling-of-pan-and-oil-heating-on-an-electric-coil-cooktop. UL 858 *Standard for Safety for Household Electric Ranges* was recently revised to include a new Section 60A that evaluates the ability of burners to not ignite cooking oil.

904.14 Aerosol Fire Extinguishing Systems

CHANGE TYPE: Modification

CHANGE SUMMARY: The installation, inspection, testing, and maintenance of aerosol fire-extinguishing systems are now addressed through applicable references to Sections 901 and 904.4 of the IBC and NFPA 2010, as well as the system's listing and manufacturer's instructions.

2018 CODE: 904.14 Aerosol Fire-extinguishing Systems. Aerosol fire-extinguishing systems shall be installed, periodically inspected, tested and maintained in accordance with Sections 901 and 904.4, NFPA 2010, and in accordance with their listing.

Such devices and appurtenances shall be listed and installed in conformance with manufacturer's instructions.

CHANGE SIGNIFICANCE: NFPA 2010 *Standard for Fixed Aerosol Fire Extinguishing Systems* was first published in 2006. Since then, the International Code Council Evaluation Service (ICC ES) has published the ICC-ES Acceptance Criteria for Fixed Condensed Aerosol Fire-Extinguishing Systems, AC432. In 2014, the ICC-ES published evaluation report ESR-3230 for an aerosol fire-extinguishing system in compliance with the 2009 and 2012 editions of the IFC as an alternative to IFC Section 904.9, Halon Fire-extinguishing Systems. The IBC now addresses the installation, inspection, testing, and maintenance of aerosol fire-extinguishing systems through applicable references to Sections 901 and 904.4 of the IBC and NFPA 2010, as well as the system's listing and manufacturer's instructions.

Condensed aerosol fire-suppression systems used as total flooding systems for the protection of Class A (surface), Class B, and Class C hazards can reduce construction, installation, and maintenance costs compared with existing fire-extinguishing systems. This technology does not use compressed gas cylinders nor pressure-rated piping. Generally, these systems are electrically operated when integrated with approved fire alarm and releasing control systems and releasing panels, or are deployed as automatic stand-alone fire-extinguishing units.

As there are no piping distribution systems required, no special storage requirements for compressed gas bottles and the ability of the flooding agent to protect areas with limited leakage, the construction costs for these systems are typically lower than for conventional chemical and gas fire-extinguishing systems requiring gas pressure.

The technology remains effective even with leakage in the space, and therefore offers an alternative technology for installations involving reconstruction, or new construction in areas with special hazards, or where total room integrity construction is problematic.

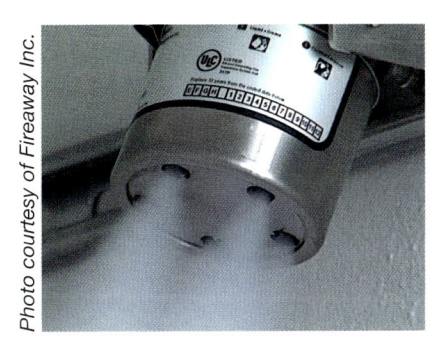

Electrically-operated aerosol fire-extinguishing device

905.3.1
Class III Standpipes

CHANGE TYPE: Modification

CHANGE SUMMARY: Standpipe system protection is now required in those buildings having four or more stories above or below grade plane regardless of the vertical distance between the floor level of the highest story and the level of the fire department vehicle access.

2018 CODE: 905.3.1 Height. Class III standpipe systems shall be installed throughout buildings where any of the following conditions exist:

1. Four or more stories above or below grade plane.
2. The floor level of the highest story is located more than 30 feet (9144 mm) above the lowest level of the fire department vehicle access. or where the.
3. The floor level of the lowest story is located more than 30 feet (9144 mm) below the highest level of fire department vehicle access.

Exceptions:

1. Class I standpipes are allowed in buildings equipped throughout with an automatic sprinkler system in accordance with Section 903.3.1.1 or 903.3.1.2.
2. Class I standpipes are allowed in Group B and E occupancies.
2.3. Class I manual standpipes are allowed in open parking garages where the highest floor is located not more than 150 feet (45 720 mm) above the lowest level of fire department vehicle access.
3.4. Class I manual dry standpipes are allowed in open parking garages that are subject to freezing temperatures, provided that the hose connections are located as required for Class II standpipes in accordance with Section 905.5.

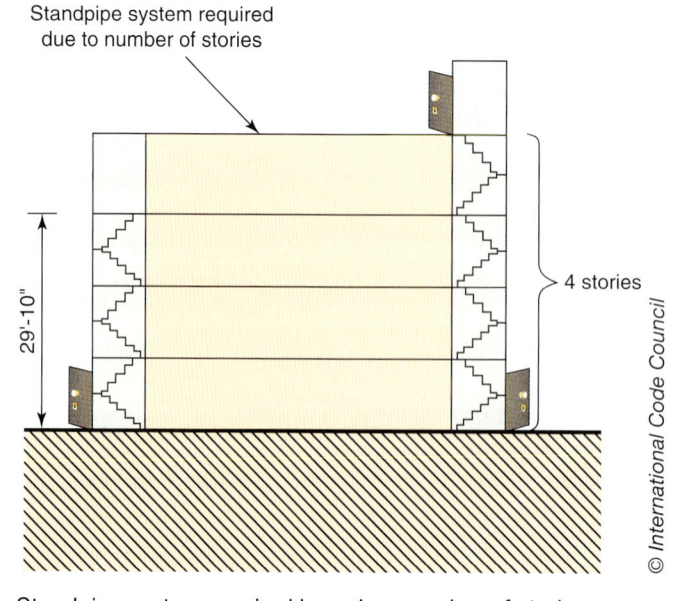

Standpipe system required based on number of stories

~~4.~~ **5.** Class I standpipes are allowed in basements equipped throughout with an automatic sprinkler system.

6. Class I standpipes are allowed in buildings where occupant-use hose lines will not be utilized by trained personnel or the fire department.

~~5.~~ **7.** In determining the lowest level of fire department vehicle access, it shall not be required to consider either of the following:

~~5.1.~~ **7.1.** Recessed loading docks for four vehicles or less.

~~5.2.~~ **7.2.** Conditions where topography makes access from the fire department vehicle to the building impractical or impossible.

CHANGE SIGNIFICANCE: A standpipe system is a system of piping, valves, and outlets that is installed exclusively for fire-fighting activities within a building. They are typically required in buildings of moderate height and greater, providing a means for trained personnel to effectively fight a fire. The primary condition for requiring standpipe protection has historically been based solely on the vertical distance between the level of fire department vehicle access and the floor level of the most remote story. A new condition requires standpipe system protection in those buildings having four or more stories above or below grade plane regardless of the vertical dimension. The base provisions mandate the installation of a Class III standpipe system where required due to height conditions. In addition, new conditions now allow for the use of Class I standpipes rather than Class III standpipes under specific conditions.

Item 1 has been added to require a standpipe in buildings that have four or more stories above or below grade plane. Previous requirements for standpipes were based solely on the height in feet between the level of fire department access and the highest or lowest floor level. This change retains the height criteria of 30 feet vertical separation, but also adds criteria that any building four or more stories above or below grade is required to have a standpipe system. The additional condition will now mandate standpipe protection in a limited number of buildings that have a vertical height of 30 feet or less between measurement limits, but have short floor-to-floor heights resulting in four or more stories.

Although the base provision mandates the installation of a Class III standpipe system where the height limits are exceeded, a number of exceptions permit a reduction to a Class I system. Hose stations for use by building occupants and hose connections for use by firefighters and other trained personnel must be provided where a Class III standpipe system is required. The hose stations need not be provided where only a Class I system is mandated.

Exception 2 has been added to address the issue of occupant-use hose lines in Group B and E occupancies. It removes the need for occupant-use hose lines in these moderate-hazard occupancies by allowing the installation of a Class I standpipe rather than a Class III system. Exception 6 has also been added allowing the installation of a Class I standpipe system in

Section 905.3.1 continues

Section 905.3.1 continued

lieu of a Class III system where the facility will not have trained personnel to utilize the occupant-use hose lines and the fire department will not utilize the hose lines.

In recent years, many fire safety and evacuation plans have all but abandoned the use of occupant-use hose lines in their training to the building occupants and employees. The primary focus of the training is evacuation. Fire behavior has changed dramatically in the past several decades due to changes in fire loading. This has created fires that develop faster, create more heat in most situations and produce greater amounts of toxic smoke. Collectively, the ability for occupants to safely and effectively utilize occupant-use hose lines without the protection of firefighting gear and respiratory protection has been greatly minimized. Where personnel will not be trained, the code now allows the elimination of the occupant-use hose lines.

905.4
Class I Standpipe Connection Locations

CHANGE TYPE: Modification

CHANGE SUMMARY: Modifications have been made regarding the location of hose connections within interior exit stairway enclosures as well as the minimum number of connections required where open breezeways and open stairs are provided.

2018 CODE: 905.4 Location of Class I standpipe hose connections. Class I standpipe hose connections shall be provided in all of the following locations:

1. In every required interior exit stairway, a hose connection shall be provided for each story above and below grade plane. Hose connections shall be located at ~~an intermediate~~ <u>the main floor</u> landing ~~between stories,~~ unless otherwise approved by the fire code official.

    Exception: <u>A single hose connection shall be permitted to be installed in the open corridor or open breezeway between open stairs that are not greater than 75 feet (22 860 mm) apart.</u>

2. No change
3. No change
4. No change
5. No change
6. No change

CHANGE SIGNIFICANCE: Where a Class I or Class III standpipe system is required, hose connections for fire department use must be provided in locations established by Section 905.4. The proper placement of standpipe

905.4 continues

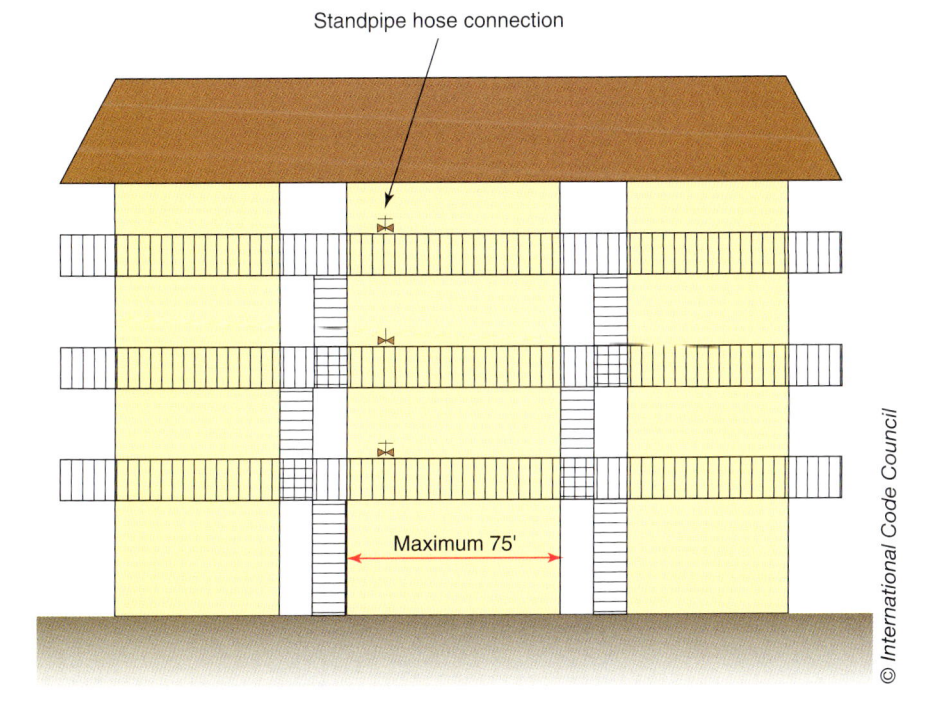

Standpipe connections in breezeways

905.4 continued

connections is necessary to best provide for firefighting operations, while also protecting fire personnel where possible. Modifications have been made regarding the location of hose connections within interior exit stairway enclosures as well as the minimum number of connections required where open breezeways and open stairs are provided.

The change to Item 1 brings the hose valve placement requirements into correlation with NFPA 14 *Standard for the Installation of Standpipe and Hose Systems*. The standard requires hose connections to be located at the main floor landing of each story, typically at the same elevation as the exit doors into the stair enclosures. Although the installation of hose valves at intermediate landings as previously required may seem more convenient for firefighter operational staging, that arrangement typically requires separate risers to be run for sprinklers and standpipes. This increases the cost and requires significantly more materials to achieve code compliance. There is no evidence to support the value of retaining the hose connections at the intermediate landings.

The new exception to Item 1 recognizes that there is no significant value to having two standpipes located at opposite ends of an open breezeway or open corridor that connects open stairs because both standpipes are essentially sharing the same environmental space. With stairways located not more than 75 feet apart, and a hose line typically consisting of 150 feet, the two hose connections would overlap each other by a distance of 75 feet. This redundancy was deemed unnecessary and therefore a second hose connection is no longer required. The intent is to provide a hose connection somewhere between the two open stairways. It is not required to locate the single connection halfway between the two stairways, as it could be located at one stairway, or closer to one stairway than the other.

907.2.1
Fire Alarms in Group A Occupancies

CHANGE TYPE: Modification

CHANGE SUMMARY: An additional criterion now mandates the installation of a manual fire alarm system where there is a Group A occupant load of more than 100 located above or below the level of exit discharge.

2018 CODE: 907.2.1 Group A. A manual fire alarm system that activates the occupant notification system in accordance with Section 907.5 shall be installed in Group A occupancies where the occupant load due to the assembly occupancy is 300 or more, <u>or where the Group A occupant load is more than 100 persons above or below the lowest level of exit discharge</u>. Group A occupancies not separated from one another in accordance with Section 707.3.10 shall be considered as a single occupancy for the purposes of applying this section. Portions of Group E occupancies occupied for assembly purposes shall be provided with a fire alarm system as required for the Group E occupancy.

> **Exception:** Manual fire alarm boxes are not required where the building is equipped throughout with an automatic sprinkler system installed in accordance with Section 903.3.1.1 and the occupant notification appliances will activate throughout the notification zones upon sprinkler water flow.

CHANGE SIGNIFICANCE: The primary hazard in an assembly space is the concentration of occupants that are at risk under fire conditions, along with the significant number of occupants that may be present. In order to provide early warning to occupants, a manual fire alarm system has historically been required in Group A occupancies where the occupant load of the assembly space is 300 or more. An additional criterion now also mandates the installation of a manual fire alarm system where there is a Group A occupant load of more than 100 located above or below the level of exit discharge.

The additional condition under which a manual fire alarm system is required in a Group A occupancy is consistent with the threshold at which the system is mandated in Group B and M occupancies. It was deemed appropriate to apply this same requirement to Group A occupancies.

A Group A occupancy located on a story other than the level of exit discharge is required to be equipped with an automatic sprinkler system, regardless of the occupant load. Now, where a Group A occupancy at other

907.2.1 continues

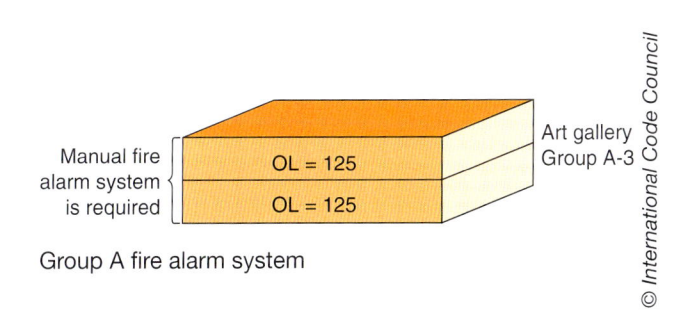

Group A fire alarm system

Manual fire alarm system is required — OL = 125, OL = 125 — Art gallery Group A-3

© International Code Council

907.2.1 continued than the lowest level of exit discharge exceeds 100 occupants, a manual fire alarm system provides an additional level of protection for those occupants located above or below the lowest discharge level. The current exception indicating that the manual fire alarm boxes are not required when the building is fully sprinklered and the occupant notification appliances will activate upon sprinkler water flow also applies to the new trigger for a manual fire alarm system.

907.2.10
Group R-4 Fire Alarm Systems

CHANGE TYPE: Deletion

CHANGE SUMMARY: The installation of a manual fire alarm system and an automatic smoke detection system is no longer required in Group R-4 occupancies.

2018 CODE: 907.2.10 Group R-4. ~~Fire alarm systems and smoke alarms shall be installed in Group R-4 occupancies as required in Sections 907.2.10.1 through 907.2.10.3.~~

907.2.10.1 Manual fire alarm system. ~~A manual fire alarm system that activates the occupant notification system in accordance with Section 907.5 shall be installed in Group R-4 occupancies.~~

~~**Exceptions:**~~

1. ~~A manual fire alarm system is not required in buildings not more than two stories in height where all individual sleeping units and contiguous attic and crawl spaces to those units are separated from each other and public or common areas by not less than 1-hour fire partitions and each individual sleeping unit has an exit directly to a public way, egress court or yard.~~

2. ~~Manual fire alarm boxes are not required throughout the building where all of the following conditions are met:~~
    - 2.1. ~~The building is equipped throughout with an automatic sprinkler system installed in accordance with Section 903.3.1.1 or 903.3.1.2.~~
    - 2.2. ~~The notification appliances will activate upon sprinkler water flow.~~
    - 2.3. ~~Not fewer than one manual fire alarm box is installed at an approved location.~~

3. ~~Manual fire alarm boxes in resident or patient sleeping areas shall not be required at exits where located at all nurses' control stations or other constantly attended staff locations, provided such stations are visible and continuously accessible and that the distances of travel required in Section 907.4.2.1 are not exceeded.~~

907.2.10.2 Automatic smoke detection system. ~~An automatic smoke detection system that activates the occupant notification system in accordance with Section 907.5 shall be installed in corridors, waiting areas open to corridors and habitable spaces other than sleeping units and kitchens.~~

~~**Exceptions:**~~

1. ~~Smoke detection in habitable spaces is not required where the facility is equipped throughout with an automatic sprinkler system installed in accordance with Section 903.3.1.1.~~

Fire alarm system not required in Group R-4

2. ~~An automatic smoke detection system is not required in buildings that do not have interior corridors serving sleeping units and where each sleeping unit has a means of egress door opening directly to an exit or to an exterior exit access that leads directly to an exit.~~

907.2.10.3 Smoke alarms. ~~Single and multiple station smoke alarms shall be installed in accordance with Section 907.2.11.~~

CHANGE SIGNIFICANCE: Group R-4 occupancies include group homes, assisted living facilities, and similar supervised residential facilities where custodial care is provided and the number of residents receiving such care is between 6 and 16, inclusive. There is an expectation that the residents are capable of self-preservation with little or no assistance from others. Because the environment is very similar to a Group R-3 occupancy, it has been deemed that the previous fire alarm requirements were overly restrictive. Therefore, the provisions requiring the installation of a manual fire alarm system and an automatic smoke detection system in Group R-4 occupancies have been deleted.

The 2015 IBC requires a manual fire alarm system and smoke detection system be installed in a Group R-4 residential care facility, which is limited to 16 or fewer residents. A Group R-4 facility is typically considered as less of a hazard than a transient Group R-1 occupancy where the occupants may not be familiar with their surroundings, but yet such an alarm system is not required in the Group R-1 occupancy unless the building is three stories or more in height. In a Group R-4 facility the residents are effectively working together similar to a single-family home rather than individual living units, and as such the requirements should be similar to those required for a Group R-3 condition.

Although deleted from Section 907.2.10, single and multiple smoke alarms continue to be required in Group R-4 occupancies as set forth in Section 907.2.10.2.

PART 4

Means of Egress

Chapter 10

- **Chapter 10** Means of Egress

The criteria set forth in Chapter 10 regulating the design of the means of egress are established as the primary method for protection of people in buildings. Both prescriptive and performance language is utilized in the chapter to provide for a basic approach in the determination of a safe exiting system for all occupancies. Chapter 10 addresses all portions of the egress system and includes design requirements as well as provisions regulating individual components. A zonal approach to egress provides a general basis for the chapter's format through regulation of the exit access, exit, and exit discharge portions of the means of egress. ■

TABLE 1004.5, 1004.8
Occupant Load Calculation in Business Use Areas

1006.2.1, TABLE 1006.2.1
Group R Spaces with One Exit or Exit Access Doorway

1006.3, 1006.3.1
Egress through Adjacent Stories

1008.2.3
Illumination of the Exit Discharge

1008.3.5, 1008.2.2
Emergency Illumination in Group I-2

1009.7.2
Protection of Exterior Areas of Assisted Rescue

1010.1.1
Size of Doors

1010.1.4.4
Locking Arrangements in Educational Occupancies

1010.1.9.8
Use of Delayed Egress Locking Systems in Group E Classrooms

1010.1.9.12
Locks on Stairway Doors

1010.3.2
Security Access Turnstiles

1013.2
Floor Level Exit Sign Location

1015.6, 1015.7
Fall Arrest for Rooftop Equipment

1017.3, 202
Measurement of Egress Travel

1023.3.1
Stairway Extensions

1023.5, 1024.6
Exit Stairway and Exit Passageway Penetrations

1025.1
Luminous Egress Path Marking in Group I Occupancies

1026.4, 1026.4.1
Refuge Areas for Horizontal Exits

1029.6, 1029.6.3, 202
Open-Air Assembly Seating

1030.1
Required Emergency Escape and Rescue Openings

Table 1004.5, 1004.8

Occupant Load Calculation in Business Use Areas

CHANGE TYPE: Modification

CHANGE SUMMARY: The method of calculating occupant load in business areas has been revised, which will typically result in reduced design occupant loads. However, higher design occupant loads can now be assigned to concentrated business areas such as telephone call centers and similar uses.

2018 CODE:

TABLE ~~1004.1.2~~ **1004.5** Maximum Floor Area Allowances Per Occupant

Function of Space	Occupant Load Factor[a]
Business areas	~~100~~150 gross
Concentrated business use areas	See Section 1004.8

(No changes to other portions of table.)

1004.8 Concentrated business use areas. The occupant load factor for concentrated business use shall be applied to telephone call centers, trading floors, electronic data processing centers and similar business use areas with a higher density of occupants than would normally be expected in a typical business occupancy environment. Where approved by the building official, the occupant load for concentrated business use areas shall be the actual occupant load, but not less than one occupant per 50 square feet (4.65 m²) of gross occupiable floor space.

CHANGE SIGNIFICANCE: Business uses have historically been viewed as having a density level of one person per 100 square feet when used in the calculation of design occupant load. It seems likely that this occupant load factor is the result of a National Bureau of Standards (NBS) [now referred to as National Institute of Standards and Technology (NIST)] study published in 1935. The occupant load factor of 100 square feet per occupant was specified for office, factory, and workroom areas. All occupant load factors were based on the gross floor area of the building, such that no deduction was permitted for corridors, closets, restrooms, or other areas of the building. Since the initial NBS study in 1935, several other studies have been conducted to determine occupant load factors for various occupancies. One common finding was that all of the subsequent studies have concluded that the factor of 100 square feet per occupant for business occupancies is conservative. Studies conducted between 1966 and 1992 have indicated that occupant load factors in business occupancies ranged from 150 to 278 square feet per occupant. A more recent project to study the appropriateness of the 100-square-feet-per-occupant factor was undertaken by the NFPA Fire Protection Research Foundation. The study was conducted by WPI undergraduate students. The recommendations of this study also supported an increase to the occupant load factor in business occupancies. Based on this information, it was deemed appropriate that the factor be increased to 150 square feet per occupant.

The NFPRF study also recommended creating a new occupant load sub-category for concentrated business use areas. New Section 1004.8 cites examples of these occupancies, including telephone call centers,

Table 1004.5, 1004.8 continues

Table 1004.5, 1004.8 continued

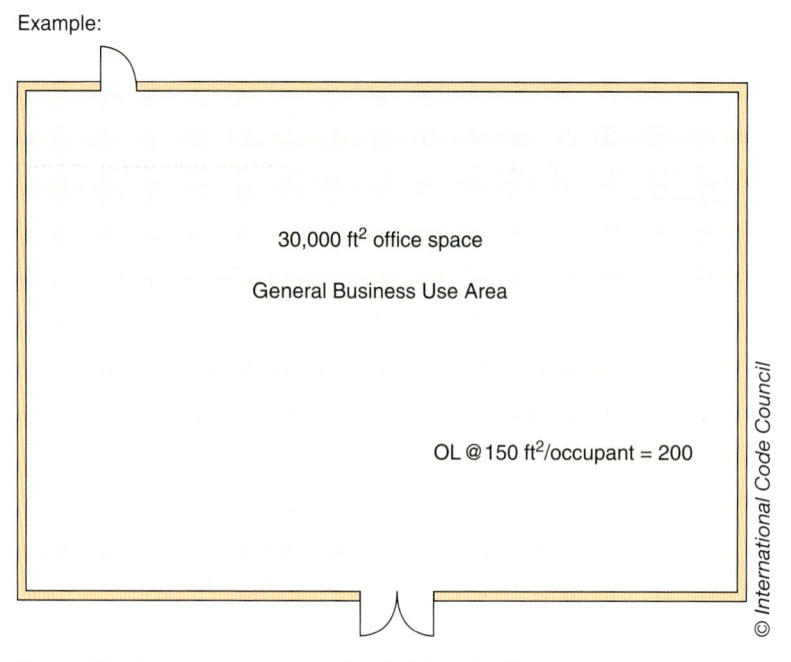

Example:

30,000 ft² office space

General Business Use Area

OL @ 150 ft²/occupant = 200

General business use occupant load determination

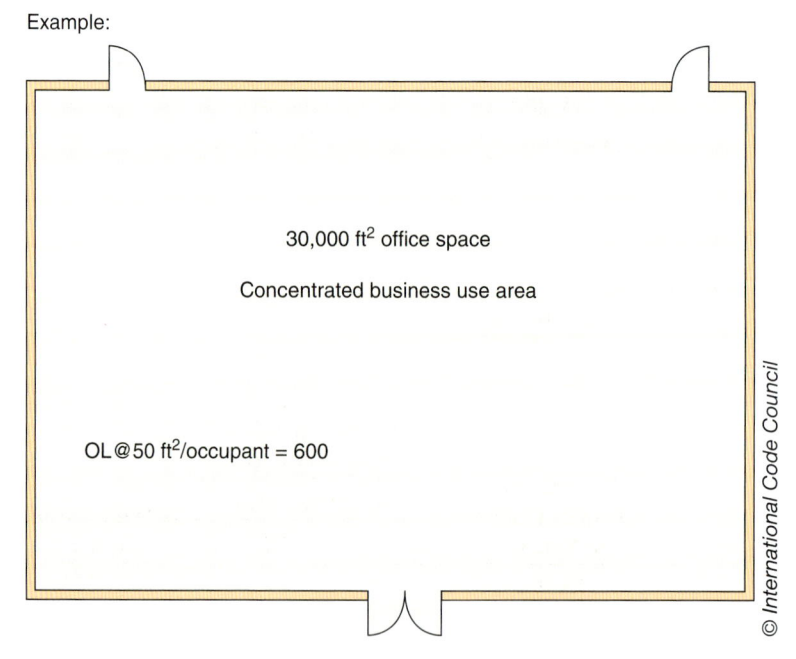

Example:

30,000 ft² office space

Concentrated business use area

OL @ 50 ft²/occupant = 600

Concentrated business use occupant load determination

trading floors, and electronic data processing centers. Essentially, the reduced factor is applicable to those business areas where a higher density of occupants would normally be expected. The actual number can be used when approved by the building official; however, the occupant load must be established at a minimum of one occupant for each 50 square feet.

For both applications of the business area occupant load calculations, the gross floor areas shall be used. Gross floor area is defined in Chapter 2 as the entire floor area, other than vent shafts and courts, within the exterior perimeter walls of the building under consideration.

1006.2.1, Table 1006.2.1

Group R Spaces with One Exit or Exit Access Doorway

CHANGE TYPE: Clarification

CHANGE SUMMARY: Allowances for single-exit Group R spaces have been reformatted and the approach to accumulating occupant loads from adjacent rooms discharging through foyers and lobbies has been clarified.

2018 CODE: 1006.2.1 Egress based on occupant load and common path of egress travel distance. Two exits or exit access doorways from any space shall be provided where the design occupant load or the common path of egress travel distance exceeds the values listed in Table 1006.2.1. <ins>The cumulative occupant load from adjacent rooms, areas or spaces shall be determined in accordance with Section 1004.2.</ins>

Exceptions:

1. ~~In Group R-2 and R-3 occupancies, one means of egress is permitted within and from individual dwelling units with a maximum occupant load of 20 where the dwelling unit is equipped throughout with an automatic sprinkler system in accordance with Section 903.3.1.1 or 903.3.1.2 and the common path of egress travel does not exceed 125 feet (38 100 mm).~~

1. <ins>The number of exits from foyers, lobbies, vestibules or similar spaces need not be based on cumulative occupant loads for areas discharging through such spaces, but the capacity of the exits from such spaces shall be based on applicable cumulative occupant loads.</ins>

2. Care suites in Group I-2 occupancies complying with Section 407.4.

TABLE 1006.2.1 Spaces With One Exit or Exit Access Doorway

| Occupancy | Maximum Occupant Load of Space | Maximum Common Path of Egress Travel Distance (feet) | | With Sprinkler System (feet) |
| | | Without Sprinkler System (feet) | | |
		OL ≤ 30	OL > 30	
R-2	~~10~~ <ins>20</ins>	NP	NP	125 [a]
R-3 [e]	~~10~~ <ins>20</ins>	NP	NP	125 [a,g]
R-4 [e]	~~10~~ <ins>20</ins>	~~75~~ <ins>NP</ins>	~~75~~ <ins>NP</ins>	125 [a,g]

(Portions of table not shown are unchanged.)

a. No change
b. No change
c. No change
d. No change
e. The ~~length of~~ common path of egress travel distance <ins>shall only apply</ins> in a Group R-3 occupancy located in a mixed occupancy building.
f. No change
g. <ins>For the travel distance limitations in Groups R-3 and R-4 equipped throughout with an automatic sprinkler system in accordance with Section 903.3.1.3, see Section 1006.2.2.6.</ins>

1006.2.1, Table 1006.2.1 continues

1006.2.1, Table 1006.2.1 continued

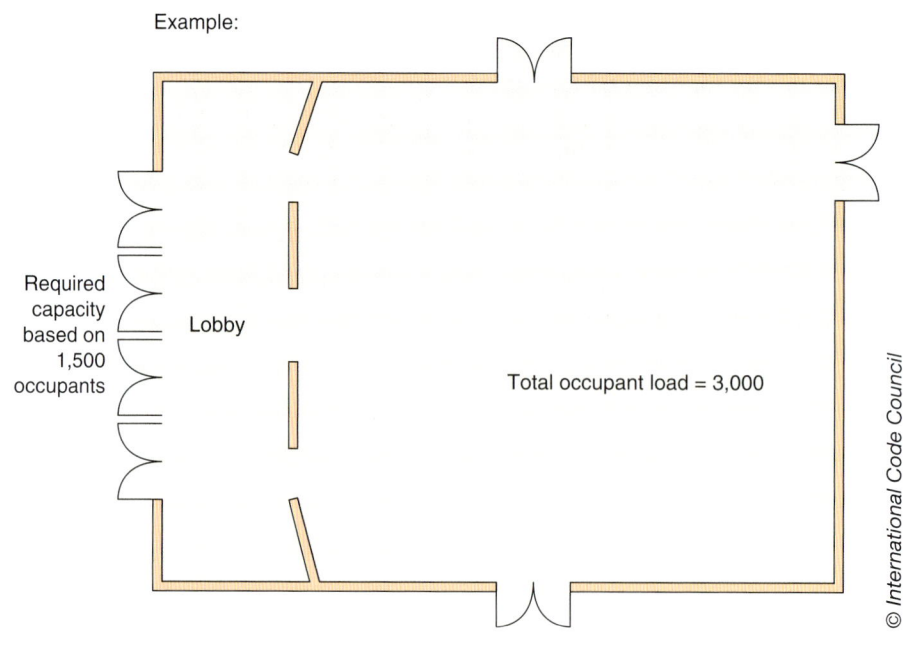

Example:

Required capacity based on 1,500 occupants — Lobby

Total occupant load = 3,000

© International Code Council

Lobby egress determination

CHANGE SIGNIFICANCE: A second exit or exit access doorway can provide an alternative route of travel for occupants of the room or area. However, it is often unreasonable to require multiple exit paths from small spaces or areas with limited occupant loads. It is also seldom beneficial because of the relatively close proximity in which such exits or exit access doorways must be located. Therefore, the code does not require a secondary egress location from all rooms, areas or spaces where they are in compliance with Table 1006.2.1. Allowances for single-exit Group R spaces have been reformatted and the approach to accumulating occupant loads from adjacent rooms discharging through foyers and lobbies have been clarified.

The previous Exception 1 to Section 1006.2.1 has been deleted. This exception stated that the maximum occupant load for one means of egress in Group R-2 or R-3 occupancies is 20 if the building is protected with an automatic sprinkler system. Because all new Group R-2 and R-3 occupancies are required to be sprinklered, this exception was always applicable. To correlate with this change, the entries for Group R-2 and R-3 occupancies in Table 1006.2.1 have been revised to acknowledge the maximum occupant load of 20 when only one means of egress is provided. The Group R-4 limit has been revised by increasing the single means of egress allowance from a maximum of 10 to a maximum of 20. As Group R-4 occupancies are limited to an occupant load of 16 residents not including staff, it is not likely that the occupant load will exceed 20.

A new Exception 1 has been added to address the situation where a lobby or foyer becomes an intervening room for egress travel. The <u>number</u> of exits from a lobby or foyer is not to be based on the cumulative occupant load; however, the <u>capacity</u>, or egress width, of the exits is to be based on the total cumulative occupant load served.

This exception is not a new approach to dealing with cumulative occupant loads but rather restates the language already established in Section 1004.2.1. The revision in Section 1006.2.1 correlates this section with Section 1004.2.1: "Design of egress path capacity shall be based on the cumulative portion of the occupant loads of all rooms, areas or spaces to that point along the path of egress travel."

Note that the main exit must still accommodate at least one-half of the required egress width when an assembly space has an occupant load greater than 300. Section 1029.2 correlates with the requirements in Section 1006.2.1 by addressing the egress capacity from the lobby rather than recalculating the number of exits from the lobby.

The addition of Footnote g to Table 1006.2.1 correlates with revisions to Table 1017.2 as discussed under that section.

1006.3, 1006.3.1
Egress through Adjacent Stories

CHANGE TYPE: Clarification

CHANGE SUMMARY: The determination of means of egress requirements has been clarified where the occupants must travel to an adjacent story to reach a complying exit or exits.

2018 CODE: 1006.3 Egress from stories or occupied roofs. The means of egress system serving any story or occupied roof shall be provided with the number of <u>separate and distinct</u> exits or access to exits based on the aggregate occupant load served in accordance with this section. ~~The path of egress travel to an exit shall not pass through more than one adjacent story.~~ <u>Where stairways serve more than one story, only the occupant load of each story considered individually shall be used in calculating the required number of exits or access to exits serving that story.</u>

1006.3.1 Adjacent story. <u>The path of egress travel to an exit shall not pass through more than one adjacent story.</u>

> **Exception:** <u>The path of egress travel to an exit shall be permitted to pass through more than one adjacent story in any of the following:</u>
>
> 1. <u>In Group R-1, R-2 or R-3 occupancies, exit access stairways and ramps connecting four stories or less serving and contained within an individual dwelling unit or sleeping unit or live/work unit.</u>
> 2. <u>Exit access stairways serving and contained within a Group R-3 congregate residence or a Group R-4 facility.</u>
> 3. <u>Exit access stairways and ramps in open parking garages that serve only the parking garage.</u>
> 4. <u>Exit access stairways and ramps serving open-air assembly seating complying with the exit access travel distance requirements of Section 1029.7.</u>
> 5. <u>Exit access stairways and ramps between the balcony, gallery or press box and the main assembly floor in occupancies such as theaters, places of religious worship, auditoriums and sports facilities.</u>

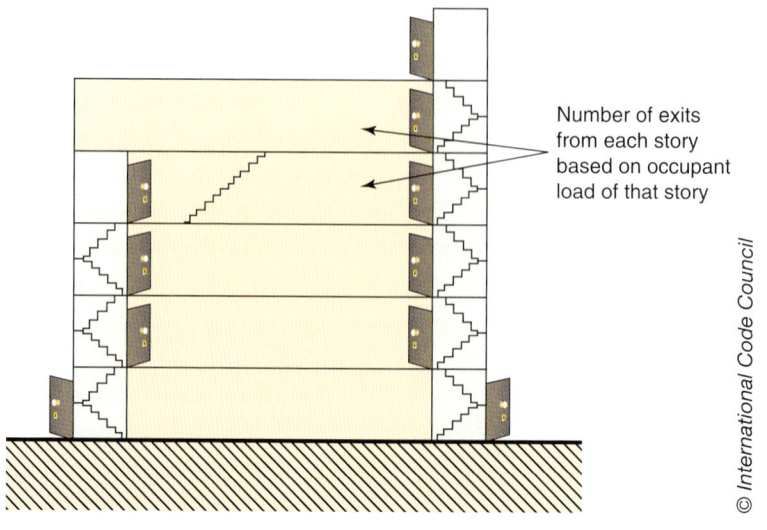

Egress travel in multi-story building

CHANGE SIGNIFICANCE: Interior exit stairways/ramps and exit access stairways/ramps often provide egress from upper stories and occupied roofs. The use of exit access stairways and ramps, typically unenclosed, is limited due to the lack of fire-resistance-rated enclosure of the vertical travel path. Therefore, the general provisions have historically mandated that egress travel to reach an exit pass through no more than one adjacent story. This requirement has been relocated, along with five previous exceptions to this limitation, to more clearly identify the limits and allowances for vertical travel not in an interior exit stairway or ramp.

In the 2015 IBC, the second sentence of Section 1006.3 states that the required number of exits must be available not more than one story above or below the floor level under consideration. That sentence has been relocated to a new Section 1006.3.1 and several applicable exceptions from Section 1019.3 have been replicated to clarify application of the new section. The five exceptions correlate with Section 1019.3 which already allows for unenclosed exit access stairways and ramps to be provided in these situations. As a result of the reformatting effort, the provisions in Sections 1019.3 and 1006.3.1 now work together to address those specific conditions under which the egress path may traverse more than one story to reach the exit, as well as the use of unenclosed stairways for those situations.

A newly introduced provision in Section 1006.3 now states that in situations where the occupants egress through an adjacent story to reach the exit, the additional occupant load is not considered when determining the required number of exits from the adjacent story. The end result is that the number of exits, or access to exits, for each story is to be based solely on the occupant load for that story. This requirement mirrors Section 1004.2.3 which has historically stated that the occupant load from an adjacent story is not added when determining the required egress width and capacity. Therefore, the number of exits, or access to exits, and the required egress capacity is based on the occupant load of each story. Keep in mind that this is specific to an adjacent story, and is treated differently than a mezzanine. The occupant load from a mezzanine is to be added to that of the room or space below into which the mezzanine travel occurs when calculating the required number of exits and egress capacity.

1008.2.3
Illumination of the Exit Discharge

CHANGE TYPE: Clarification

CHANGE SUMMARY: The introduction of illumination provisions specific to the exit discharge portion of the means of egress clarifies the extent of the illumination requirement. In addition, new language recognizes a long-held allowance for the use of safe dispersal areas and the necessary illumination where such areas are provided.

2018 CODE: 1008.2.3 Exit discharge. Illumination shall be provided along the path of travel for the exit discharge from each exit to the public way.

> **Exception:** Illumination shall not be required where the path of the exit discharge meets both of the following requirements:
>
> 1. The path of exit discharge is illuminated from the exit to a safe dispersal area complying with Section 1028.5.
> 2. A dispersal area shall be illuminated to a level not less than 1 foot candle (11 lux) at the walking surface.

CHANGE SIGNIFICANCE: In order for the egress system to afford a safe path of travel and for the building occupant to be able to negotiate the system efficiently, it is necessary that the entire egress system be provided with a certain minimum amount of illumination. Without such lighting, it would be impossible for building occupants to identify and follow the appropriate path of travel. The lack of adequate illumination would also be the cause of various other concerns, such as an increase in evacuation time, a greater potential for injuries during the egress process, and most probably an increased level of panic to those individuals trying to exit the building. General illumination has always been required throughout the means of egress system, which would include the exit discharge portion. This requirement has been further emphasized through the introduction

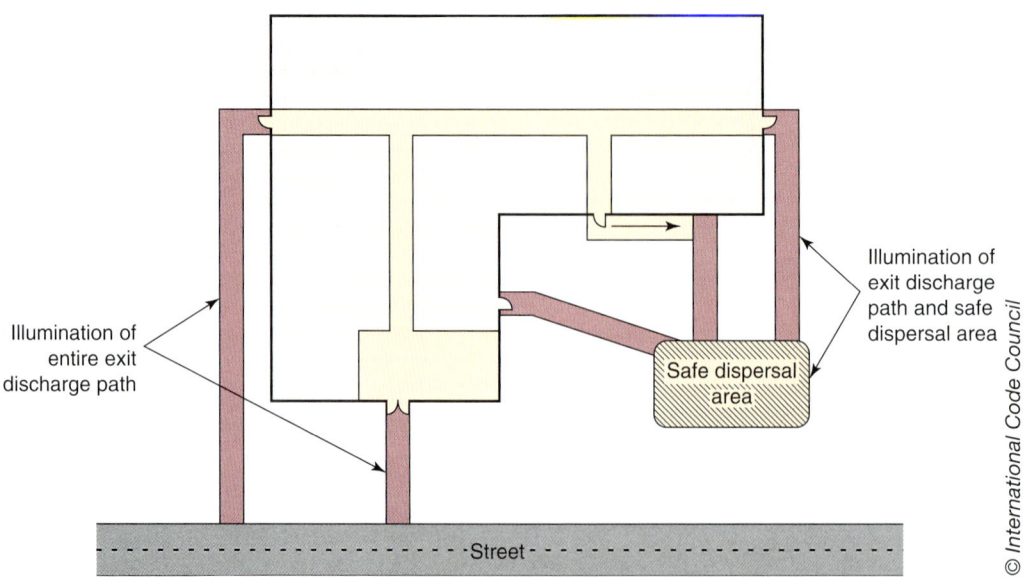

Exterior egress illumination

of illumination provisions specific to the exit discharge portion of the means of egress. In addition, new language recognizes a long-held allowance for the use of safe dispersal areas and the necessary illumination where such areas are provided.

Section 1008.1 mandates that illumination throughout the means of egress. Although the exit discharge is considered as a portion of the means of egress, the new provisions clearly specify that the required illumination must be provided for the entire exit discharge path to the public way. There are conditions under which the exit discharge is extensive and the use of a safe dispersal area is an acceptable alternative. Through a reference to Section 1028.5, a safe dispersal area must be located at least 50 feet from the building and provide adequate area to accommodate the anticipated occupant load. A minimum level of 1 footcandle is required to, and within, the safe dispersal area.

Note that this section does not require emergency illumination be provided for the exit discharge path or the safe dispersal area. Exterior emergency illumination is only required at exterior landings at exit doors as stated in Section 1008.3.2.

[Handwritten note: NOT EM LIGHTING JUST SUN/EXT. LIGHTS]

1008.3.5, 1008.2.2

Emergency Illumination in Group I-2

CHANGE TYPE: Modification

CHANGE SUMMARY: In Group I-2 occupancies, the required minimum illumination level of 0.2 footcandle must now be available upon failure of a single lamp in a multi-lamp lighting unit.

2018 CODE: **1008.3.5 Illumination level under emergency power.** Emergency lighting facilities shall be arranged to provide initial illumination that is not less than an average of 1 footcandle (11 lux) and a minimum at any point of 0.1 footcandle (1 lux) measured along the path of egress at floor level. Illumination levels shall be permitted to decline to 0.6 footcandle (6 lux) average and a minimum at any point of 0.06 footcandle (0.6 lux) at the end of the emergency lighting time duration. A maximum-to-minimum illumination uniformity ratio of 40 to 1 shall not be exceeded. In Group I-2 occupancies, failure of ~~any~~ a single ~~lighting unit~~ lamp in a luminaire shall not reduce the illumination level to less than 0.2 foot-candle (2.2 lux).

1008.2.2 ~~Exit discharge~~ Group I-2. In Group I-2 occupancies where two or more exits are required, on the exterior landings required by Section 1010.1.6, means of egress illumination levels for the exit discharge shall be provided such that failure of ~~any~~ a single ~~lighting unit~~ lamp in a luminaire shall not reduce the illumination level on that landing to less than 1 foot-candle (11 lux).

CHANGE SIGNIFICANCE: Given the possible critical nature of some of the patients and residents of Group I-2 hospitals and nursing homes and the need to move some of the people with life-sustaining equipment, there is an additional requirement that the lighting along the exit path will always have some redundancy in the fixtures providing that illumination. Previous language required that the minimum illumination level of

Emergency lighting unit

0.2 footcandles must be provided upon failure of any single lighting unit. Article 100 of the 2017 *National Electrical Code*® defines a luminaire as a "lighting unit consisting of a lamp or lamps." The required minimum illumination level must now be available upon failure of a single lamp in a multi-lamp lighting unit.

The evaluation of the emergency lighting system will now consider one of the lamps has failed, rather than both lamps in a luminaire with two lamps. A similar revision occurred in Section 1008.2.2 in regard to required illumination levels under normal building power. This mandate is applicable to the exit discharge serving Group I-2 occupancies where two or more exits are required.

1009.7.2
Protection of Exterior Areas of Assisted Rescue

CHANGE TYPE: Modification

CHANGE SUMMARY: The fire-resistance-rated exterior wall with protected openings separation between a required exterior area of assisted rescue and the interior of the building is no longer mandated, provided the building is protected with an automatic sprinkler system.

2018 CODE: 1009.7.2 Separation. Exterior walls separating the exterior area of assisted rescue from the interior of the building shall have a minimum fire-resistance rating of 1 hour, rated for exposure to fire from the inside. The fire-resistance-rated exterior wall construction shall extend horizontally <u>not less than</u> 10 feet (3048 mm) beyond the landing on either side of the landing or equivalent fire-resistance-rated construction is permitted to extend out perpendicular to the exterior wall <u>not less than</u> 4 feet (1220 mm) ~~minimum~~ on the side of the landing. The fire-resistance-rated construction shall extend vertically from the ground to a point 10 feet (3048 mm) above the floor level of the area for assisted rescue or to the roof line, whichever is lower. Openings within such fire-resistance-rated exterior walls shall be protected in accordance with Section 716.

> **Exception:** <u>The fire-resistance rating and opening protectives are not required in the exterior wall where the building is equipped throughout with an automatic sprinkler system installed in accordance with Section 903.3.1.1 or 903.3.1.2.</u>

CHANGE SIGNIFICANCE: An exterior area of assisted rescue must be provided where the exit discharge path does not consist of an accessible path from an exit at the level of exit discharge completely to a public way. The exterior area of assisted rescue provides a location for the mobility-impaired person to wait for assistance. As a general requirement, an area of assisted rescue is to be protected from the interior of the building by a minimum 1-hour fire-resistance-rated exterior wall with opening protectives. If the building is equipped with an automatic sprinkler system designed to NFPA 13 or 13R, the required fire separation and opening protection are no longer required.

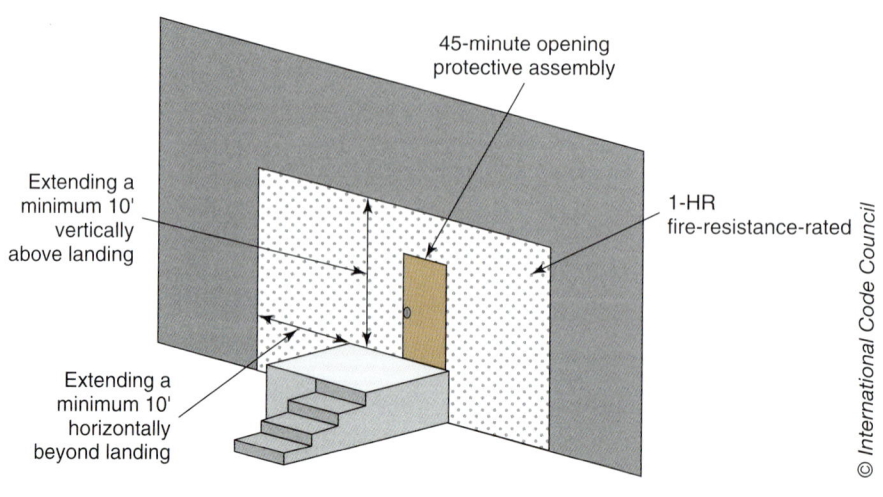

Exterior area of assisted rescue in nonsprinklered application

Section 1009.3.3, Exception 2 allows for the elimination of areas of refuge in stairways and, where applicable, at elevators if the building is fully sprinklered. The new exception to Section 1009.7.2 is based on the area of refuge concept in that if the person is adequately protected inside the building because it was sprinklered, now that the person is outside the building, the level of protection should be equivalent at the least.

This exception only allows the elimination of the fire-resistance-rated separation if the building is fully sprinklered. It does not eliminate the need to provide the exterior area of assisted rescue. The exterior area of assisted rescue must still be sized to provide one wheelchair space for every 200 occupants, or portion thereof. It must also continue to be at least 50 percent open to the outside air.

1010.1.1
Size of Doors

CHANGE TYPE: Clarification

CHANGE SUMMARY: Provisions addressing limits to the width and height of door openings have been selectively reformatted and revised as necessary to correlate with the technical accessibility requirements of ICC A117.1.

2018 CODE: 1010.1.1 Size of doors. The required capacity of each door opening shall be sufficient for the occupant load thereof and shall provide a minimum clear <u>opening</u> width of 32 inches (813 mm). ~~Clear openings~~ <u>The clear opening width</u> of doorways with swinging doors shall be measured between the face of the door and the stop, with the door open 90 degrees (1.57 rad). Where this section requires a minimum clear <u>opening</u> width of 32 inches (813 mm) and a door opening includes two door leaves without a mullion, one leaf shall provide a <u>minimum</u> clear opening width of 32 inches (813 mm). <u>In Group I-2, doors serving as means of egress doors where used for the movement of beds shall provide a minimum clear opening width of 41½ inches (1054 mm).</u> The maximum width of a swinging door leaf shall be 48 inches (1219 mm) nominal. ~~Means of egress doors in a Group I-2 occupancy used for the movement of beds shall provide a clear width not less than 41½ inches (1054 mm).~~ The <u>minimum clear opening</u> height of door openings shall be not less than 80 inches (2032 mm).

Exceptions:

1. ~~The~~ <u>In Group R-2 and R-3 dwelling and sleeping units that are not required to be an Accessible unit, Type A unit or Type B unit, the</u> minimum and maximum width shall not apply to door openings that are not part of the required means of egress ~~in Group R-2 and R-3 occupancies~~.

2. ~~Door~~ <u>In Group I-3, door</u> openings to resident sleeping units ~~in Group I-3 occupancies~~ <u>that are not required to be an Accessible unit</u> shall have a <u>minimum</u> clear <u>opening</u> width ~~of not less than~~ 28 inches (711 mm).

3. Door openings to storage closets less than 10 square feet (0.93 m²) in area shall not be limited by the minimum <u>clear opening</u> width.

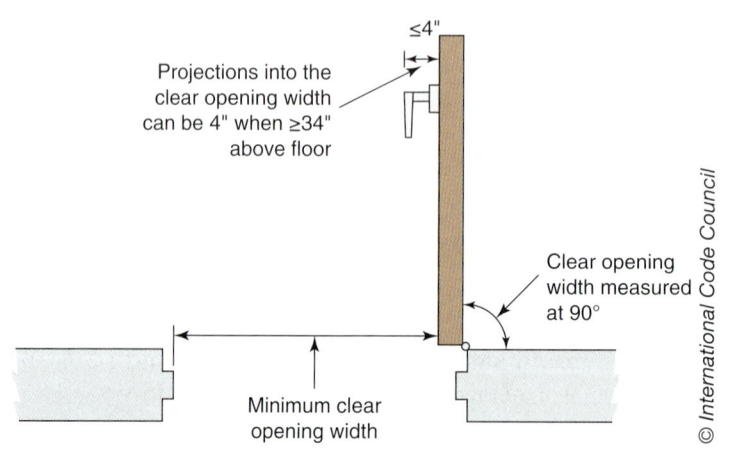

Egress door opening width

4. ~~Width~~ The width of door leaves in revolving doors that comply with Section 1010.1.4.1 shall not be limited.

5. The maximum width of door leaves in power-operated doors that comply with Section 1010.1.4.2 shall not be limited.

6. Door openings within a dwelling unit or sleeping unit shall ~~be not less than~~ have a minimum clear opening height of 78 inches (1981 mm) ~~in height~~.

~~6.~~ 7. ~~Exterior door openings in dwelling units and sleeping units, other than the required exit door, shall be not less than 76 inches (1930 mm) in height.~~ In dwelling and sleeping units that are not required to be Accessible, Type A or Type B units, exterior door openings other than the required exit door shall have a minimum clear opening height of 76 inches (1930 mm).

~~7.~~ 8. In ~~other than Group R-1 occupancies, the minimum widths shall not apply to interior egress doors within a dwelling unit or sleeping unit that is not required to be an Accessible unit, Type A unit or Type B unit.~~ Groups I-1, R-2, R-3 and R-4, in dwelling and sleeping units that are not required to be Accessible, Type A or Type B units, the minimum clear opening widths shall not apply to interior egress doors.

~~8.~~ 9. Door openings required to be accessible within Type B units intended for user passage shall have a minimum clear opening width of 31.75 inches (806 mm).

~~9.~~ 10. Doors to walk-in freezers and coolers less than 1,000 square feet (93 m²) in area shall have a maximum width of 60 inches (1524 mm) nominal.

~~10.~~ 11. ~~In Group R-1 dwelling units or sleeping units not required to be Accessible units, the~~ The minimum clear opening width shall not apply to doors for nonaccessible shower or saunas compartments.

12. The minimum clear opening width shall not apply to the doors for nonaccessible toilet stalls.

CHANGE SIGNIFICANCE: Doors are regulated for both width and height for two primary purposes: 1) their ability to provide clear and efficient egress for use during emergency purposes, and 2) their function as components of the building circulation system. In both cases, it is very important that the doors be able to accommodate individuals with disabilities. Provisions addressing limits to the width and height of door openings have been selectively reformatted and revised as necessary to correlate with the technical accessibility requirements of ICC A117.1.

This reformat of the text now provides consistency throughout the section. The same terminology is used in each condition (e.g., minimum clear opening width/height). In addition, the scoping appears first within each specific requirement (Group I-1, Group R-2, etc.).

A phrase referring to nonaccessible units has been added to Exceptions 1, 2, 7, and 8, providing consistency with ICC A117.1, the ADA Standards, and FHA. Exception 8 has been further modified to recognize that dwelling units and sleeping units in Group I-2 and I-3 occupancies have specific criteria elsewhere in this section. Additionally, because Group R-1 units are not permitted to apply the requirements of Exception

1010.1.1 continues

1010.1.1 continued

8 under the *ADA Standards for Accessible Design*, the exception has been limited to Groups I-1, R-2, R-3, and R-4 for consistency purposes.

Exception 9 has been revised to be consistent with the language in ICC A117.1 for Type B dwelling units. Revisions to Exception 11 state that the exception is not limited to nonaccessible Group R-1 dwelling units, but rather it more specifically applies to all showers and sauna compartments that are not required to be accessible. In addition, Exception 12 has been provided to address a similar issue regarding doors on toilet stalls. With a minimum required stall width of 30 inches mandated by the *International Plumbing Code*, the exception addresses a potential conflict between requirements.

1010.1.4.4
Locking Arrangements in Educational Occupancies

CHANGE TYPE: Addition

CHANGE SUMMARY: Guidance has been provided to allow for enhanced security measures on educational classroom egress doors and yet still continue to comply with applicable means of egress requirements.

2018 CODE: <u>**1010.1.4.4 Locking arrangements in educational occupancies.**</u> <u>In Group E and Group B educational occupancies, egress doors from classrooms, offices and other occupied rooms shall be permitted to be provided with locking arrangements designed to keep intruders from entering the room where all of the following conditions are met:</u>

1. <u>The door shall be capable of being unlocked from outside the room with a key or other approved means.</u>
2. <u>The door shall be openable from within the room in accordance with Section 1010.1.9.</u>
3. <u>Modifications shall not be made to listed panic hardware, fire door hardware or door closers.</u>

<u>**1010.1.4.4.1 Remote operation of locks.**</u> <u>Remote operation of locks complying with Section 1010.1.4.4 shall be permitted.</u>

CHANGE SIGNIFICANCE: A high priority in educational facilities is the safety of occupants while in classrooms and other occupied spaces during the event of a threatening situation. It is important that the IBC provides criteria which balance the challenges of providing protection for students and teachers in the classroom and at the same time provide for free and immediate egress. Guidance has now been provided to allow for enhanced security measures on educational classroom egress doors and yet still continue to comply with applicable means of egress requirements.

Oftentimes locks or devices are added to doors to provide security, but they do not allow free egress, or comply with the single operation requirement. The new provisions provide guidance for combining security while maintaining safe egress capabilities.

1010.1.4.4 continues

Security hardware

Photo courtesy of Schlage Lock Co., LLC (part of Allegion plc)

1010.1.4.4 continued

Door locksets with some type of "classroom security function" are readily available at a comparable cost to the traditional "classroom function" door locksets. The most common configuration of a classroom security function lockset is the ability to lock the door from inside the classroom with a key preventing entry to the classroom; and for egress, the door may be easily opened from inside the classroom without a key by a single action on the lever handle. On the outside of the classroom, consistent with tradition, the door may also be locked and unlocked with a key. Many of the traditional locksets required the instructor to leave the classroom and lock the door with a key from the hallway or exterior side, then reenter the classroom for a defend-in-place strategy. This action places instructors at risk by forcing them to leave the classroom and become exposed. The classroom security function eliminates the need to leave the classroom to lock the door and it still allows unrestricted egress from inside the classroom.

Additionally, this language requires that the door shall be unlockable from outside the classroom. This allows for school personnel, law enforcement and emergency responders to obtain entry even after the door is locked from the inside. This can be accomplished at the door with a key, or other approved means, including remotely as permitted in Section 1010.4.4.1.

1010.1.9.8
Use of Delayed Egress Locking Systems in Group E Classrooms

CHANGE TYPE: Modification

CHANGE SUMMARY: The allowance for the use of delayed egress locking systems has been expanded to also include egress doors serving Group E classrooms with an occupant load of less than 50, as well as secondary exits or exit access doors serving courtrooms.

2018 CODE: ~~1010.1.9.7~~ <u>1010.1.9.8</u> **Delayed egress.** Delayed egress locking systems shall be permitted to be installed on doors serving ~~any occupancy except Group A, E and H~~ <u>the following occupancies</u> in buildings that are equipped throughout with an automatic sprinkler system in accordance with Section 903.3.1.1 or an approved automatic smoke or heat detection system installed in accordance with Section 907.

<u>1. Group B, F, I, M, R, S and U occupancies</u>

<u>2. Group E classrooms with an occupant load of less than 50.</u>

Exception: <u>Delayed egress locking systems shall be permitted to be installed on exit or exit access doors, other than the main exit or exit access door, serving a courtroom in buildings that are equipped throughout with an automatic sprinkler system in accordance with Section 903.3.1.1.</u>

1010.1.9.8 continues

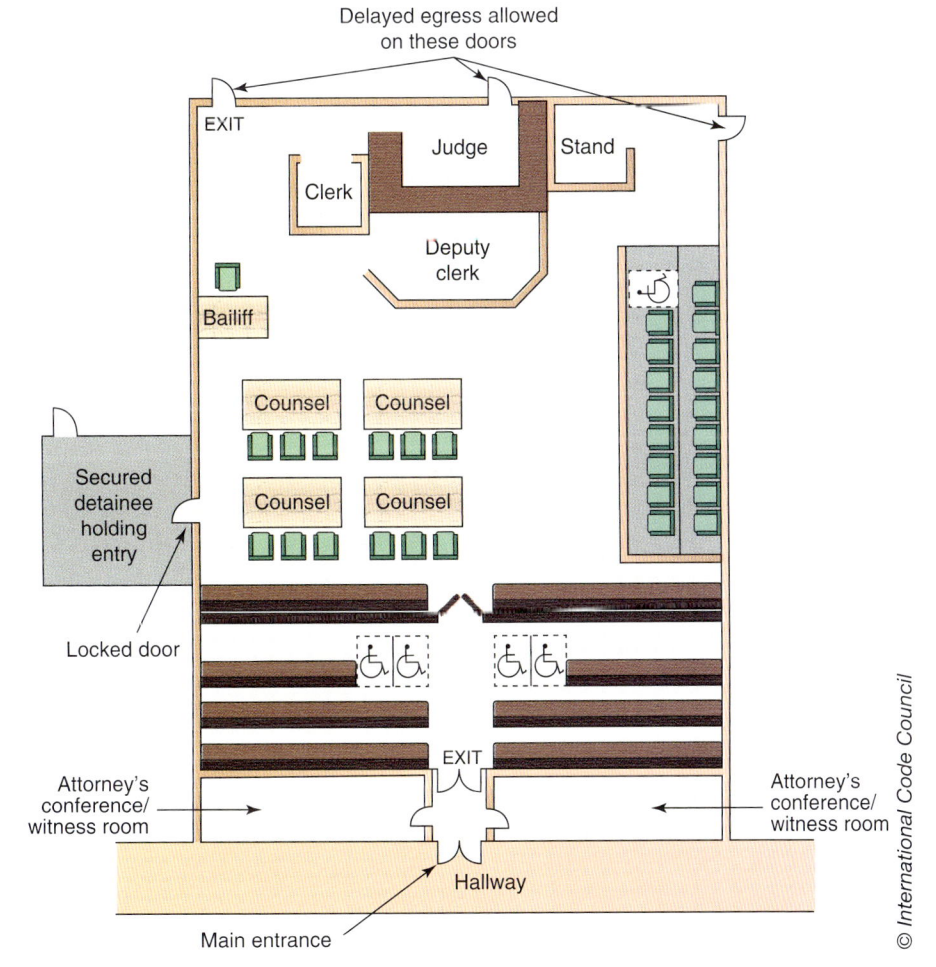

Courtroom egress example

1010.1.9.8 continued

1010.1.9.8.1 Delayed egress locking system. The delayed egress locking system shall be installed and operated in accordance with all of the following:

1. The delay electronics of the delayed egress locking system shall deactivate upon actuation of the automatic sprinkler system or automatic fire detection system, allowing immediate, free egress.
2. The delay electronics of the delayed egress locking system shall deactivate upon loss of power controlling the lock or lock mechanism, allowing immediate free egress.
3. The delayed egress locking system shall have the capability of being deactivated at the fire command center and other approved locations.
4. An attempt to egress shall initiate an irreversible process that shall allow such egress in not more than 15 seconds when a physical effort to exit is applied to the egress side door hardware for not more than 3 seconds. Initiation of the irreversible process shall activate an audible signal in the vicinity of the door. Once the delay electronics have been deactivated, rearming the delay electronics shall be by manual means only.

   Exception: Where approved, a delay of not more than 30 seconds is permitted on a delayed egress door.

5. The egress path from any point shall not pass through more than one delayed egress locking system.

   Exceptions:
   1. In Group I-2 or I-3 occupancies, the egress path from any point in the building shall pass through not more than two delayed egress locking systems provided the combined delay does not exceed 30 seconds.
   2. In Group I-1 or I-4 occupancies, the egress path from any point in the building shall pass through not more than two delayed egress locking systems provided the combined delay does not exceed 30 seconds and the building is equipped throughout with an automatic sprinkler system in accordance with Section 903.3.1.1.

6. A sign shall be provided on the door and shall be located above and within 12 inches (305 mm) of the door exit hardware:
   6.1. For doors that swing in the direction of egress, the sign shall read: PUSH UNTIL ALARM SOUNDS. DOOR CAN BE OPENED IN 15 [30] SECONDS.
   6.2. For doors that swing in the opposite direction of egress, the sign shall read: PULL UNTIL ALARM SOUNDS. DOOR CAN BE OPENED IN 15 [30] SECONDS.
   6.3. The sign shall comply with the visual character requirements in ICC A117.1.

Exception: Where approved, in Group I occupancies, the installation of a sign is not required where care recipients who because of clinical needs require restraint or containment as part of the function of the treatment area.

7. Emergency lighting shall be provided on the egress side of the door.
8. The delayed egress locking system units shall be listed in accordance with UL 294.

CHANGE SIGNIFICANCE: The acceptable use of delayed egress devices was introduced in the code years ago to resolve the problem of an exit door being illegally blocked by building operators desperate to stop the theft of merchandise through unsupervised, secondary exits. Institutional and residential occupancies have been more recently included because it is perceived that they also have security issues that must be addressed while maintaining viable exit systems. The allowance for the use of delayed egress locking systems has now been expanded to also include egress doors serving Group E classrooms with an occupant load of less than 50, as well as secondary exits or exit access doors serving courtrooms.

Previously, if a courtroom was classified as a Group B occupancy because the occupant load was below 50, then delayed egress locking devices were allowed by the code. The new exception to Section 1010.1.9.8 allows for delayed egress devices to also be installed on egress doors within a courtroom classified as Group A-3. This exception only permits delayed egress on the exit doors or exit access doors which are not the main exit. This typically will be the doors on the judge's side of the bar. Note that for use in Group A-3 courtrooms, a fire sprinkler system is required throughout the building. This differs a bit from the other acceptable occupancies where either a fire sprinkler system or a fire detection system is required if delayed egress is provided. A fire detection system cannot be used in the Group A courtroom as a compliant condition to allow for delayed egress.

Delayed egress is also now allowed in Group E classrooms, but only when the occupant load is less than 50. Delayed egress is often used to prevent wandering or elopement of younger children. It provides additional supervisory support while maintaining a fully functioning means of egress.

1010.1.9.12
Locks on Stairway Doors

CHANGE TYPE: Modification

CHANGE SUMMARY: Previously limited to only those stairways serving four or fewer stories, the allowance for stairway doors to be locked on the stairway side until simultaneously unlocked from a signal by emergency personnel is now applicable to all multi story conditions which are not considered as high-rise buildings.

2018 CODE: ~~1010.1.9.11~~ <u>1010.1.9.12</u> **Stairway doors.** Interior stairway means of egress doors shall be openable from both sides without the use of a key or special knowledge or effort.

Exceptions:

1. Stairway discharge doors shall be openable from the egress side and shall only be locked from the opposite side.
2. This section shall not apply to doors arranged in accordance with Section 403.5.3.
3. ~~In stairways serving not more than four stories,~~ <u>Stairway exit</u> doors are permitted to be locked from the side opposite the egress side, provided <u>that</u> they are openable from the egress side and capable of being unlocked simultaneously without unlatching upon a signal from the fire command center, if present, or a signal by emergency personnel from a single location inside the main entrance to the building.
4. Stairway exit doors shall be openable from the egress side and shall only be locked from the opposite side in Group B, F, M and S occupancies where the only interior access to the tenant space is from a single exit stairway where permitted in Section ~~1006.3.2~~ <u>1006.3.3</u>.

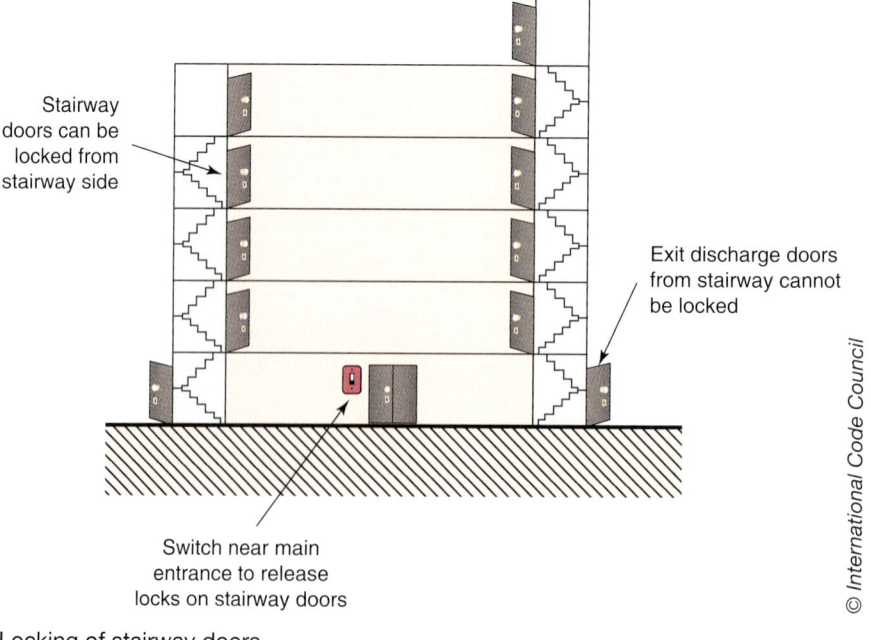

Locking of stairway doors

5. Stairway exit doors shall be openable from the egress side and shall only be locked from the opposite side in Group R-2 occupancies where the only interior access to the dwelling unit is from a single exit stairway where permitted in Section ~~1006.3.2~~ 1006.3.3.

CHANGE SIGNIFICANCE: The general requirement for interior stairway doors is much broader than the general requirement for all means of egress doors. Although the general mandate for all means of egress doors is that they be openable from the egress side without the use of a key, special effort, or special knowledge, interior stairway doors must be openable from both sides under the same conditions. This allows for immediate access from the stairway enclosure to the adjacent floor area for emergency responders. In addition, in the unlikely event that the stairwell becomes untenable during evacuation procedures, occupants may reenter a floor level as an alternative means of egress. Exception 3 to this general requirement provides an allowance for stairway doors to be locked on the stairway side until simultaneously unlocked from a signal by emergency personnel. Previously limited to only those stairways serving four or fewer stories, this exception is now applicable to all multi story conditions which are not considered as high-rise buildings.

In a high-rise building, stairway doors are permitted to be locked from the side opposite egress under the provisions of Section 403.5.3. For buildings other than high-rises, 2015 Section 1010.1.9.11 similarly regulates the locking of stairway doors where the stairway serves up to four stories. Although the high-rise allowance could be applied to non-high-rise buildings, the stairway communication requirement for high-rise conditions would apply in such cases. The revised scoping of the exception now permits locking from the stairway side without the presence of a telephone or two-way communications system.

Much like the requirements for high-rise buildings, the doors must be capable of being unlocked simultaneously without unlatching. The unlocking action must occur when signaled by emergency personnel from a single location inside the main entrance to the building. Where the building has a fire command center, the signal may also originate from that location.

1010.3.2
Security Access Turnstiles

CHANGE TYPE: Addition

CHANGE SUMMARY: New conditions of use are now provided to the building official with criteria to evaluate security access turnstiles that are located in a manner to obstruct a means of egress.

2018 CODE: 1010.3 Turnstiles and similar devices. Turnstiles or similar devices that restrict travel to one direction shall not be placed so as to obstruct any required means of egress, except where permitted in accordance with Sections 1010.3.1, 1010.3.2 and 1010.3.3.

~~Exception:~~ **1010.3.1 Capacity.** Each turnstile or similar device shall be credited with a capacity based on not more than a 50-person occupant load where all of the following provisions are met:

1. Each device shall turn free in the direction of egress travel when primary power is lost and on the manual release by an employee in the area.
2. Such devices are not given credit for more than 50 percent of the required egress capacity or width.
3. Each device is not more than 39 inches (991 mm) high.
4. Each device has not less than 16½ inches (419 mm) clear width at and below a height of 39 inches (991 mm) and not less than 22 inches (559 mm) clear width at heights above 39 inches (991 mm).

1010.3.1.1 Clear width. Where located as part of an accessible route, turnstiles shall have not less than 36 inches (914 mm) clear at and below a height of 34 inches (864 mm), not less than 32 inches (813 mm) clear width between 34 inches (864 mm) and 80 inches (2032 mm) and shall consist of a mechanism other than a revolving device.

Security access devices

1010.3.2 Security access turnstiles. Security access turnstiles that inhibit travel in the direction of egress utilizing a physical barrier shall be permitted to be considered as a component of the means of egress, provided that all the following criteria are met:

1. The building is protected throughout by an automatic sprinkler system in accordance with Section 903.3.1.1.
2. Each security access turnstile lane configuration has a minimum clear passage width of 22 inches (559 mm).
3. Any security access turnstile lane configuration providing a clear passage width of less than 32 inches (810 mm) shall be credited with a maximum egress capacity of 50 persons.
4. Any security access turnstile lane configuration providing a clear passage width of 32 inches (810 mm) or more shall be credited with a maximum egress capacity as calculated in accordance with Section 1005.
5. Each secured physical barrier shall automatically retract or swing to an unobstructed open position in the direction of egress, under each of the following conditions:
    5.1. Upon loss of power to the turnstile or any part of the access control system that secures the physical barrier.
    5.2. Upon actuation of a readily accessible and clearly identified manual release device that results in direct interruption of power to each secured physical barrier, after which such barriers remain in the open position for not less than 30 seconds. The manual release device shall be positioned at one of the following locations:
        5.2.1. On the egress side of each security access turnstile lane.
        5.2.2. At an approved location where it can be actuated by an employee assigned to the area at all times that the building is occupied.
    5.3. Upon actuation of the building fire alarm system, if provided, after which the physical barrier remains in the open position until the fire alarm system is manually reset.
        Exception: Actuation of a manual fire alarm box.
    5.4. Upon actuation of the building automatic sprinkler or fire detection system, after which the physical barrier remains in the open position until the fire alarm system is manually reset.

~~1010.3.1~~ **1010.3.3 High turnstile.** Turnstiles more than 39 inches (991 mm) high shall meet the requirements for revolving doors or the requirements of Section 1010.3.2 for security access turnstiles.

~~1010.3.2~~ **1010.3.4 Additional door.** Where serving an occupant load greater than 300, each turnstile that is not portable shall have a side-hinged swinging door that conforms to Section 1010.1 within 50 feet (15 240 mm).

Exception: A side-hinged swinging door is not required at security access turnstiles that comply with Section 1010.3.2.

1010.3.2 continues

1010.3.2 continued

CHANGE SIGNIFICANCE: As turnstiles within the means of egress system typically create a considerable degree of obstruction to efficient use of the system, they are strictly regulated by the code. Manufacturers of turnstile devices have expanded into the security access control market and currently have products that have physical barrier leaves that restrict access into and out of buildings. These devices can vary in height and sophistication to address building security concerns that may not meet safety requirements related to the means of egress. Typically, these turnstile devices are located at building entrances and elevator lobbies. The current requirements for turnstiles apply historically to the "three arm" waist-high turnstiles for entertainment or transportation venues and do not apply to the new installations. New conditions of use are now provided to the building official with criteria to evaluate these new modern security access turnstiles.

To be considered for installation in the path of egress travel, a turnstile or similar device must provide a clear egress width of at least 22 inches. Such devices between 22 inches and 32 inches can only be considered to accommodate a maximum of 50 occupants. Where the device provides at least 32 inches of clear egress width, the maximum egress capacity of the turnstile is calculated based upon the criteria of Section 1005.

Turnstiles and similar devices which inhibit travel in the direction of egress are only allowed if the building is protected with an approved, supervised sprinkler system designed to the requirements of NFPA 13, and the devices are capable of being retracted or opened automatically or manually by a security guard or similar employee. Automatic operation must occur upon actuation of the fire sprinkler system required in Item 1, or fire alarm system, if one is provided. The manual override for the turnstile access is similar to that required for delayed egress locks.

1013.2 Floor Level Exit Sign Location

CHANGE TYPE: Modification

CHANGE SUMMARY: The permitted location for low-level exit signs selectively required in Group R-1 occupancies has been expanded to now allow the bottom of such sign to be mounted up to 18 inches above the floor.

2018 CODE: 1013.2 ~~Floor-level~~ Low-level exit signs in Group R-1. Where exit signs are required in Group R-1 occupancies by Section 1013.1, additional low-level exit signs shall be provided in all areas serving guest rooms in Group R-1 occupancies and shall comply with Section 1013.5.

The bottom of the sign shall be not less than 10 inches (254 mm) nor more than ~~12 inches (305 mm)~~ 18 inches (455 mm) above the floor level. The sign shall be flush mounted to the door or wall. Where mounted on the wall, the edge of the sign shall be within 4 inches (102 mm) of the door frame on the latch side.

CHANGE SIGNIFICANCE: To help guide occupants of Group R-1 guest rooms to the exits during emergency conditions, additional exit signs are required within the egress system serving the guest rooms. Limiting the application to the egress system serving the guest rooms of hotels and other Group R-1 occupancies recognizes the transient nature of building's use, the often-delayed response to emergency conditions, and the typical low ceiling height in the corridors.

In the 2015 IBC, only a 2-inch tolerance was established for where the bottom of required low-level exit signs must be located. This 2-inch allowance was often challenging for designers and property owners due to field conditions or desired interior finish and trim. For example, several high-end resort properties have installed 12-inch-tall base boards in the exit access corridors of their hotels. The requirement for the bottom of the sign to be located within 10 to 12 inches above the floor level creates issues for these and similar facilities.

The bottom of the required low-level exit signs is now permitted to be located between 10 and 18 inches of the floor level. The additional 6 inches now available provides sufficient flexibility for designers and owners without adversely impacting the level of life safety of the occupants of the Group R-1 occupancies because the low-level exit signs will still be visible below the smoke layer from a fire (in the zone in which the occupants would presumably be crawling).

NFPA 101 *Life Safety Code* Section 7.10.1.6 permits the bottom of low-level exit signs to be installed between 6 and 18 inches above the floor level. Therefore, there is another life safety standard that permits the bottom of low-level exit signs to be installed up to 18 inches above the floor level. Although NFPA 101 permits the bottom of the low-level exit signs to be just 6 inches above the floor level, the IBC retains the requirement for a minimum of 10 inches above the floor level because ICC A117.1 Section 404.2.9 requires door surfaces within 10 inches of the floor to be a smooth surface for the full width of the door. As low-level exit signs can be installed either on the wall or the door, it is important that the low-end limit be such that it does not conflict with accessibility requirements.

Low-level exit sign

1015.6, 1015.7
Fall Arrest for Rooftop Equipment

CHANGE TYPE: Modification

CHANGE SUMMARY: The prescriptive provisions addressing the installation of personal fall arrest/restraint anchorage where mechanical equipment or roof hatches are located close to a roof edge have now been deleted with simply a reference to the ANSI/ASSE Z 359.1 standard.

2018 CODE: 1015.6 Mechanical equipment, systems and devices. Guards shall be provided where various components that require service are located within 10 feet (3048 mm) of a roof edge or open side of a walking surface and such edge or open side is located more than 30 inches (762 mm) above the floor, roof or grade below. The guard shall extend not less than 30 inches (762 mm) beyond each end of such components. The guard shall be constructed so as to prevent the passage of a sphere 21 inches (533 mm) in diameter.

> **Exception:** Guards are not required where ~~permanent~~ personal fall arrest/restraint anchorage connector devices that comply with ANSI/ASSE Z 359.1 are ~~affixed for use during the entire roof covering lifetime. The devices shall be reevaluated for possible replacement when the entire roof covering is replaced. The devices shall be placed not more than 10 feet (3048 mm) on center along hip and ridge lines and placed not less than 10 feet (3048 mm) from the roof edge or open side of the walking surface~~ installed.

1015.7 Roof access. Guards shall be provided where the roof hatch opening is located within 10 feet (3048 mm) of a roof edge or open side of a walking surface and such edge or open side is located more than 30 inches (762 mm) above the floor, roof or grade below. The guard shall be constructed so as to prevent the passage of a sphere 21 inches (533 mm) in diameter.

Fall arrest/restraint anchorage

Exception: Guards are not required where ~~permanent~~ personal fall arrest/restraint anchorage connector devices that comply with ANSI/ASSE Z 359.1 are ~~affixed for use during the entire roof covering lifetime. The devices shall be reevaluated for possible replacement when the entire roof covering is replaced. The devices shall be placed not more than 10 feet (3048 mm) on center along hip and ridge lines and placed not less than 10 feet (3048 mm) from the roof edge or open side of the walking surface~~ installed.

CHANGE SIGNIFICANCE: Guard protection is typically required at open sides of walking surfaces, stairways, mezzanines, and other elevated areas where a significant fall could occur. Where mechanical equipment requiring service or a roof hatch opening is located adjacent to a roof edge or open side of a walking surface where the elevation change exceeds 30 inches to the surface below, a guard must be provided for fall protection purposes. An exception was introduced in the 2015 IBC allowing for the omission of such required guards where fall arrest/restraint anchorage connector devices are installed. In addition to referencing the requirements of ANSI/ASSE Z 359.1, the 2015 provisions addressed device reevaluation triggered by roof covering replacement and the placement locations of such devices. Through the removal of the language requiring placement of anchors along hip or ridge lines and along roof edges, the provisions now simply refer to ANSI/ASSE Z359.1 *Safety Requirements for Personal Fall Arrest Systems, Subsystems and Components, Part of the Fall Protection Code* for compliance criteria.

Section 102.4.1 specifies that when a conflict occurs between the code and a referenced standard, the code language applies. Previously, the anchors had to be installed at intervals of 10 feet based on the IBC criteria, even if the standard did not require the same installation method. Removing the criteria in the code allows the standard to govern for all installations and eliminate any confusion. In addition, the revision allows the anchors for each building to be designed based on the actual roof system and equipment location. The standard provides guidance on anchor placement and installation which can be applied on a case-by-case basis to fit the specific activities that may occur on each individual roof.

1017.3, 202
Measurement of Egress Travel

CHANGE TYPE: Clarification

CHANGE SUMMARY: Additional language clarifies that the common path of egress travel limitations must be applied to each room or space on every story.

2018 CODE: 1017.3 Measurement. Exit access travel distance shall be measured from the most remote point ~~within a story~~ <u>of each room, area or space</u> along the natural and unobstructed path of horizontal and vertical egress travel to the entrance to an exit.

> **Exception:** In open parking garages, exit access travel distance is permitted to be measured to the closest riser of an exit access stairway or the closest slope of an exit access ramp.

SECTION 202 DEFINITIONS

COMMON PATH OF EGRESS TRAVEL. That portion of ~~the~~ exit access travel distance measured from the most remote point ~~within a story~~ <u>of each room, area or space</u> to that point where the occupants have separate <u>and distinct</u> access to two exits or exit access doorways.

CHANGE SIGNIFICANCE: The means of egress shall be accessible in terms of its arrangement so that the distance of travel from an occupied point in the building to an exit is not excessive. Maximum distances permitted to exits are established from any occupiable part of a building. Such "travel distance" is measured to the door of an exit, such as an interior exit stairway or an exterior door at the discharge level. It has been clarified that every room, space, or area shall be considered in the evaluation of travel distance, not just from the remote point of the story. In addition, similar clarification has been provided regarding the evaluation of "common paths of egress travel."

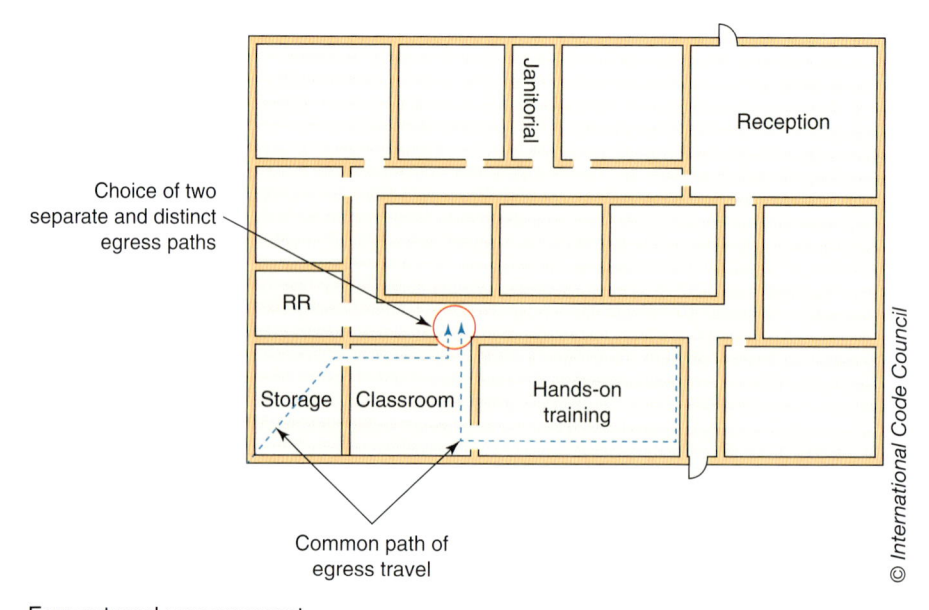

Egress travel measurement

The provisions addressing the measurement of travel distance have been revised along with the definition of "common path of egress travel" to clarify that each room or space must be individually evaluated. If applied literally, the 2015 IBC could be interpreted such that the common path of egress travel need only be considered from one point (the most remote) on a given story.

Additionally, the deletion of the definition reference to a single story allows for the common path of travel to continue to an adjacent level, as provided in Section 1006.3. Common path of travel requirements could potentially apply to a multi level design condition. The definition is further refined by specifying that common path of egress travel is measured to the point where separate and distinct egress paths are available. The reference to "separate and distinct" access to exits or exit access doorways clarifies that the egress paths must be independent.

1023.3.1
Stairway Extensions

CHANGE TYPE: Modification

CHANGE SUMMARY: Fire-resistance-rated separation is not required between an interior exit stairway and its exit passageway extension where both the stair enclosure and exit passageway are pressurized.

2018 CODE: 1023.3.1 Extension. Where interior exit stairways and ramps are extended to an exit discharge or a public way by an exit passageway, the interior exit stairway and ramp shall be separated from the exit passageway by a fire barrier constructed in accordance with Section 707 or a horizontal assembly constructed in accordance with Section 711, or both. The fire-resistance rating shall be not less than that required for the interior exit stairway and ramp. A fire door assembly complying with Section 716.5 shall be installed in the fire barrier to provide a means of egress from the interior exit stairway and ramp to the exit passageway. Openings in the fire barrier other than the fire door assembly are prohibited. Penetrations of the fire barrier are prohibited.

Exceptions:

1. Penetrations of the fire barrier in accordance with Section 1023.5 shall be permitted.

2. Separation between an interior exit stairway or ramp and the exit passageway extension shall not be required where there are no openings into the exit passageway extension.

3. <u>Separation between an interior exit stairway or ramp and the exit passageway extension shall not be required when the interior exit stairway and the exit passageway extension are pressurized in accordance with Section 909.20.5.</u>

CHANGE SIGNIFICANCE: Egress in an interior exit stairway not located at the building's perimeter typically is extended through an exit passageway to reach the exterior. The exit passageway provides a level of occupant protection equivalent to that of the interior exit stairway. As a general rule, the interior exit stairway must be separated from the exit passageway through the use of fire-resistance-rated construction with

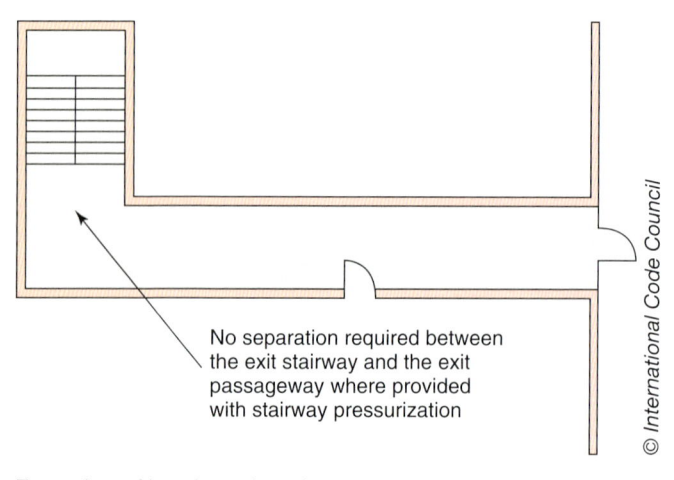

No separation required between the exit stairway and the exit passageway where provided with stairway pressurization

© International Code Council

Extension of interior exit stairway

protected openings. Where both the interior exit stairway and the exit passageway are pressurized in accordance with Section 909.20.5, the separation is no longer required.

In a high-rise building or an underground building, interior exit stairways are required to be smokeproof enclosures. Exception 2 of Section 1023.11.1 already permits the elimination of the separation between the smokeproof enclosure and the exit passageway where the exit passageway is pressurized in the same manner as the enclosure. The new exception to Section 1023.3.1 extends this allowance to all interior exit stairways that are extended by an exit passageway.

Where a stair enclosure is pressurized, any accompanying exit passageway is also typically required to be pressurized because it is a continuation of the pressurized stair enclosure. A single mechanical system is often provided to pressurize the exit stairway and exit passageway. Technical compliance with the previous provisions would require separate systems, if a separation is required to be maintained. In addition, the introduction of a door and fire barrier between the exit passageway and the stair enclosure creates an obstruction to airflow which inhibits the combined pressurization of the stairway and passageway.

The addition of the fire-resistance-rated separation between the stairway and passageway does not provide any added level of safety, and could actually impede egress. With the separation not required, the door and wall are not required. This creates a single atmosphere for the exit stairway and exit passageway.

1023.5, 1024.6

Exit Stairway and Exit Passageway Penetrations

CHANGE TYPE: Modification

CHANGE SUMMARY: Security system and two-way communication system components are now specifically permitted to penetrate the fire-resistant-rated enclosure of exit passageways, interior exit stairways, and interior exit ramps.

2018 CODE: 1023.5 Penetrations. Penetrations into or through interior exit stairways and ramps are prohibited except for <u>the following</u>:

1. Equipment and ductwork necessary for independent ventilation or pressurization.
2. ~~sprinkler piping, standpipes~~ <u>Fire protection systems.</u>
3. <u>Security systems.</u>
4. <u>Two-way communication systems.</u>
5. Electrical raceway for fire department communication systems. ~~and~~
6. Electrical raceway serving the interior exit stairway and ramp and terminating at a steel box not exceeding 16 square inches (0.010 m²).

Such penetrations shall be protected in accordance with Section 714. There shall not be penetrations or communication openings, whether protected or not, between adjacent interior exit stairways and ramps.

Exception: Membrane penetrations shall be permitted on the outside of the interior exit stairway and ramp. Such penetrations shall be protected in accordance with Section 714.4.2.

1024.6 Penetrations. Penetrations into or through an exit passageway are prohibited except for <u>the following</u>:

1. Equipment and ductwork necessary for independent ventilation or pressurization.
2. ~~sprinkler piping, standpipes~~ <u>Fire protection systems.</u>
3. <u>Security systems.</u>
4. <u>Two-way communication systems.</u>
5. Electrical raceway for fire department communication systems. ~~and~~
6. Electrical raceway serving the interior exit stairway and ramp and terminating at a steel box not exceeding 16 square inches (0.010 m²).

Such penetrations shall be protected in accordance with Section 714. There shall not be penetrations or communicating openings, whether protected or not, between adjacent exit passageways.

Exception: Membrane penetrations shall be permitted on the outside of the exit passageway. Such penetrations shall be protected in accordance with Section 714.4.2.

Stair enclosure penetrations

1023.5, 1024.6 ■ Exit Stairway and Exit Passageway

CHANGE SIGNIFICANCE: Because the enclosure of interior exit stairways, interior exit ramps, and exit passageways is so fundamental to the safety of building occupants and their ability to safely exit during a fire emergency, the code is careful to protect the integrity of the enclosures in every way possible. Therefore, penetrations into such enclosures are prohibited unless necessary to service or protect the exit component. Acceptable penetrations identified in the IBC have historically included sprinkler piping, standpipes, and electrical conduits serving the enclosures. Two additional items have been added to the list, security systems and two-way communication systems.

Building security systems, including cameras, in stairway enclosures are becoming more prevalent. Their components pose a concern to the integrity of the fire-resistive enclosure due to their penetration of the enclosure walls. However, if properly protected, a limited number of penetrations for security systems will not result in an unacceptable level of safety. It was deemed important to make it clear that such penetrations are acceptable and sometimes required. As an example, NFPA 101 *Life Safety Code* requires stairway video monitoring in high-rise buildings with an occupant load of 4,000 or more persons.

In addition, the specified penetrations now include those related to two-way communication systems that are required in areas of refuge for accessibility purposes. The inclusion of these items now clearly allows for these systems to be provided in the exit enclosure to provide for safety and security of the building while still maintaining the integrity of the enclosure.

1025.1
Luminous Egress Path Marking in Group I Occupancies

CHANGE TYPE: Modification

CHANGE SUMMARY: Luminous egress path marking is no longer required in high-rise buildings classified as Group I-2, I-3, or I-4 occupancies.

2018 CODE: 1025.1 General. Approved luminous egress path markings delineating the exit path shall be provided in high-rise buildings of Group A, B, E, ~~I~~ I-1, M, or R-1 occupancies in accordance with this section.

> **Exception:** Luminous egress path markings shall not be required on the level of exit discharge in lobbies that serve as part of the exit path in accordance with Section 1028.1, Exception 1.

CHANGE SIGNIFICANCE: The use of luminous egress path markings in specified high-rise buildings is intended to provide for the visibility of stair treads and handrails in interior exit stairways under emergency conditions where both the building's primary power supply and the emergency power system fail. The use of photoluminescent or self-illuminating materials to delineate the exit path has historically been required in interior exit stairways and exit passageways of high-rise buildings housing Group A, B, E, I, M, and R-1 occupancies. Such markings are no longer required in those high-rise buildings classified as Groups I-2, I-3, or I-4.

The 2015 IBC mandates luminous egress path markings in all of the Group I classifications. This requirement has been revised for Group I to only require luminous egress path markings in Group I-1 occupancies. Groups I-2 and I-3 have been removed because hospitals, nursing homes, jails, and detention facilities have trained staff that operate with a defend-in-place strategy. The emergency generators are continually monitored and maintained, so the chance of the emergency egress lighting required in the means of egress failing is extremely minimal. For the luminous egress path marking to be utilized, both the normal power for means of egress lighting and the emergency generator have to fail.

Group I-4 occupancies were also removed from the list of required classifications. For the requirement for luminous path marking to apply, the high-rise building would need to be a day care facility used for custodial care. It was determined that there was a very limited chance that such a use would occur in a high-rise condition.

Luminous egress path markings

1026.4, 1026.4.1

Refuge Areas for Horizontal Exits

CHANGE TYPE: Modification

CHANGE SUMMARY: The method for determining the minimum required refuge area size where a horizontal exit has been provided has been modified to allow for a more appropriate determination of the occupant load assigned to the refuge area.

2018 CODE: 1026.4 Refuge area. The refuge area of a *horizontal exit* shall be a space occupied by the same tenant or a public area and each such refuge area shall be adequate to accommodate the original *occupant load* of the refuge area plus the *occupant load* anticipated from the adjoining compartment. The anticipated *occupant load* from the adjoining compartment shall be based on the capacity of the *horizontal exit* doors entering the refuge area or the total occupant load of the adjoining compartment, whichever is less.

1026.4.1 Capacity. The capacity of the refuge area shall be computed based on a *net floor area* allowance of 3 square feet (0.2787 m²) for each occupant to be accommodated therein. Where the horizontal exit also forms a smoke compartment, the capacity of the refuge area for Group I-1, I-2 and I-3 occupancies and Group B ambulatory care facilities shall comply with Sections 407.5.3, 408.6.2, 420.6.1 and 422.3.2 as applicable.

> **Exceptions:** The net floor area allowable per occupant shall be as follows for the indicated occupancies:
> 1. Six square feet (0.6 m²) per occupant for occupancies in Group I-3.
> 2. Fifteen square feet (1.4 m²) per occupant for ambulatory occupancies in Group I-2.
> 3. Thirty square feet (2.8 m²) per occupant for nonambulatory occupancies in Group I-2.

1026.4, 1026.4.1 continues

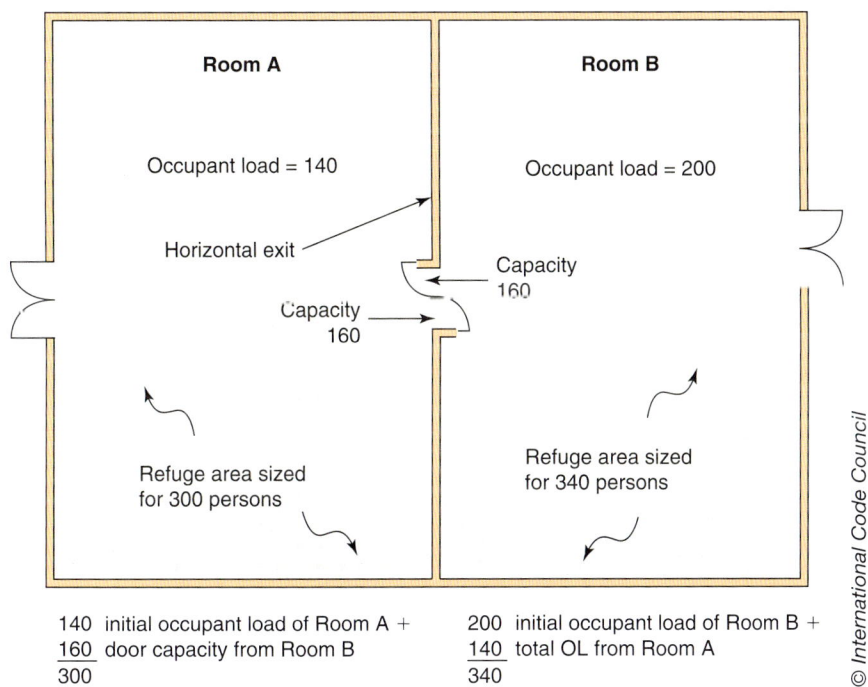

Calculation of refuge area occupant load

Room A: Occupant load = 140, Horizontal exit, Capacity 160, Refuge area sized for 300 persons
Room B: Occupant load = 200, Capacity 160, Refuge area sized for 340 persons

140 initial occupant load of Room A +
160 door capacity from Room B
300

200 initial occupant load of Room B +
140 total OL from Room A
340

1026.4, 1026.4.1 continued

CHANGE SIGNIFICANCE: A horizontal exit consists essentially of separating a story into portions by dividing it with construction having a fire-resistance rating. The construction of one or more horizontal exit walls divides the story into multiple compartments. The concept of the horizontal exit is to permit each of these fire compartments to serve as a refuge area for occupants in one or more of the fire compartments in the event of a fire emergency. As a refuge area, each compartment must provide sufficient space for the individuals to be protected in place until such time further evacuation is possible or the fire has been suppressed. The method for determining the minimum required refuge area size has been modified to allow for a more appropriate determination of the occupant load assigned to the refuge area.

Where one or more horizontal exits are established to serve as a required portion of the means of egress, it is important to determine what portion of the occupant load of the original space is to be added to the refuge area occupant load in the calculation of the cumulative occupant load that is necessary for sizing the required refuge area. In the 2015 IBC, the capacity of the horizontal exit doors is to be added to the original occupant load of the refuge area. This approach often results in an assigned occupant load significantly higher than the actual occupant load of the adjoining compartment. Section 1026.4 has been modified to indicate that the anticipated occupant load assigned for refuge purposes is based upon the capacity of the horizontal exit, but need not exceed the total occupant load of the adjoining compartment.

A single egress door is required to provide a minimum clear width of 32 inches which, at 0.2 inches per person, can serve up to 160 occupants. Where the occupant load of the space is more than 160, then the horizontal exit door capacity becomes the limiting factor. Where the occupant load of the space is less than 160, then the actual occupant load is the factor to be used for sizing the refuge area. Note that the entire occupant load is to be assigned, even where the horizontal exit is only one exit of multiple exits available from the space. The entire occupant load will be used, up to the capacity limit of the horizontal exit, rather than dividing the occupant load among the available exits or exit access doorways. The primary concept of requiring multiple exit points is so that when one egress path is blocked, there is another option. In the situation where the other egress path is blocked, the entire occupant load will be using the horizontal exit. Sizing the refuge area based on the total occupant load allows for that situation.

Section 1026.4.1 has also been revised by relocating the exceptions into the text and replacing the required floor area factors with a reference to the code sections where the factors can be found. Since the defend-in-place strategy is now included in the code, these items are no longer exceptions but are accepted and required methods of providing egress in Group I-1, I-2, and I-3 occupancies and Group B ambulatory care facilities. This reformatting provides coordination of the requirements, as the individual provisions may change in those sections dealing specifically with each occupancy.

1029.6, 1029.6.3, 202
Open-Air Assembly Seating

CHANGE TYPE: Clarification

CHANGE SUMMARY: The various assembly seating methods have been clarified through the introduction of a new definition for open-air assembly seating and an expanded definition for smoke-protected assembly seating.

2018 CODE: 1029.6 Capacity of aisle for assembly. The required capacity of aisles shall be not less than that determined in accordance with Section 1029.6.1 where smoke-protected assembly seating is not provided, ~~and with~~ Section 1029.6.2 ~~or 1029.6.3~~ where smoke-protected assembly seating is provided <u>and Section 1029.6.3 where open-air assembly seating is provided</u>.

1029.6.3 ~~Outdoor smoke-protected~~ <u>Open-air</u> **assembly seating.** ~~The~~ <u>In open-air assembly seating, the</u> required capacity in inches (mm) of aisles shall be not less than the total occupant load served by the egress element multiplied by 0.08 (2.0 mm) where egress is by stepped aisle and multiplied by 0.06 (1.52 mm) where egress is by level aisles and ramped aisles.

Exception: The required capacity in inches (mm) of aisles shall be permitted to comply with Section 1029.6.2 for the number of seats in the ~~outdoor smoke-protected~~ <u>open-air</u> assembly seating where Section 1029.6.2 permits less capacity.

SECTION 202 DEFINITIONS

<u>**OPEN-AIR ASSEMBLY SEATING.** Seating served by means of egress that is not subject to smoke accumulation within or under a structure and is open to the atmosphere.</u>

1029.6, 1029.6.3, 202 continues

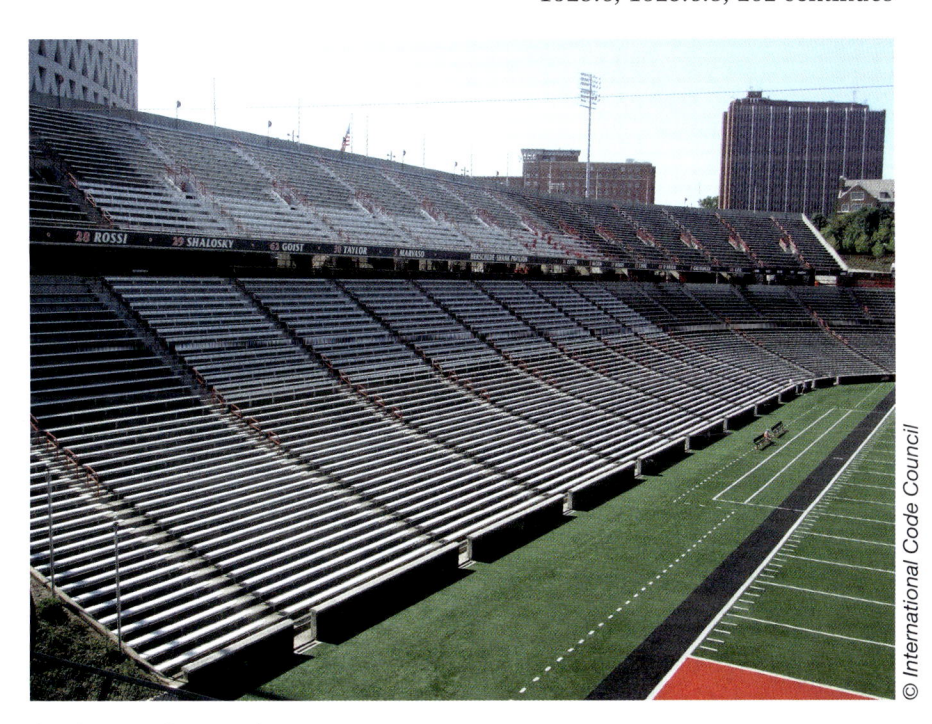

Outdoor stadium seating

1029.6, 1029.6.3, 202 continued

SMOKE-PROTECTED ASSEMBLY SEATING. Seating served by means of egress that is not subject to smoke accumulation within or under a structure <u>for a specified design time by means of passive design or by mechanical ventilation</u>.

CHANGE SIGNIFICANCE: Historically there have been two distinct methods for determining assembly aisle capacities and travel distance limitations, those assembly areas provided with smoke-protected seating areas and those without. It has been recognized that where an essentially smoke-free means of egress system can be maintained, the required limit on egress time can be extended for the occupants under emergency conditions. In addition, greater benefits are provided where the smoke protection is available due to the assembly seating facilities being outdoors. Clarification of the various assembly seating methods has been made through the introduction of a new definition for open-air assembly seating and an expanded definition for smoke-protected assembly seating.

The 2015 IBC uses the terms "smoke-protected assembly seating" and "outdoor smoke-protected assembly seating." The definition of smoke-protected assembly seating has been revised by adding performance language to better scope its application. The additional language recognizes that travel will be slowed in a manner that extends the egress time for an established period. In addition, either a passive system or mechanical ventilation can be provided.

A definition has also been added for open-air assembly seating clarifying that the seating area is to be open to the atmosphere and should not be subject to the accumulation of smoke where within or under a structure. The main distinction between these definitions is that open-air assembly seating is open to the atmosphere. These two definitions, and coordination of the terms, distinguish between the two types of systems that provide smoke protection for assembly seating.

1030.1
Required Emergency Escape and Rescue Openings

CHANGE TYPE: Clarification

CHANGE SUMMARY: The occupancies where emergency openings are required have been clarified and the minimum number of required openings in a residential basement has been revised.

2018 CODE: 1030.1 General. In addition to the means of egress required by this chapter, ~~provisions shall be made for~~ emergency escape and rescue openings <u>shall be provided</u> in <u>the following occupancies:</u>

1. Group R-2 occupancies <u>located</u> in ~~accordance~~ <u>stories</u> with <u>only one exit or access to only one exit as permitted by</u> Tables ~~1006.3.2(1)~~ <u>1006.3.3(1)</u> and ~~1006.3.2(2)~~ <u>1006.3.3(2)</u>. ~~and~~
2. Group R-3 <u>and R-4</u> occupancies.

Basements and sleeping rooms below the fourth story above grade plane shall have ~~at least~~ <u>not fewer than</u> one exterior emergency escape and rescue opening in accordance with this section. Where basements contain one or more sleeping rooms, emergency escape and rescue openings shall be required in each sleeping room, but shall not be required in adjoining areas of the basement. Such openings shall open directly into a public way or to a yard or court that opens to a public way.

Exceptions:

1. Basements with a ceiling height of less than 80 inches (2032 mm) shall not be required to have emergency escape and rescue openings.
2. Emergency escape and rescue openings are not required from basements or sleeping rooms that have an exit door or exit access door that opens directly into a public way or to a yard, court or exterior exit balcony that opens to a public way.

1030.1 continues

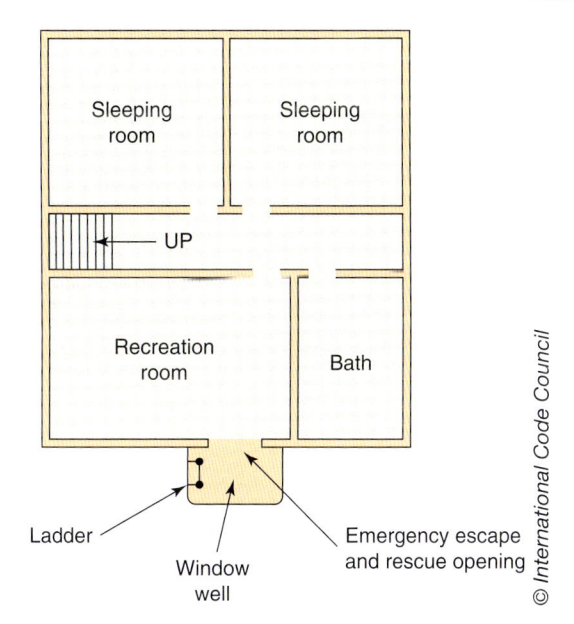

Sprinklered Group R-3 basement

1030.1 continued

3. Basements without habitable spaces and having not more than 200 square feet (18.6 m²) in floor area shall not be required to have emergency escape and rescue openings.

4. <u>Within individual dwelling and sleeping units in Groups R-2 and R-3, where the building is equipped throughout with an automatic sprinkler system installed in accordance with Section 903.3.1.1, 903.3.1.2 or 903.3.1.3, sleeping rooms in basements shall not be required to have emergency escape and rescue openings provided that the basement has one of the following:</u>

   4.1. <u>One means of egress and one emergency escape and rescue opening.</u>

   4.2. <u>Two means of egress.</u>

CHANGE SIGNIFICANCE: Because so many fire deaths occur as the result of occupants of residential buildings being asleep at the time of a fire, the IBC selectively requires that basements and all sleeping rooms below the fourth story have windows or doors that may be used for emergency escape or rescue. While residents are asleep, a fire can spread before the residents are aware of the problem and normal exit channels will most likely be blocked. The occupancies where such openings are required have been clarified, and the minimum number of required openings in a residential basement has been revised.

Several revisions occurred in the charging language in Section 1030.1.

- The section has been revised by creating a list identifying the requirements for Group R-2 and Group R-3 occupancies separately.

- For a Group R-2 occupancy, it has been clarified that escape and rescue openings are only required on those stories having only one complying exit. A single exit would be permitted under the allowances established in Section 1006.3.3. Although this requirement is intended by the 2015 IBC, the inclusion of scoping language in Section 1030.1 no longer requires the code user to view Section 1006.3.2 to determine applicability.

- The application to Group R-4 occupancies has also been clarified by adding it to the scoping list. As a general rule, Group R-4 occupancies are required to comply with the design and construction requirements for Group R-3 occupancies. Therefore, its addition to the scoping provisions only provides further clarification.

Exception 4 has also been added to address dwelling and sleeping units in Group R-2 and R-3 buildings equipped with an automatic sprinkler system. The new exception states that basements in sprinklered buildings of Groups R-2 or R-3 are allowed to provide either two means of egress, or one means of egress and one emergency escape and rescue opening. It should be noted that the escape and rescue opening is not required in each sleeping room, but rather just provided somewhere in the basement. Therefore, in a sprinklered condition, compliance could be achieved by providing a single complying exit and a single escape and rescue opening from the basement level.

PART 5
Accessibility

Chapter 11

- Chapter 11 Accessibility

Chapter 11 is intended to address the accessibility and usability of buildings and their elements to persons having physical disabilities. The provisions within the chapter are generally considered as scoping requirements that state what and where accessibility is required or how many accessible features or elements must be provided. The technical requirements, addressing how accessibility is to be accomplished, are found in ICC A117.1, as referenced by Chapter 11. The concept of the code is to initially mandate that all buildings and building elements be accessible and then to reduce the required accessibility where logical and reasonable. ■

1103.2.14
Access to Walk-In Coolers and Freezers

1109.2.1.2
Fixtures in Family or Assisted-Use Toilet Rooms

1109.15
Access to Gaming Machines and Gaming Tables

1110.4.13
Access to Play Areas for Children

1103.2.14
Access to Walk-In Coolers and Freezers

CHANGE TYPE: Modification

CHANGE SUMMARY: Revised conditions have now been placed on the use of walk-in coolers and freezers exempted from accessibility provisions by requiring them to be accessed from only employee work areas and limiting the scope to only pieces of equipment.

2018 CODE: 1103.2.14 Walk-in coolers and freezers. Walk-in coolers and ~~freezers intended for~~ freezer equipment accessed only from employee ~~use only~~ work areas is not required to comply with this chapter.

CHANGE SIGNIFICANCE: In general, access to individuals with disabilities is required in all buildings. There are, however, certain conditions under which sites, buildings, facilities, and elements are exempt from the provisions where specifically addressed. One such condition addresses walk-in coolers and freezers. In the past, these spaces were not required to comply with the IBC for accessibility purposes provided they were intended only for employee use. Revised conditions have now been placed on the use of exempt walk-in coolers and freezers by requiring them to be accessed from only employee work areas and limiting the scope to only pieces of equipment.

The previous scope of the allowance included all walk-in coolers and freezers, which could conceivably include giant coolers where employees spend all of their working hours. It could also include buildings or large areas where forklifts are utilized. By revising the scope of the allowance to cooler and freezer equipment, the intent has been clarified to more appropriately address those types of equipment that are found adjacent to

Walk-in cooler

a restaurant's commercial kitchen and similar types of uses. The expressed intent was not to include in the allowance a walk-in refrigerated room that was part of a facility such as a meat packing plant.

Another aspect of the new scoping language is the requirement that the walk-in cooler or freezer be accessed from only work areas. This will no longer allow the accessibility exemption to apply to freezers and coolers that are found in most large warehouse stores. Because these spaces are located in the public sales area and not accessed directly from an employee work area, they would not be granted the exemption from accessibility.

1109.2.1.2

Fixtures in Family or Assisted-Use Toilet Rooms

CHANGE TYPE: Modification

CHANGE SUMMARY: Family or assisted-use toilet rooms may now also contain a child-height water closet and lavatory in order to provide a higher level of accommodation.

2018 CODE: 1109.2.1.2 Family or assisted-use toilet rooms. Family or assisted-use toilet rooms shall include only one water closet and only one lavatory. A family or assisted-use bathing room in accordance with Section 1109.2.1.3 shall be considered <u>to be</u> a family or assisted-use toilet room.

> **Exception:** <u>The following additional fixtures shall be permitted in a family or assisted-use toilet room:</u>
> 1. A urinal ~~is permitted to be provided in addition to the water closet in a family or assisted-use toilet room~~.
> 2. <u>A child-height water closet.</u>
> 3. <u>A child-height lavatory.</u>

CHANGE SIGNIFICANCE: Family or assisted-use toilet rooms are selectively required in assembly and mercantile occupancies in order to provide a more comprehensive choice of facilities for users. Although the primary issue relative to such toilet rooms from an accessibility perspective is that some people with disabilities may require the assistance of persons of the opposite sex, there are a variety of other benefits to persons

Signage at family toilet room

with disabilities and caregivers that can be achieved through the presence of such facilities. As a general rule, a family or assisted-use toilet room may only contain a single water closet, urinal, and lavatory. It is now permissible for such rooms to also contain a child-height water closet and lavatory in order to provide a higher level of accommodation.

In recent editions of the IBC, toilet facilities sized for children are considered as accessible elements in those buildings where significant children's use is expected. The application could include buildings such as schools, recreational venues, day-care facilities, and children's museums. The dimensional criteria for child-sized water closets and lavatories are set forth in ICC A117.1. It may be appropriate that two types of water closets and lavatories, one type for adults and one type for children, be provided in family or assisted-use toilet rooms to achieve a higher level of accommodation. The additional facilities are not mandated by the code, but rather just as an acceptable condition where desired.

1109.15
Access to Gaming Machines and Gaming Tables

CHANGE TYPE: Modification

CHANGE SUMMARY: A more practical approach to the appropriate distribution of accessible gaming machines and gaming tables in casinos and other gaming facilities has been established and new definitions provide guidance in the application of the revised provisions.

2018 CODE: 1109.15 Gaming machines and gaming tables. <u>At least</u> two percent <u>of the total</u>, but not fewer than one, of each <u>gaming machine type</u> ~~of~~ <u>and</u> gaming table <u>type</u> ~~provided~~ shall be accessible ~~and provided with a front approach. Two percent of~~. <u>Where multiple gaming areas occur, accessible</u> gaming machines <u>and</u> ~~provided shall be accessible and provided with a front approach. Accessible~~ gaming <u>tables</u> shall be distributed throughout ~~the different types of gaming machines provided~~.

SECTION 202 DEFINITIONS

<u>**GAMING.**</u> <u>To deal, operate, carry on, conduct, maintain or expose for play any game played with cards, dice, equipment or any mechanical, electromechanical or electronic device or machine for money, property, checks, credit or any representative of value except wherein occurring at a private home or as operated by a charitable or educational organization.</u>

<u>**GAMING AREA.**</u> <u>Single or multiple areas of a building or facility where gaming machines or tables are present and gaming occurs, including but not limited to: primary casino gaming areas, VIP gaming areas, high-roller gaming areas, bar tops, lobbies, dedicated rooms or spaces such as in retail or restaurant establishments, sports books, and tournament areas.</u>

Casino gaming machines

Casino gaming tables

GAMING MACHINE TYPE. Categorization of gaming machines per type of game played on them, including, but not limited to: slot machines, video poker and video keno.

GAMING TABLE TYPE. Categorization of gaming tables per the type of game played on them, including, but not limited to: baccarat, bingo, blackjack/21, craps, pai-gow, poker and roulette.

CHANGE SIGNIFICANCE: Provisions for the accessibility of gaming machines and gaming tables were first established in the 2015 edition of the IBC. The provisions have been considered somewhat vague as the terms are not specifically defined and the required distribution of the machines based on type can be interpreted in a variety of ways. The original scope did not take into consideration the various gaming locations, designations, and needs of the gaming industry. As a result, new definitions have been provided and a more practical approach to the appropriate distribution of accessible gaming machines and gaming tables has been provided.

Although most gaming activities occur in larger casinos, there are also many smaller non-casino establishments where limited types of gaming are available. These types of facilities include bars, convenience stores, and restaurants. The revised provisions allow the accessibility requirements to be scaled for both small and large venues, and to be applied effectively regardless of the type of venue where the gaming is present.

The definitions of the terms "gaming machine type" and "gaming table type" recognize that while there are multiple different games available, they generally fall within two broad categories. Gaming machines include slots, video poker, and video keno. Gaming tables represent baccarat, bingo, craps, poker, and roulette. While there are a wide variety of different games, they all essentially fall into one of these two categories.

1109.15 continues

1109.15 continued   As such, they should be regulated for accessibility, particularly for distribution purposes, based upon the general category into which they resemble.

The revised text also considers the unique anthropometric design of most existing gaming machines, while also providing a reasonable level of access that does not require a wholesale redesign of the machine itself. By removing the requirement for "front approach" at the gaming machines, the new considerations allow nearly all upright-type machines to be accessible provided they do not have a fixed chair or other obstruction in front of them. In addition, gaming tables are typically standardized, requiring no special consideration with respect to their approach and clearance requirements.

1110.4.13
Access to Play Areas for Children

CHANGE TYPE: Clarification

CHANGE SUMMARY: Access to children's play areas is now specifically required where those areas contain play components.

2018 CODE: 1110.4.13 Play areas. Play areas containing play components designed and constructed for children shall be located on an accessible route.

CHANGE SIGNIFICANCE: The 2015 IBC introduced scoping provisions regarding the accessibility of recreational facilities so that persons with mobility impairments can participate to the best of their ability. The extent of the provisions is not to change any essential aspects of that specific recreational activity, but rather to allow increased participation in a variety of activities such as court sports, boating, exercise, and swimming. Access to children's play areas is now specifically required where those play areas contain play components.

Although the general provisions of Section 1110.3 mandate that recreational facilities be accessible, Section 1110.4 further scopes the extent of such accessibility for specific recreational activities. In order to make it clear that accessibility to playgrounds and similar play areas with play components intended for children's use is required, this type of recreational feature is now specifically addressed. This requirement is also set forth in the *2010 ADA Standards for Accessible Design*. It should be emphasized that the requirement is only applicable to the route to the play area and not to each individual play component. There is also no requirement for the play components themselves to be accessible.

Children's playground

PART 6

Building Envelope, Structural Systems, and Construction Materials

Chapters 12 through 26

- **Chapter 12** Interior Environment
- **Chapter 13** Energy Efficiency
  No changes addressed
- **Chapter 14** Exterior Walls
- **Chapter 15** Roof Assemblies and Rooftop Structures
- **Chapter 16** Structural Design
- **Chapter 17** Special Inspections and Tests
- **Chapter 18** Soils and Foundations
- **Chapter 19** Concrete
- **Chapter 20** Aluminum
  No changes addressed
- **Chapter 21** Masonry
  No changes addressed
- **Chapter 22** Steel
- **Chapter 23** Wood
- **Chapter 24** Glass and Glazing
- **Chapter 25** Gypsum Board, Gypsum Panel Products, and Plaster
- **Chapter 26** Plastic

The interior environment provisions of Chapter 12 include requirements for lighting, ventilation, and sound transmission. Chapter 13 provides a reference to the *International Energy Conservation Code* for provisions governing energy efficiency. Regulations governing the building envelope are located in Chapters 14 and 15, addressing exterior wall coverings and roof coverings, respectively. Structural systems are regulated through the structural design provisions of Chapter 16, whereas structural testing and special inspections are addressed in Chapter 17. The provisions of Chapter 18 apply to soils and foundation systems. The requirements for materials of construction, both structural and non-structural, are located in Chapters 19 through 26. Structural materials regulated by the code include concrete, lightweight metals, masonry, steel, and wood. Glass and glazing, gypsum board, plaster, and plastics are included as regulated non-structural materials. ■

1206.2, 1206.3
Engineering Analysis of Sound Transmission

TABLE 1404.2
Weather Covering Minimum Thickness

1404.18
Polypropylene Siding

1504.3.3
Metal Roof Shingles

1507.1
Underlayment

1507.18
Building Integrated Photovoltaic Panels

1603.1
Construction Documents

1604.3.7
Deflection of Glass Framing

1604.5.1
Multiple Occupancies

1604.10
Storm Shelters

TABLE 1607.1
Deck Live Load

TABLE 1607.1
Live Load Reduction

1607.15.2
Minimum Fire Load

1609
Wind Loads

1613
Earthquake Loads

1613.2.1
Seismic Maps

1615, 1604.5
Tsunami Loads

1704.6
Structural Observation

1705.5.2
Metal-Plate-Connected Wood Trusses

1705.12.1, 1705.13.1
Seismic Force-Resisting Systems

1705.12.6
Fire Sprinkler Clearance

1804.4
Site Grading

1807.2
Retaining Walls

1810.3.8.3
Precast Prestressed Piles

1901.2
Seismic Loads for Precast Concrete Diaphragms

2207.1
SJI Standard

2209.2
Cantilevered Steel Storage Racks

2211
Cold-Formed Steel Light-Frame Construction

2303.2.2
Fire-Retardant-Treated Wood

2303.6
Nails and Staples

TABLE 2304.9.3.2
Mechanically Laminated Decking

TABLE 2304.10.1
Ring Shank Nails

2304.10.5
Fasteners in Treated Wood

2304.11
Heavy-Timber Construction

2304.12.2.5, 2304.12.2.6
Supporting Members for Permeable Floors and Roofs

TABLE 2308.4.1.1(1)
Header and Girder Spans—Exterior Walls

TABLE 2308.4.1.1(2)
Header and Girder Spans—Interior Walls

2308.5.5.1
Openings in Exterior Bearing Walls

2407.1
Structural Glass Baluster Panels

2510.6
Water-Resistive Barrier

2603.13
Cladding Attachment over Foam Sheathing to Wood Framing

193

1206.2, 1206.3

Engineering Analysis of Sound Transmission

CHANGE TYPE: Modification

CHANGE SUMMARY: A performance-based alternative approach for meeting the required sound transmission class ratings for unit separation walls and floor-ceiling assemblies in residential buildings has been introduced which allows for the use of an engineering analysis based upon a comparison to previously tested assemblies.

2018 CODE: ~~1207.2~~ <u>1206.2</u> **Airborne sound.** Walls, partitions and floor-ceiling assemblies separating dwelling units and sleeping units from each other or from public or service areas shall have a sound transmission class of not less than 50, or not less than 45 if field tested, for airborne noise where tested in accordance with <u>ASTM E 90. Alternatively, the sound transmission class of walls, partitions and floor/ceiling assemblies shall be established by engineering analysis based on a comparison of walls, partitions and floor-ceiling assemblies having sound transmission class ratings as determined by the test procedures set forth in</u> ASTM E 90. Penetrations or openings in construction assemblies for piping; electrical devices; recessed cabinets; bathtubs; soffits; or heating, ventilating or exhaust ducts shall be sealed, lined, insulation or otherwise treated to maintain the required ratings. This requirement shall not apply to entrance doors; however, such doors shall be tight fitting to the frame and sill.

~~1207.3~~ <u>1206.3</u> **Structural-borne sound.** Floor-ceiling assemblies between dwelling units and sleeping units or between a dwelling unit or sleeping unit and a public or service area within the structure shall have an impact insulation class rating of not less than 50, or not less than 45 if field tested, where tested in accordance with ASTM E 492. <u>Alternatively, the impact insulation class of floor-ceiling assemblies shall be established by engineering analysis based on a comparison of floor-ceiling assemblies having impact insulation class ratings as determined by the test procedures set forth in ASTM E 492.</u>

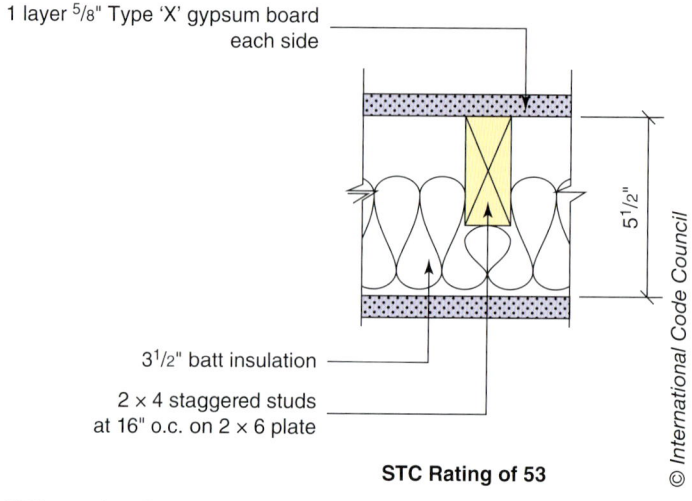

STC-rated wall assembly

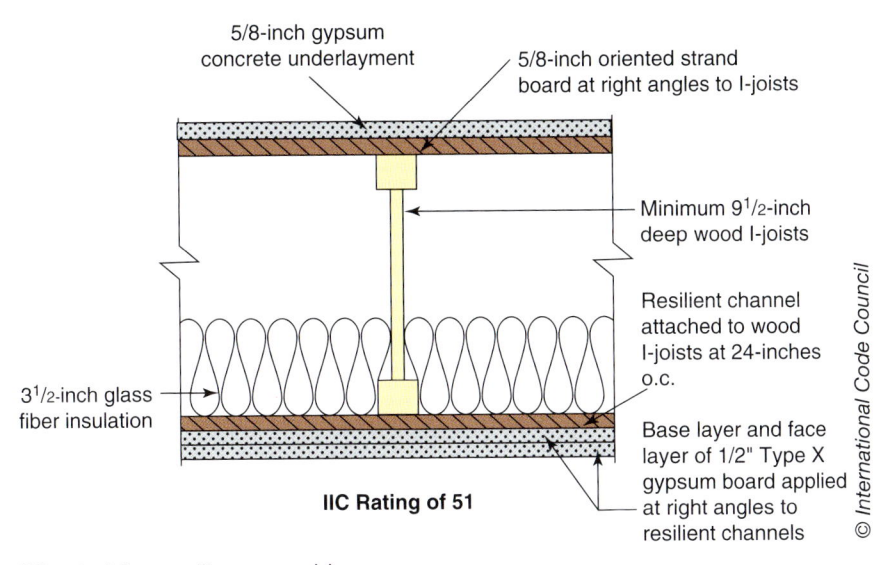

IIC-rated floor-ceiling assembly

CHANGE SIGNIFICANCE: For the well-being of the occupants, it is important that individual dwelling units and sleeping units be provided with sound transmission separation from the remainder of the building. Sound control construction is mandated for both the walls (for airborne sound) and the floor-ceiling assemblies (for impact-sound control). Compliance has historically been determined through the testing of each individual wall and floor-ceiling assembly to the requirements of ASTM E 90 or ASTM E 492, for airborne noise or impact sound, respectively. A performance-based alternative approach for meeting the required sound transmission class ratings has been introduced which allows for the use of an engineering analysis based upon a comparison to previously tested assemblies.

Section 703.3 sets forth a number of different methods or procedures that can be used to establish fire-resistance ratings of building materials and assemblies. Method #4 allows for an engineering analysis based on a comparison of building element, component, or assembly designs having fire-resistance ratings as determined by the specified test procedures. The new performance alternative for evaluating sound transmission is similar to the approach established in Method #4 of Section 703.3 for fire-resistance ratings. The sound transmission class of both walls and floor-ceiling assemblies can now be determined by an engineering analysis based on a comparison with previously tested assemblies.

Table 1404.2

Weather Covering Minimum Thickness

CHANGE TYPE: Modification

CHANGE SUMMARY: The minimum required thickness of masonry and stone veneer weather coverings has been updated to align with current industry standards.

2018 CODE:

TABLE ~~1405.2~~ <u>1404.2</u> Minimum Thickness of Weather Coverings

Covering Type	Minimum Thickness (inches)
Adhered masonry veneer	~~0.25~~
• <u>Architectural cast stone</u>	<u>0.75</u>
• <u>Other</u>	<u>0.25</u>
Anchored masonry veneer	~~2.625~~
• <u>Stone (natural)</u>	<u>2.0</u>
• <u>Architectural cast stone</u>	<u>1.25</u>
• <u>Other</u>	<u>2.625</u>
~~Stone (cast artificial, anchored)~~	~~1.5~~
~~Stone (natural)~~	~~2.0~~

(Portions of table and footnotes not shown remain unchanged.)

CHANGE SIGNIFICANCE: Table 1404.2 addresses two types of masonry veneer systems: anchored masonry and adhered masonry. Anchored masonry veneer describes masonry attached with ties or anchors. Adhered masonry veneer applies to masonry bonded using mortar or other approved adhesive material to a wood-sheathed or concrete wall. Adhered masonry is a thinner material than anchored masonry with an adhesive bond that is relatively weak compared to an anchor tied into the supporting wall behind the masonry veneer.

In previous editions of the IBC, anchored masonry veneer was required to have a minimum thickness of 2.625 inches while the minimum required thickness of adhered masonry veneer was established at 0.25 inches. In the

Decorative adhered masonry veneer

2018 IBC, the table for minimum veneer thickness has been reorganized to clarify which minimum thickness requirements apply to specific products on the market depending upon whether they are used as an anchored or adhered veneer.

Additionally, changes in terminology have been added. The term "stone cast artificial" is replaced with "architectural cast stone" to be consistent with industry practice. The *International Residential Code* (IRC) has historically allowed a minimum nominal thickness of anchored masonry veneer to be 2 inches. Changes in the 2018 IBC clarify that anchored stone shall have a minimum thickness of 2 inches. The minimum thickness of anchored architectural cast stone has been reduced slightly from 1.5 inches to 1.25 inches, in order to be consistent with industry practices and recommendations from the Cast Stone Institute. All other anchored masonry continues to have a required minimum thickness of 2.625 inches.

A minimum thickness requirement of 0.75 inches for adhered architectural cast stone products has also been added. All other types of adhered veneer continue to have a minimum required thickness of 0.25 inches. The thinner limit is not appropriate for architectural cast stone due to production, transportation, and installation constraints. The new minimum thickness of 0.75 inches is consistent with industry practices and recommendations from the Cast Stone Institute.

1404.18
Polypropylene Siding

CHANGE TYPE: Modification

CHANGE SUMMARY: Polypropylene siding is now specifically permitted for use on exterior walls of any type of construction when other provisions of the *International Building Code* allow its use.

2018 CODE: ~~1405.18~~**1404.18 Polypropylene siding.** Polypropylene siding conforming to the requirements of this section and complying with Section ~~1404.12~~ 1403.12 shall be limited to exterior walls ~~of Type VB construction~~ located in areas where the wind speed specified in Chapter 16 does not exceed 100 miles per hour (45 m/s) and the building height is less than or equal to 40 feet (12 192 mm) in Exposure C. Where construction is located in areas where the basic wind speed exceeds 100 miles per hour (45 m/s), or building heights are in excess of 40 feet (12 192 mm), tests or calculations indicating compliance with Chapter 16 shall be submitted. Polypropylene siding shall be installed in accordance with the manufacturer's instructions. Polypropylene siding shall be secured to the building so as to provide weather protection for the exterior walls of the building.

CHANGE SIGNIFICANCE: In the 2015 IBC, polypropylene siding was only permitted on exterior walls of Type VB buildings. Type VB construction has the fewest limits on fire resistance for materials used to construct a building. Additionally, Section 1406.2.1 allowed combustible materials to be used as exterior wall coverings in Type I through IV construction if the covering passed NFPA 268 testing requirements, met surface area and height limits, and met the radiant heat energy limits in Table 1406.2.1.1.2.

In the 2018 IBC, polypropylene siding is now allowed on exterior walls in Type I through V construction if the siding meets the limits of Section 1405.1.1 and all other applicable limits in the *International Building Code.* Other limits in Section 1404.18 for polypropylene siding have been maintained including a maximum height of construction and maximum wind speed. For buildings that exceed these limits, tests or calculations showing adequacy of the siding must be submitted with the construction documents.

Polypropylene siding on a residential building

1504.3.3
Metal Roof Shingles

CHANGE TYPE: Addition

CHANGE SUMMARY: Metal roof shingles are now addressed separately from other metal panel roof systems with reference made to applicable standards for the labeling and testing of wind resistance for the shingles.

2018 CODE: 1504.3.3 Metal roof shingles. Metal roof shingles applied to a solid or closely fitted deck shall be tested in accordance with ASTM D 3161, FM 4474, UL 580, or UL 1897. Metal roof shingles tested in accordance with ASTM D 3161 shall meet the classification requirements of Table 1504.1.1 for the appropriate maximum basic wind speed and the metal shingle packaging shall bear a label to indicate compliance with ASTM D 3161 and the required classification in Table 1504.1.1.

Metal roof shingles

CHANGE SIGNIFICANCE: In the 2015 edition of the *International Building Code* (IBC), Section 1504.3 deals with the wind resistance of non-ballasted, non-asphalt shingled roofs. In Section 1504.3.1, roof systems, including metal panel roof systems, may be tested for wind uplift resistance using FM 4474, UL 580, or UL 1897. Additionally, in Section 1504.3.2, metal panel roof systems functioning as both roof deck and roof covering may use the standards FM 4474, UL 580, or ASTM E 1592 to test wind uplift resistance. Metal roof shingles are not listed as a separate roofing system, but are best located under the structural metal panel roof systems.

In the 2018 IBC, new Section 1504.3.3 addresses metal roof shingles separately from metal roof panel systems in Sections 1504.3.1 and 1504.3.2. Metal roof shingles are not similar in all respects to asphalt shingles, addressed in Section 1504.1.1, nor to metal panel roof systems, addressed in Sections 1504.3.1 and 1504.3.2.

To address the similarities between asphalt and metal shingles, ASTM D 3161, a fan-induced wind test originally developed for asphalt shingles, has been added as an option for testing metal roof shingles. The 2015 edition of ASTM D 3161 is no longer constrained to asphalt shingles, but expanded to evaluate the wind resistance of discontinuous, air-permeable, steep-slope roofing products. Inclusion of this standard as a compliance path alleviates some difficulties experienced due to current language in UL 1897 and UL 580, which test in a non-air-permeable fashion. This change adds the standard as an option for testing while maintaining the three standards for metal roof panel systems as well so manufacturers meeting the standards referenced in the 2015 IBC will not need to retest their products.

1507.1 Underlayment

CHANGE TYPE: Clarification

CHANGE SUMMARY: Underlayment and ice barrier requirements have been relocated from sections describing each type of roofing material and placed into one new section describing the type, attachment, and application of underlayment.

2018 CODE: 1507.1 Scope. Roof coverings shall be applied in accordance with the applicable provisions of this section and the manufacturer's installation instructions.

<u>**1507.1.1 Underlayment.** Underlayment for asphalt shingles, clay and concrete tile, metal roof shingles, mineral surfaced roll roofing, slate and slate-type shingles, wood shingles, wood shakes, metal roof panels and photovoltaic shingles shall conform to the applicable standards listed in this chapter. Underlayment materials required to comply with ASTM D 226, D 1970, D 4869 and D 6757 shall bear a label indicating compliance to the standard designation and, if applicable, type classification indicated in Table 1507.1.1(1). Underlayment shall be applied in accordance with Table 1507.1.1(2). Underlayment shall be attached in accordance with Table 1507.1.1(3).</u>

> **Exceptions:**
>
> 1. <u>As an alternative, self-adhering polymer modified bitumen underlayment complying with ASTM D 1970 and installed in accordance with the manufacturer's installation instructions</u>

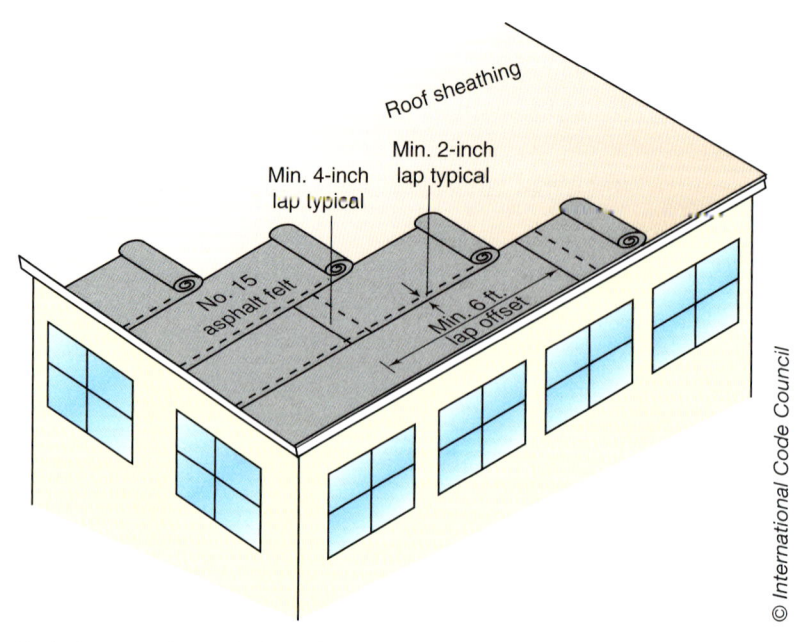

Underlayment for roof pitch of 4:12 or greater

for the deck material, roof ventilation configuration and climate exposure for the roof covering to be installed shall be permitted.

2. As an alternative, a minimum 4-inch (102 mm) wide strip of self-adhering polymer modified bitumen membrane complying with ASTM D 1970 and installed in accordance with the manufacturer's installation instructions for the deck material shall be applied over all joints in the roof decking. An approved underlayment for the applicable roof covering for design wind speeds less than 120 mph (54 m/s) shall be applied over the 4-inch (102 mm) wide membrane strips.

3. As an alternative, two layers of underlayment complying with ASTM D 226 Type II or ASTM D 4869 Type IV shall be permitted to be installed as follows: Apply a 19-inch (483 mm) strip of underlayment parallel with the eave. Starting at the eave, apply 36-inch (914 mm) wide strips of underlayment felt, overlapping successive sheets 19 inches (483 mm). The underlayment shall be attached with corrosion-resistant fasteners in a grid pattern of 12 inches (305 mm) between side laps with a 6-inch (152 mm) spacing at side and end laps. End laps shall be 4 inches (102 mm) and shall be offset by 6 feet (1829 mm).

   Underlayment shall be attached using metal or plastic cap nails with a nominal cap diameter of not less than 1 inch (25.4 mm). Metal caps shall have a thickness of not less than 32-gage [0.0134 inch (0.34 mm)] sheet metal. Power-driven metal caps shall have a thickness of not less than 0.010 inch (0.25 mm). Thickness of the outside edge of plastic caps shall be not less than 0.035 inch (0.89 mm). The cap nail shank shall be not less than 0.083 inch (2.11 mm) for ring shank cap nails and 0.091 inch (2.31 mm) for smooth shank cap nails. The cap nail shank shall have a length sufficient to penetrate through the roof sheathing or not less than 0.75 inch (19.1 mm) into the roof sheathing.

4. Structural metal panels that do not require a substrate or underlayment.

1507.1.2 Ice barriers. In areas where there has been a history of ice forming along the eaves causing a backup of water, an ice barrier shall be installed for asphalt shingles, metal roof shingles, mineral-surfaced roll roofing, slate and slate-type shingles, wood shingles, and wood shakes. The ice barrier shall consists of not less than two layers of underlayment cemented together, or a self-adhering polymer modified bitumen sheet shall be used in place of normal underlayment and extend from the lowest edges of all roof surfaces to a point not less than 24 inches (610 mm) inside the exterior wall line of the building.

> **Exception:** Detached accessory structures that do not contain conditioned floor area.

1507.1 continues

1507.1 continued

TABLE 1507.1.1(1)   Underlayment Types

Roof Covering	Section	Maximum Basic Design Wind Speed, V < 140 mph	Maximum Basic Design Wind Speed, V ≥ 140 mph
Asphalt shingles	1507.2	ASTM D 226 Type I or II ASTM D 4869 Type I, II, III, or IV ASTM D 6757	ASTM D 226 Type II ASTM D 4869 Type IV ASTM D 6757
Clay and concrete tiles	1507.3	ASTM D 226 Type II ASTM D 2626 Type I ASTM D 6380 Class M mineral surfaced roll roofing	ASTM D 226 Type II ASTM D 2626 Type I ASTM D 6380 Class M mineral surfaced roll roofing
Metal panels	1507.4	Manufacturer's instructions	ASTM D 226 Type II ASTM D 4869 Type IV
Metal roof shingles	1507.5	ASTM D 226 Type I or II ASTM D 4869 Type I, II, III, or IV	ASTM D 226 Type II ASTM D 4869 Type IV
Mineral-surfaced roll roofing	1507.6	ASTM D 226 Type I or II ASTM D 4869 Type I, II, III, or IV	ASTM D 226 Type II ASTM D 4869 Type IV
Slate shingles	1507.7	ASTM D 226 Type II ASTM D 4869 Type III or IV	ASTM D 226 Type II ASTM D 4869 Type IV
Wood shingles	1507.8	ASTM D 226 Type I or II ASTM D 4869 Type I, II, III, or IV	ASTM D 226 Type II ASTM D 4869 Type IV
Wood shakes	1507.9	ASTM D 226 Type I or II ASTM D 4869 Type I, II, III, or IV	ASTM D 226 Type II ASTM D 4869 Type IV
Photovoltaic shingles	1507.17	ASTM D 226 Type I or II ASTM D 4869 Type I, II, III, or IV ASTM D 6757	ASTM D 226 Type II ASTM D 4869 Type IV ASTM D 6757

TABLE 1507.1.1(2)   Underlayment Application

Roof Covering	Section	Maximum Basic Design Wind Speed, V < 140 mph	Maximum Basic Design Wind Speed, V ≥ 140 mph
Asphalt shingles	1507.2	For roof slopes from two units vertical in 12 units horizontal (2:12), up to four units vertical in 12 units horizontal (4:12), underlayment shall be two layers applied as follows: Apply a 19-inch (483 mm) strip of underlayment felt parallel to and starting at the eaves. Starting at the eave, apply 36-inch-wide (914 mm) sheets of underlayment, overlapping successive sheets 19 inches (483 mm). End laps shall be 4 inches (102 mm) and shall be offset by 6 feet (1829 mm). Distortions in the underlayment shall not interfere with the ability of the shingles to seal. For roof slopes of four units vertical in 12 units horizontal (4:12) or greater, underlayment shall be one layer applied as follows: Underlayment shall be applied shingle fashion, parallel to and starting from the eave and lapped 2 inches (51 mm). Distortions in the underlayment shall not interfere with the ability of the shingles to seal. End laps shall be 4 inches (102 mm) and shall be offset by 6 feet (1829 mm).	Same as Maximum Basic Design Wind Speed, V < 140 mph except all laps shall be not less than 4 inches (102 mm).

Significant Changes to the IBC 2018 Edition   1507.1 ■ Underlayment   **203**

Roof Covering	Section	Maximum Basic Design Wind Speed, V < 140 mph	Maximum Basic Design Wind Speed, V ≥ 140 mph
Clay and concrete tile	1507.3	For roof slopes from two and one-half units vertical in 12 units horizontal (2½:12), up to four units vertical in 12 units horizontal (4:12), underlayment shall be not fewer than two layers underlayment applied as follows: Starting at the eave, a 19-inch (483 mm) strip of underlayment shall be applied parallel with the eave. Starting at the eave, a 36-inch-wide (914 mm) strips of underlayment felt shall be applied, overlapping successive sheets 19 inches (483 mm). End laps shall be 4 inches (102 mm) and shall be offset by 6 feet (1829 mm). For roof slopes of four units vertical in 12 units horizontal (4:12) or greater, underlayment shall be one layer applied as follows: Underlayment shall be applied shingle fashion, parallel to and starting from the eave and lapped 2 inches (51 mm), End laps shall be 4 inches (102 mm) and shall be offset by 6 feet (1829 mm).	Same as Maximum Basic Design Wind Speed, V < 140 mph except all laps shall be not less than 4 inches (102 mm).
Metal roof panels	1507.4	Apply in accordance with the manufacturer's installation instructions.	For roof slopes from two units vertical in 12 units horizontal (2:12), up to four units vertical in 12 units horizontal (4:12), underlayment shall be two layers applied as follows: Apply a 19-inch (483 mm) strip of underlayment felt parallel to and starting at the eaves. Starting at the eave, apply 36-inch-wide (914 mm) sheets of underlayment, overlapping successive sheets 19 inches (483 mm). End laps shall be 4 inches (102 mm) and shall be offset by 6 feet (1829 mm). For roof slopes of four units vertical in 12 units horizontal (4:12) or greater, underlayment shall be one layer applied as follows: Underlayment shall be applied shingle fashion, parallel to and starting from the eave and lapped 4 inches (102 mm). End laps shall be 4 inches (102 mm) and shall be offset by 6 feet (1829 mm).
Metal roof shingles	1507.5		
Mineral-surfaced roll roofing	1507.6		
Slate shingles	1507.7		
Wood shakes	1507.8		
Wood shingles	1507.9		
Photovoltaic shingles	1507.17	For roof slopes from three units vertical in 12 units horizontal (3:12), up to four units vertical in 12 units horizontal (4:12), underlayment shall be two layers applied as follows: Apply a 19-inch (483 mm) strip of underlayment felt parallel to and starting at the eaves. Starting at the eave, apply 36-inch-wide (914 mm) sheets of underlayment, overlapping successive sheets 19 inches (483 mm). End laps shall be 4 inches (102 mm) and shall be offset by 6 feet (1829 mm). Distortions in the underlayment shall not interfere with the ability of the shingles to seal. For roof slopes of four units vertical in 12 units horizontal (4:12) or greater, underlayment shall be one layer applied as follows: Underlayment shall be applied shingle fashion, parallel to and starting from the eave and lapped 2 inches (51 mm). Distortions in the underlayment shall not interfere with the ability of the shingles to seal. End laps shall be 4 inches (102 mm) and shall be offset by 6 feet (1829 mm).	Same as Maximum Basic Design Wind Speed, V < 140 mph except all laps shall be not less than 4 inches (102 mm).

For SI: 1 inch = 25.4 mm, 1 foot = 304.8 mm; 1 mile per hour = 0.447 m/s.

1507.1 continues

1507.1 continued

TABLE 1507.1.1(3) Underlayment Attachment

Roof Covering	Section	Maximum Basic Design Wind Speed, V < 140 mph	Maximum Basic Design Wind Speed, V ≥ 140 mph
Asphalt shingles	1507.2		The underlayment shall be attached with corrosion-resistant fasteners in a grid pattern of 12 inches (305 mm) between side laps with a 6-inch (152 mm) spacing at side and end laps.
Clay and concrete tile	1507.3	Fastened sufficiently to hold in place	Underlayment shall be attached using metal or plastic cap nails or cap staples with a nominal cap diameter of not less than 1 inch. (25.4 mm) Metal caps shall have a thickness of not less than 32-gage [0.0134 inch (0.34 mm)] sheet metal. Power-driven metal caps shall have a minimum thickness of 0.010 inch (0.25 mm). Minimum thickness of the outside edge of plastic caps shall be 0.035 inch (0.89 mm). The cap nail shank shall be not less than 0.083 inch (2.11 mm) for ring shank cap nails and 0.091 inch (2.31 mm) for smooth shank cap nails. Staples shall be not less than 21 gage [0.032 inch (0.81 mm)]. The cap nail shank and cap staple legs shall have a length sufficient to penetrate through the roof sheathing or not less than 0.75 inch (19.1 mm) into the roof sheathing.
Photovoltaic shingles	1507.17		
Metal roof panels	1507.4		The underlayment shall be attached with corrosion-resistant fasteners in a grid pattern of 12 inches (305 mm) between side laps with a 6-inch (152 mm) spacing at side and end laps.
Metal roof shingles	1507.5	Manufacturer's installation instructions	Underlayment shall be attached using metal or plastic cap nails or cap staples with a nominal cap diameter of not less than 1 inch (25.4 mm). Metal caps shall have a thickness of not less than 32-gage [0.0134 inch (0.34 mm)] sheet metal. Power-driven metal caps shall have a minimum thickness of 0.010 inch (0.25 mm). Minimum thickness of the outside edge of plastic caps shall be 0.035 inch (0.89 mm). The cap nail shank shall be not less than 0.083 inch (2.11 mm) for ring shank cap nails and 0.091 inch (2.31 mm) for smooth shank cap nails. Staples shall be not less than 21 gage [0.032 inch (0.81 mm)]. The cap nail shank and cap staple legs shall have a length sufficient to penetrate through the roof sheathing or not less than 0.75 inch (19.1 mm) into the roof sheathing.
Mineral-surfaced roll roofing	1507.6		
Slate shingles	1507.7		
Wood shingles	1507.8		
Wood shakes	1507.9		

(*As this code change affected substantial portions of Section 1507, the entire code change text is too extensive to be included here. Changes may be seen in Section 1507 of the* 2018 IBC *or refer to code change S27 in the* Complete Revision History to the 2018 I-Codes *for the complete text and history of the change.*)

CHANGE SIGNIFICANCE: In the 2015 IBC, underlayment provisions are specified individually for each type of roof covering. Many of the roof-covering provisions contain similar and overlapping requirements for underlayment type, application, and attachment. Additionally, self-adhering membrane is permitted as an alternative to the underlayment provisions for high wind.

In the 2018 IBC, Section 1507 has been reorganized to move underlayment provisions from individual roofing material sections to a single section addressing underlayment and ice barriers for all roof-covering materials. Three new tables have been added that address underlayment type, application, and attachment respectively. Consolidating the underlayment requirements into a single section makes the provisions more user-friendly and highlights the key differences between the requirements for underlayment for the different types of roof coverings addressed by the IBC.

Additionally, the wind speed threshold that triggers enhanced underlayment provisions has been revised. The threshold changes from $V_{asd} = 120$ mph to $V_{ult} = 140$ mph making the IBC and *International Residential Code* (IRC) consistent. The original code change that placed this trigger at 120 mph was developed to correspond with the wind speed maps in the 2009 IBC and ASCE 7-05. New maps in ASCE 7-10 shifted the contours closer to the coast for the entire hurricane-prone region, which resulted in a reduction of the geographic area required to comply with the enhanced underlayment provisions. The threshold was originally chosen based upon a geographic location on the wind speed map rather than a particular design limitation. This change in the wind speed threshold maintains the intent of the provision.

In Section 1507.1.1 Exceptions, use of ASTM D 1970, self-adhering membrane as an underlayment, has also been clarified to require use of the manufacturer's installation instructions to install self-adhering membrane, providing an equivalent level of water intrusion prevention in regions with high winds.

In Section 1507.1.1 Exception 2, minimum four-inch strips of self-adhering membrane may be applied over all joints in the roof decking to seal the joints. Then a single layer of underlayment is applied over the membrane strips. This alternate method is equivalent to the application prescribed in Table 1507.1.1(2). Note, the proponent's intent with this exception was to allow underlayment applicable for wind speeds less than $V_{asd} = 120$ mph (now $V_{ult} = 140$ mph). The description of the type of design wind speed is missing from this exception and the wind speed has not been updated.

In Section 1507.1.1 Exception 3, a double layer of underlayment has been added for prevention of water penetration when the primary roof covering is lost due to high winds. Water penetration is well documented from post-hurricane damage assessments where hurricane winds were strong enough to blow off the primary roof covering, but not strong enough to blow off roof sheathing. In such instances, significant property damage and extended occupant displacement occurs due to water intrusion. The damage is particularly common in inland areas, where hurricane-strength winds occur and building codes and standards are not as stringent as in coastal jurisdictions.

Tests conducted at the Insurance Institute for Business and Home Safety (IBHS) Research Facility have found the double layer of underlayment, new Exception 3, performs similar to self-adhering polymer-modified bitumen underlayment. As a result, this system of underlayment application and attachment is now recognized by the Fortified Program (IBHS) for creating a sealed roof deck. While this system is currently required at the eave for roof slopes between 2:12 and 4:12, it provides an equal level of water penetration protection for roof slopes greater than a 4:12 pitch.

1507.18
Building-Integrated Photovoltaic Panels

CHANGE TYPE: Addition

CHANGE SUMMARY: Building-integrated photovoltaic panel systems have specific requirements as a roof-covering material in the *International Building Code*.

2018 CODE: 1507.18 Building-integrated photovoltaic roof panels. The installation of building-integrated photovoltaic (BIPV) roof panels shall comply with the provisions of this section.

1507.18.1 Deck requirements. BIPV roof panels shall be applied to a solid or closely-fitted deck, except where the roof covering is specifically designed to be applied over spaced sheathing.

1507.18.2 Deck slope. BIPV roof panels shall be used only on roof slopes of two units vertical in 12 units horizontal (2:12) or greater.

1507.18.3 Underlayment. Underlayment shall comply with ASTM D 226, ASTM D 4869 or ASTM D 6757.

1507.18.4 Underlayment application. Underlayment shall be applied shingle fashion, parallel to and starting from the eave, lapped 2 inches (51 mm) and fastened sufficiently to hold in place.

1507.18.4.1 High wind attachment. Underlayment applied in areas subject to high winds [V_{asd} greater than 110 mph (49 m/s) as determined in accordance with Section 1609.3.1] shall be applied in accordance with the manufacturer's instructions. Fasteners shall be applied along the overlap at not more than 36 inches (914 mm) on center. Underlayment installed where V_{asd} is not less than 120 mph (54 m/s) shall comply

BIPV panels

with ASTM D 226, Type III, ASTM D 4869, Type IV, or ASTM D 6757. The underlayment shall be attached in a grid pattern of 12 inches (305 mm) between side laps with an 6-inch (152 mm) spacing at the side laps. The underlayment shall be applied in accordance with Section 1507.1.1 except all laps shall be not less than 4 inches (102 mm). Underlayment shall be attached using cap nails or cap staples. Caps shall be metal or plastic with a nominal head diameter of not less than 1 inch (25.4 mm). Metal caps shall have a thickness of not less than 0.010 inch (0.25 mm). Power driven metal caps shall have a thickness of not less than 0.010 inch (0.25 mm). Thickness of the outside edge of plastic caps shall be not less than 0.035 inch (0.89 mm). The cap nail shank shall be not less than 0.083 inch (2.11 mm) for ring shank cap nails and 0.091 inch (2.31 mm) for smooth shank cap nails. Staple gage shall be not less than 21 gage [0.032 inch (0.81 mm)]. Cap nail shank and cap staple legs shall have a length sufficient to penetrate through the roof sheathing or a minimum of 0.75 inch (19.1 mm) into the roof sheathing.

> **Exception:** As an alternative, adhered underlayment complying with ASTM D 1970 shall be permitted.

1507.18.4.2 Ice barrier. In areas where there has been a history of ice forming along the eaves causing a back-up of water, an ice barrier that consists of at least two layers of underlayment cemented together or of a self-adhering polymer modified bitumen sheet shall be used instead of normal underlayment and extend from the lowest edges of all roof surfaces to a point not less than 24 inches (610 mm) inside the exterior wall line of the building.

> **Exception:** Detached accessory structures that contain no conditioned floor area.

1507.18.5 Material standards. BIPV roof panels shall be listed and labeled in accordance with UL 1703.

1507.18.6 Attachment. BIPV roof panels shall be attached in accordance with the manufacturer's installation instructions.

1507.18.7 Wind resistance. BIPV roof panels shall be tested in accordance with UL 1897. BIPV roof panel packaging shall bear a label to indicate compliance with UL 1897.

SECTION 202 DEFINITIONS

BUILDING-INTEGRATED PHOTOVOLTAIC ROOF PANEL (BIPV ROOF PANEL). A photovoltaic panel that functions as a component of the building envelope.

CHANGE SIGNIFICANCE: A new definition for BIPV roof panels has been added to Chapter 2 to identify the panels as a unique building-integrated photovoltaic product. Building-integrated photovoltaic (BIPV) roof panels form part of the building envelope and are subject to the requirements for roof coverings. BIPV panels are larger than typical

1507.18 continues

1507.18 continued

BIPV shingles. The panels are a thin-film layer and can be the typical rectangular panel shape or have a rounded shape. There are now clay roof tile-shaped BIPVs available.

Section 1507.18 details proper application of BIPV roof panels. As the BIPV panels act as a roof covering, wind resistance must be determined by UL 1897, *Uplift Tests for Roof Covering Systems*. This standard is already referenced in the IBC for other roofing products including built-up modified bitumen, single-ply, and metal panel roof systems in Section 1504.3.

1603.1 Construction Documents

CHANGE TYPE: Modification

CHANGE SUMMARY: The construction document requirements for environmental and special loads have been updated for rain, snow and wind forces and their components.

2018 CODE: 1603.1 General. Construction documents shall show the size, section and relative locations of structural members with floor levels, column centers and offsets dimensioned. The design loads and other information pertinent to the structural design required by Sections 1603.1.1 through ~~1603.1.8~~ 1603.1.9 shall be indicated on the construction documents.

> **Exception:** Construction documents for buildings constructed in accordance with the conventional light-frame construction provisions of Section 2308 shall indicate the following structural design information:
>
> 1. Floor and roof ~~doad and~~ live loads.
> 2. Ground snow load, P_g.
> 3. ~~Ultimate~~ Basic design wind speed, V_{ult}, ~~(3-second gust)~~, miles per hour (mph) (km/hr) and ~~nominal~~ allowable stress design wind speed, V_{asd}, as determined in accordance with Section 1609.3.1 and wind exposure.
> 4. Seismic design category and site class.
> 5. Flood design data, if located in flood hazard areas established in Section 1612.3.
> 6. Design load-bearing values of soils.
> 7. Rain load data.

1603.1 continues

Flood waters in a canal

1603.1 continued

Snow on sloped roof

1603.1.3 Roof snow load data. The ground snow load, P_g, shall be indicated. In areas where the ground snow load, P_g, exceeds 10 pounds per square foot (psf) (0.479 kN/m²), the following additional information shall also be provided, regardless of whether snow loads govern the design of the roof:

1. Flat-roof snow load, P_f.
2. Snow exposure factor, C_e.
3. Snow load importance factor, I_s.
4. Thermal factor, C_t.
5. <u>Slope factor(s), C_s</u>
~~5.~~ <u>6.</u> Drift surcharge load(s), P_d, where the sum of P_d and P_f exceeds 20 psf (0.96 kN/m²).
~~6.~~ <u>7.</u> Width of snow drift(s), w.

1603.1.4 Wind design data. The following information related to wind loads shall be shown, regardless of whether wind loads govern the design of the lateral force-resisting system of the structure:

1. ~~Ultimate~~ <u>Basic</u> design wind speed, V_{ult}, ~~(3-second gust),~~ miles per hour (km/hr) and ~~nominal~~ <u>allowable stress</u> design wind speed, V_{asd}, as determined in accordance with Section 1609.3.1.

(No changes to Items 2 – 5)

1603.1.8 Special loads. Special loads that are applicable to the design of the building, structure or portions thereof, including but not limited to the loads of machinery or equipment, and that are greater than specified floor and roof loads shall be ~~indicated along with the specified section of this code that addresses the special loading condition~~ specified by their descriptions and locations.

1603.1.9 Roof rain load data. Rain intensity, i (in/hr) (cm/hr), shall be shown regardless of whether rain loads govern the design.

CHANGE SIGNIFICANCE: The construction document requirements for snow and rain loads have been updated to include critical design forces and their components. The roof slope factor, C_s, has been added to the required list of snow load data and factors. The roof slope factor is occasionally overlooked in designs. Including the factor in the list of required factors and loads to be listed in the submittal documents adds a reminder for engineers to calculate and use the value. Inclusion creates a line item to check during plan review. Requiring declaration of the value simplifies checking roof snow load calculations for the building department.

Rain load data is added to the list of design loads required on construction documents. Having the rain load intensity listed allows for a quicker check by the building department of the rain load versus snow load values to determine the controlling environmental load. Values for snow and rain loads are to be determined using the 2016 edition of *ASCE 7, Minimum Design Loads and Associated Criteria for Buildings and Other Structures* (ASCE 7-16).

Special loads are clarified while wind load terminology is updated to match ASCE 7.

1604.3.7
Deflection of Glass Framing

CHANGE TYPE: Addition

CHANGE SUMMARY: Limits to the deflection of framing which supports glazing have been added to Section 1604.3.

2018 CODE: <u>**1604.3.7 Framing supporting glass.** The deflection of framing members supporting glass subjected to 0.6 times the 'component and cladding' wind loads shall not exceed either of the following:</u>

> <u>1. 1/175 of the length of span of the framing member, for framing members having a length not more than 13 feet 6 inches (4115 mm).</u>
>
> <u>2. 1/240 of the length of span of the framing member + 1/4 inch (6.4 mm), for framing members having a length greater than 13 feet 6 inches (4115 mm).</u>

CHANGE SIGNIFICANCE: The deflection limit given in Section 2403 is appropriate for glass design and is similar to the limit in Section 1604.3. But Chapter 24 has not historically addressed the deflection of the framing members surrounding glazing panels, in particular deflection over the entire length of the frame's span.

New Section 1604.3.7 clarifies the allowable deflection of exterior wall framing members supporting the glazing on the basis of the framing member spans. The change addresses serviceability issues by adding specific deflection limits, based on wind loads. The deflection limit is based upon criteria in the American Architectural Manufacturers Association (AAMA) publication TIR-A11, *Maximum Allowable Deflection of Framing Systems for Building Cladding Components at Design Wind Loads*, the industry standard for fenestration. Deflection limits in the AAMA TIR-A11 publication are based on using full allowable stress design wind loads with mean recurrence intervals (MRI) of 50 or 100 years. The limits do not allow a reduction to this design load for deflection analysis as was previously permitted by footnote f in the 2015 IBC Table 1604.3.

Glazing in metal frame

1604.5.1 Multiple Occupancies

CHANGE TYPE: Addition

CHANGE SUMMARY: The provisions addressing multiple occupancies within a structure now include an exception exempting buildings in their entirety from needing to qualify as Risk Category IV buildings when a storm shelter is part of the structure.

2018 CODE: 1604.5.1 Multiple occupancies. Where a building or structure is occupied by two or more occupancies not included in the same risk category, it shall be assigned the classification of the highest risk category corresponding to the various occupancies. Where buildings or structures have two or more portions that are structurally separated, each portion shall be separately classified. Where a separated portion of a building or structure provides required access to, required egress from or shares life safety components with another portion having a higher risk category, both portions shall be assigned to the higher risk category.

> **Exception:** Where a storm shelter designed and constructed in accordance with ICC 500 is provided in a building, structure or portion thereof normally occupied for other purposes, the risk category for the normal occupancy of the building shall apply unless the storm shelter is a designated emergency shelter in accordance with Table 1604.5.

CHANGE SIGNIFICANCE: Risk categories are assigned to buildings to account for consequences and risks to human life in the event of a structural failure. The intent is to assign higher risk categories, and hence higher design criteria, to buildings or structures that, if they experience a failure, would inhibit the availability of essential community services necessary to cope with the emergency situation and therefore have grave consequences to either the building occupants or the population around the building that rely upon the provided services.

1604.5.1 continues

Building with multiple occupancies on separate floors

1604.5.1 continued

Community storm shelters are defined in the IBC and ICC-500, *Standard for the Design and Construction of Storm Shelters*, as shelters that either serve a nonresidential use or serve dwelling units and provide a capacity exceeding 16 persons. The standard confirms that the area of a building that has been constructed to the ICC 500 criteria has been specifically designed and constructed to provide life-safety protection for people seeking refuge from a high wind event.

ICC 500-compliant storm shelters are designed and constructed to account for extreme wind loads and have specific requirements for structural stability, vertical and horizontal load transfer, and egress that meet or exceed the basic requirements of the building code for property protection. Even if the storm shelter is not structurally separated from the host building, ICC 500 details the strength requirements for the members of the host building that connect to the storm shelter. Issues related to protection of occupants due to building collapse have been considered and do not need to be addressed for the other portions of the facility.

A storm shelter is a self-contained and defined space within the building that does not rely upon other portions of the building to provide life-safety protection from high winds, floods, or structural collapse. Hardening the other portions of the building that are outside the storm shelter or increasing the risk category for portions of the building that may be used to egress the space is not necessary. The statements in Section 1604.5.1 regarding egress are to be applied when a building or portion thereof is being used to provide long-term, post-disaster response capabilities the loss of which would have considerable consequences to the community outside the occupied building. Section 1604.5.1 does not apply to ICC 500-compliant storm shelters.

The intent of the storm shelter is to provide short-term life safety in the event of a severe storm when the host building cannot. This allows a building owner to provide a storm shelter in one portion of the structure while designing the entire structure to meet the risk category provisions required for the primary use. The new exception in the 2018 IBC clarifies this intent. While the storm shelter will be designed to meet the requirements of ICC 500, the rest of the building need only meet the requirements for the risk categories of its various occupancies.

1604.10 Storm Shelters

CHANGE TYPE: Addition

CHANGE SUMMARY: The development of loads for storm shelters is to be based on ICC 500 which provides wind speeds for tornado and hurricane shelter design using ASCE 7 load combinations.

2018 CODE: 1604.10 Loads on storm shelters. Loads and load combinations on storm shelters shall be determined in accordance with ICC 500.

CHANGE SIGNIFICANCE: The *International Building Code* (IBC) Section 423, Storm Shelters, includes wind load criteria for storm shelters by reference to ICC 500, *ICC/NSSA Standard for the Design and Construction of Storm Shelters*, which was first referenced in the 2009 edition of the IBC. Chapter 16 of the IBC, which defines wind loads for buildings and structures, does not include specific criteria for storm shelters. Although wind loads are not specified, Table 1604.5 does regulate hurricane and other emergency shelters as Risk Category IV.

Since Chapter 16 does not include other requirements for storm shelter wind loads, Risk Category IV (RC IV) wind speeds may be interpreted as appropriate for storm shelters. Unfortunately, RC IV wind speeds are too low for a hurricane or tornado when considering an emergency shelter. RC IV wind speeds use a mean recurrence interval (MRI) of 3,000 years, an infrequent event. Meanwhile, ICC 500, the standard specifically for emergency or storm shelters, uses wind speeds based on a MRI of 10,000 years for hurricane storm shelters and a MRI of approximately 20,000 to 1,000,000 years for tornado storm shelters, highly infrequent events. By decreasing the probability that the storm may happen, the wind speed

1604.10 continues

Traditional protection from tornadoes

1604.10 continued

assumed for designing the building greatly increases. We know from wind speed measurements that wind speeds in and around a large tornado can exceed 250 mph. A straight-line wind, caused by a thunderstorm or seasonal storm, rarely has winds exceeding 70 mph. By adding a specific reference in Section 1604.10, wind design for shelters is consistent with the ICC 500 standard which contains appropriate wind speeds for a shelter required to remain standing and protect people inside during a tornado.

The new code section also allows development of load combinations based solely on ICC 500 which has load combination equations mirroring ASCE 7, *Minimum Design Loads and Associated Criteria for Buildings and Other Structures*. This limits the need to simultaneously search for design criteria in the IBC, ASCE 7, and ICC 500.

Table 1607.1 Deck Live Load

CHANGE TYPE: Modification

CHANGE SUMMARY: Table 1607.1 is now consistent with provisions in the 2010 and 2016 editions of ASCE 7 for minimum uniformly distributed live loads on decks and balconies by increasing the deck live load to one and one-half times the live load of the area served.

2018 CODE:

TABLE 1607.1 Minimum Uniformly Distributed Live Loads, L_0, and Minimum Concentrated Live Loads

Occupancy or Use	Uniform (psf)	Concentrated (pounds)
5. Balconies and decks[h]	<u>1.5 times the live load for the area served, not required to exceed 100</u> ~~Same as occupancy served~~	—

h. See Section 1604.8.3 for decks attached to exterior walls.

CHANGE SIGNIFICANCE: The *2018 International Building Code* (IBC) Table 1607.1 is now consistent with the provisions for live loads of decks and balconies in the 2010 and 2016 editions of *ASCE 7 Minimum Design Loads and Associated Criteria for Buildings and Other Structures* (ASCE 7-10 and ASCE 7-16). Live loads on decks have been increased to one and one-half times the live load for the area served, but not greater than 100 psf. Balconies and decks are recognized as often having different loading patterns than the interior of a building. A deck is often subjected to concentrated line loads from people congregating along the edge of the deck. This loading condition is acknowledged in ASCE 7 as an increase of the live load for the area served, up to the loading requirement for assembly occupancies.

Concern regarding deck failures, degradation over time, and overloading of decks has encouraged this increase of the design live load to be greater than the live load of the interior rooms served by the deck. As an upper limit to the live load, given that balconies and decks are often used as places of assembly, it is reasonable that the required live load need not exceed the specified uniform live load required in assembly areas.

Typical deck in a commercial building

Table 1607.1
Live Load Reduction

CHANGE TYPE: Modification

CHANGE SUMMARY: Table 1607.1 now clarifies where heavy live loads of 100 psf or greater may be reduced.

2018 CODE:

TABLE 1607.1 Minimum Uniformly Distributed Live Loads, L_0, and Minimum Concentrated Live Loads

Occupancy or Use	Uniform (psf)	Concentrated (pounds)
3. Armories and drill rooms	150$^{\underline{mn}}$	—
4. Assembly areas		
Fixed seats (fastened to floor)	60^m	
Follow spot, projections and control rooms	50	
Lobbies	100^m	—
Movable seats	100^m	
Stage floors	150$^{\underline{mn}}$	
Platforms (assembly)	100^m	
Other assembly areas	100^m	
14. Garages (passenger vehicles only)	40$^{m\underline{o}}$	Note a
Trucks and buses	See Section 1607.7	
19. Libraries		
Corridors above first floor	80	1,000
Reading rooms	60	1,000
Stack rooms	150$^{b,\underline{mn}}$	1,000
20. Manufacturing		
Heavy	250$^{\underline{mn}}$	3,000
Light	125$^{\underline{mn}}$	2,000
24. Recreational uses:		
Bowling alleys, poolrooms and similar uses	75^m	
Dance halls and ballrooms	100^m	
Gymnasiums	100^m	
Ice skating rink	250$^{\underline{mn}}$	—
Reviewing stands, grandstands and bleachers	100^{c,m}	
Roller skating rink	100^m	
Stadiums and arenas with fixed seats (fastened to floor)	60^{c,m}	

Occupancy or Use	Uniform (psf)	Concentrated (pounds)
26. Roofs		
Occupiable roofs:		
Roof gardens	100	
Assembly areas	100^m	
All other similar areas	Note l	Note l
29. Sidewalks, vehicular driveways and yards, subject to trucking	250$^{d,\,\underline{mn}}$	8,000^e

(Footnotes a-k not included for brevity.)

l. Areas of occupiable roofs, other than roof gardens and assembly areas, shall be designed for appropriate loads as approved by the building official. Unoccupied landscaped areas of roofs shall be designed in accordance with Section ~~1607.12.3~~ <u>1607.13.3</u>.

m. Live load reduction is not permitted ~~unless specific exceptions of Section 1607.10 apply~~.

n. <u>Live load reduction is only permitted in accordance with Section 1607.11.1.2 or Item 1 of Section 1607.11.2.</u>

o. <u>Live load reduction is only permitted in accordance with Section 1607.11.1.3 or Item 2 of Section 1607.11.2.</u>

Theater stage

CHANGE SIGNIFICANCE: The 2015 IBC Table 1607.1 restricts the use of the live load reduction equations in IBC Sections 1607.10.1 (basic) and 1607.10.2 (alternate), unless the specific exceptions of Section 1607.10 apply. This clause has caused confusion for both engineers and building officials. Section 1607.10.1 states that, except for uniform live loads at roofs, all other minimum uniformly distributed live loads are permitted to be reduced. Section 1607.10.2 Item 3 reads, for live loads not exceeding 100 psf the design live load for any structural member supporting

Table 1607.1 continues

Table 1607.1 continued

Library bookshelves

Source: Rhododendrites from Wikimedia Commons

150 square feet or more is permitted to be reduced. These two provisions, which conflict with Table 1607.1, have led to confusion about whether live loads greater than 100 psf may be reduced.

Both building officials and engineers have on occasion maintained that the provisions do allow any live load to be reduced. The intent of the multiple provisions is that heavy live loads, greater than 100 psf, and the live loads of passenger vehicle garages be reduced only when a member supporting the large live load also supports two or more floors. Lastly, Footnote m in Table 1607.1 was introduced in the 2012 IBC to align the 2012 IBC with ASCE 7-10. Assembly loads are not allowed to be reduced in ASCE 7-10.

In the 2018 IBC, in order to clear up the confusion, two new footnotes have been added to the table, and footnote m was modified. For reference to live loads greater than 100 psf, new footnote n covers the exceptions stated for heavy live loads. The footnote refers the user to exceptions in Section 1607.11.1.2 and Section 1607.11.2 Item 1. New footnote o does the same for passenger vehicle garage live loads by referencing an exception in Section 1607.11.1.3 and Section 1607.11.2 Item 2.

Footnote m has been modified to completely prohibit any live load reduction for the remaining items covered by IBC Table 1607.1, footnote m. These clarifications make the restrictions on the application of live load reductions clear for users of the IBC.

1607.15.2
Minimum Live Load for Fire Walls

CHANGE TYPE: Addition

CHANGE SUMMARY: The minimum lateral load that fire walls are required to resist has been established at five pounds per square foot.

2018 CODE: 1607.15.2 Fire walls. In order to meet the structural stability requirements of Section 706.2 where the structure on either side of the wall has collapsed, fire walls and their supports shall be designed to withstand a minimum horizontal allowable stress load of 5 psf (0.240 kN/m^2).

CHANGE SIGNIFICANCE: In the 2015 IBC, no horizontal fire wall load criteria exist for walls not using the "deemed to comply" NFPA option. A fire wall is a code defined term to describe a fire-resistance-rated wall having protected openings, which restricts the spread of fire and extends continuously from the foundation to or through the roof, with sufficient structural stability under fire conditions to allow collapse of construction on either side without collapse of the wall.

For the 2018 IBC, the minimum lateral loading that fire walls are required to resist to meet the structural stability requirements of Section 706.2 is five pounds per square foot (psf). This load assumes the structure on one side of a wall has collapsed and can no longer provide support. This lateral load (horizontal) is required for fire walls in NFPA 221, *Standard for High Challenge Fire Walls, Fire Walls, and Fire Barrier Walls*.

Jurisdictions vary in application of interior wall loads and potential loads after a fire. Some jurisdictions require consideration of wind loads; others only require a five psf lateral load. To create consistency, Section 1607.15.2 explicitly requires a five psf horizontal allowable stress design (ASD) load.

Fires may remove portions of exterior walls

PART 6 ■ Building Envelope, Structural Systems, and Construction Materials

1609
Wind Loads

CHANGE TYPE: Modification

CHANGE SUMMARY: Section 1609 now has updated wind speed maps, including maps for the state of Hawaii. Terminology for describing wind speeds has been changed again, with ultimate design wind speeds now called basic design wind speeds.

2018 CODE: Section 1602.1 ~~Definitions:~~ **Notations.**

V_{asd} = ~~Nominal~~ Allowable stress design wind speed ~~(3-second gust)~~, miles per hour (mph) (km/hr) where applicable.

V_{ult} = ~~Ultimate~~ Basic design wind speeds ~~(3-second gust)~~, miles per hour (mph) (km/hr) determined from Figures 1609.3(1)~~, 1609.3(2), 1609.3(3)~~ through 1609.3(8) or ASCE 7.

1609.1.1.1 Applicability. The provisions of ICC 600 are applicable only to buildings located within Exposure B or C as defined in Section 1609.4. The provisions of ICC 600, AWC WFCM and AISI S230 shall not

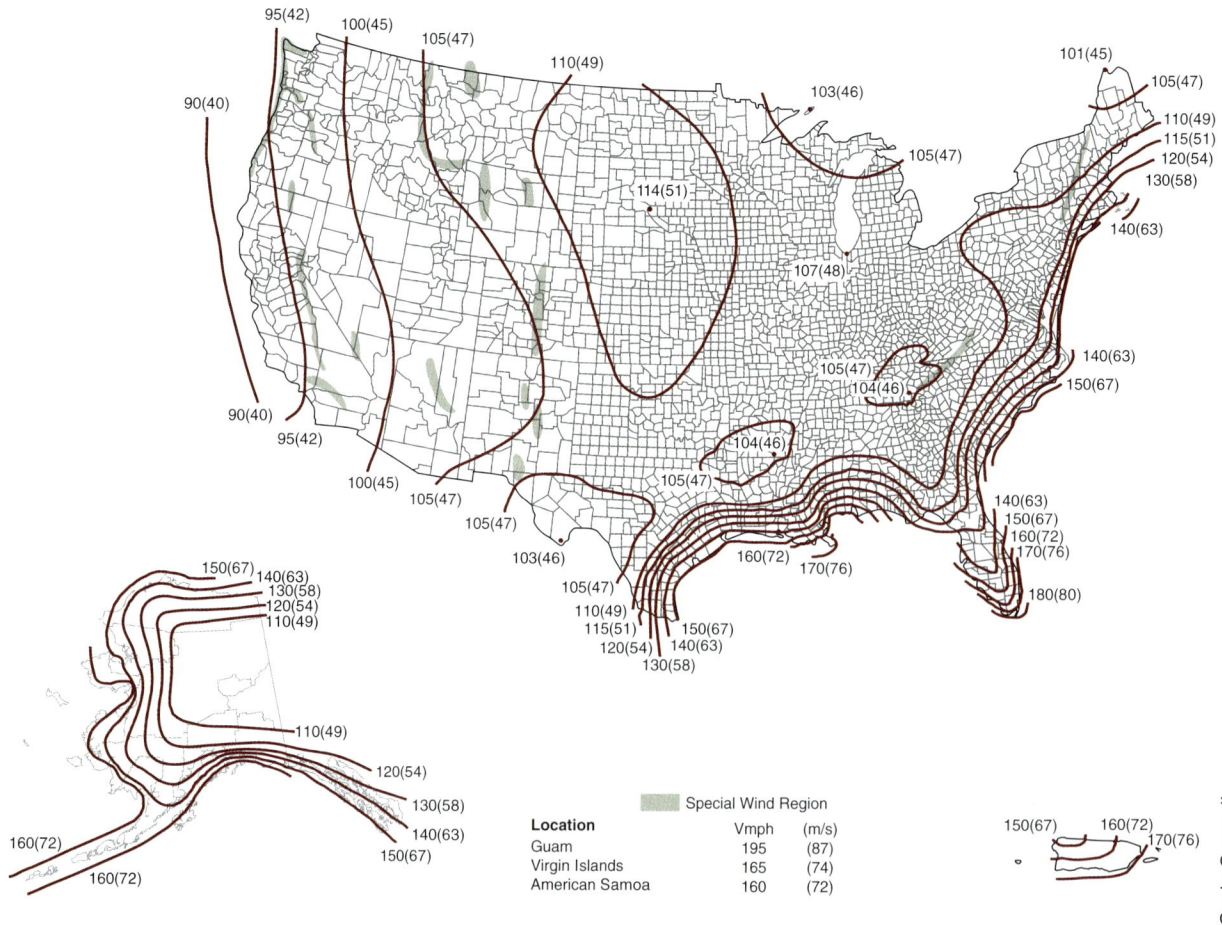

Location	Vmph	(m/s)
Guam	195	(87)
Virgin Islands	165	(74)
American Samoa	160	(72)

Notes:
1. Values are nominal design 3-second gust wind speeds in miles per hour (m/s) at 33 ft (10 m) above ground for Exposure C category.
2. Linear interpolation is permitted between contours. Point values are provided to aid with interpolation.
3. Islands, coastal areas, and land boundaries outside the last contour shall use the last wind speed contour.
4. Mountainous terrain, gorges, ocean promontories, and special wind regions shall be examined for unusual wind conditions.
5. Wind speeds correspond to approximately a 7% probability of exceedance in 50 years (Annual Exceedance Probability = 0.00143, MRI = 700 Years).
6. Location-specific basic wind speeds shall be permitted to be determined using www.atcouncil.org/windspeed.

Basic Design Wind Speeds, V, For Risk Category II Buildings and Other Structures

Notes:
1. Values are nominal design 3-second gust wind speeds in miles per hour (m/s) at 33 ft (10 m) above ground for Exposure C category.
2. Linear interpolation is permitted between contours. Point values are provided to aid with interpolation.
3. Islands, coastal areas, and land boundaries outside the last contour shall use the last wind speed contour.
4. Mountainous terrain, gorges, ocean promontories, and special wind regions shall be examined for unusual wind conditions.
5. Wind speeds correspond to approximately a 3% probability of exceedance in 50 years (Annual Exceedance Probability = 0.000588, MRI = 1700 Years).
6. Location-specific basic wind speeds shall be permitted to be determined using www.atcouncil.org/windspeed

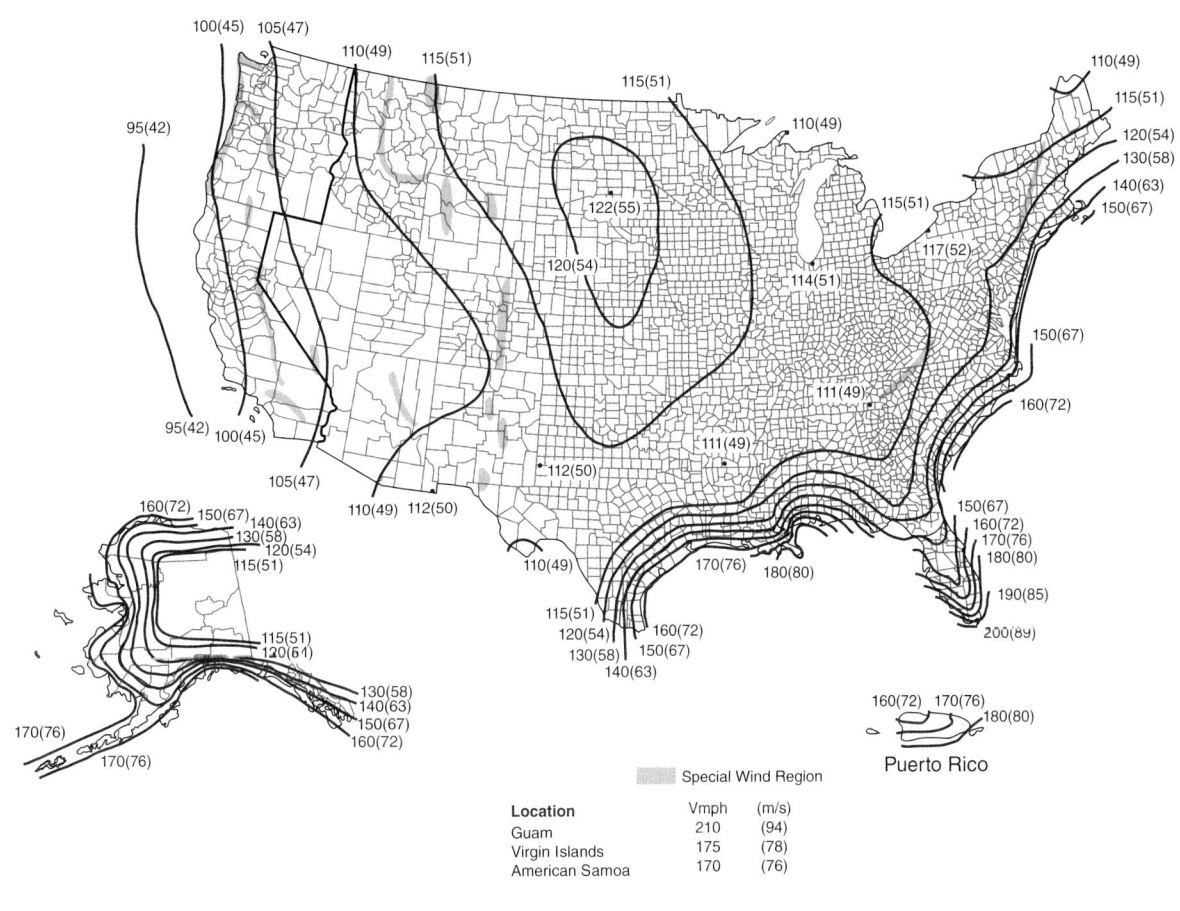

Basic Design Wind Speeds, V, For Risk Category III Buildings and Other Structures

apply to buildings sited on the upper half of an isolated hill, ridge or escarpment meeting <u>all of</u> the following conditions:

1. The hill, ridge or escarpment is 60 feet (18 288 mm) or higher if located in Exposure B or 30 feet (9144 mm) or higher if located in Exposure C;
2. The maximum average slope of the hill exceeds 10 percent; and
3. The hill, ridge or escarpment is unobstructed upwind by other such topographic features for a distance from the high point of 50 times the height of the hill or ~~1 mile~~ <u>2 miles</u> (~~1.61~~<u>3.22</u> km), whichever is greater.

1609.3 ~~Ultimate~~ <u>Basic</u> design wind speed. The ~~ultimate~~ <u>basic</u> design wind speed, V_{ult}, in mph, for the determination of the wind loads shall be determined by Figures 1609.3(1)~~, 1609.3(2) and 1609.3(3)~~ <u>through (8)</u>. The ~~ultimate~~ <u>basic</u> design wind speed, V_{ult}, for use in the design of Risk Category II buildings and structures shall be obtained from Figure 1609.3(1) <u>and 1609.3(5)</u>. The ~~ultimate~~ <u>basic</u> design wind speed, V_{ult}, for use in the design of Risk Category III <u>buildings</u> and <u>structures shall be obtained from</u>

1609 continues

1609 continued

Figures 1609.3(2) and 1609.3(6). The basic design wind speed, V, for use in the design of Risk Category IV buildings and structures shall be obtained from Figures ~~1609.3(2)~~ 1609.3(3) and 1609.3(7). The ~~ultimate~~ basic design wind speed, V_{ult}, for use in the design of Risk Category I buildings and structures shall be obtained from Figures ~~1609.3(3)~~ 1609.3(4) and 1609.3(8). The ~~ultimate~~ basic design wind speed, V_{ult}, for the special wind regions indicated near mountainous terrain and near gorges shall be in accordance with local jurisdiction requirements. The ~~ultimate~~ basic design wind speeds, V_{ult}, determined by the local jurisdiction shall be in accordance with ~~Section 26.5.1~~ Chapter 26 of ASCE 7.

In nonhurricane-prone regions, when the ~~ultimate~~ basic design wind speed, V_{ult}, is estimated from regional climatic data, the ~~ultimate~~ basic design wind speed, V_{ult}, shall be determined in accordance with ~~Section 26.5.3~~ Chapter 26 of ASCE 7.

(Hawaiian wind speed maps have many complex isolines and have not been included in this section. See the 2018 IBC for Figures 1609.3(5) through (8) which contain wind maps for the Hawaiian Islands.)

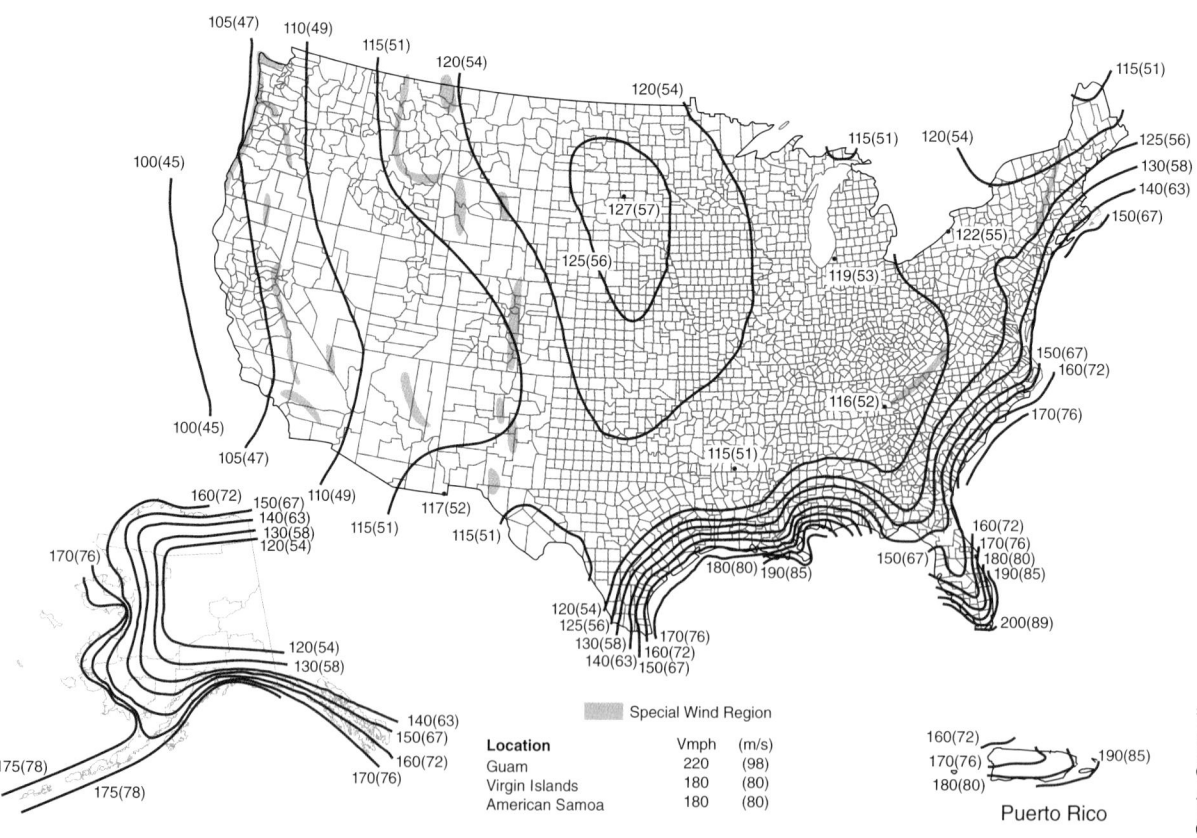

Location	Vmph	(m/s)
Guam	220	(98)
Virgin Islands	180	(80)
American Samoa	180	(80)

Notes:
1. Values are nominal design 3-second gust wind speeds in miles per hour (m/s) at 33 ft (10 m) above ground for Exposure C category.
2. Linear interpolation is permitted between contours. Point values are provided to aid with interpolation.
3. Islands, coastal areas, and land boundaries outside the last controur shall use the last wind speed contour.
4. Mountainous terrain, gorges, ocean promontories, and special wind regions shall be examined for unusual wind conditions.
5. Wind speeds correspond to approximately a 1.6% probability of exceedance in 50 years (Annual Exceedance Probability = 0.00033, MRI = 3000 Years).
6. Location-specific basic wind speeds shall be permitted to be determined using www.atcouncil.org/windspeed.

Basic Design Wind Speeds, V, For Risk Category IV Buildings and Other Structures

Additionally, as this code change affected substantial portions of Section 1609, the entire code change text is too extensive to be included here. Changes may be seen in Section 1609 of the 2018 IBC or refer to code change S56 in the Complete Revision History to the 2018 I-Codes for the complete text and history of the change.)

CHANGE SIGNIFICANCE: This code change aligns the 2018 IBC with changes to provisions in the 2016 edition of *ASCE 7 Minimum Design Loads and Associated Criteria for Buildings and Other Structures* (ASCE 7-16).

The code changes harmonize terminology between the IBC and structural loads standard. Some of the changes include deleting the word "ultimate" in favor of the term "basic" and deletion of the subscript "ult" from the variable "V" for wind speed. It is thought that use of the term ultimate with wind speeds is no longer necessary as users have had time using ASCE 7-10 and either the 2012 or 2015 IBC; all of which use the ultimate design wind speed terminology. Similarly, the word "nominal" is deleted and the term "allowable stress" is added to express where use of older terminology still exists. These changes allow for consistency with terminology used in ASCE 7-16.

The increase in the minimum distance without obstructions from one mile to two miles for buildings on the top of a hill, ridge, or escarpment in Section 1609.1.1.1, Applicability, has been done to correct a discrepancy between the code and ASCE 7. The load standard requires consideration of topography when a building is on the upper half of a hill and winds are unobstructed for at least two miles in the direction under consideration. Limits for the use of other optional standards to determine loading do not consider buildings on the top of a hill unprotected by other hills or ridges. By changing the minimum distance to two miles without obstructions, more buildings will be allowed to use ICC 600, AWC WFCM, or AISI 230 as an alternate loading standard.

The design wind speed maps in Figures 1609.3(1) through 1609.3(8) have been updated to reflect the maps adopted into ASCE 7-16. During the development of the ASCE 7-16 standard, the ASCE 7 Wind Load Subcommittee made substantial revisions to the wind speed maps contained within the standard, and the number of maps increased to eight maps. These revisions include the development of separate maps for Risk Category III and IV structures; reconstruction of the special wind regions within the maps, correction for known deficiencies in the wind speed contours; and modification of the basic wind speed based on updated climatic and weather data. New hurricane contours in the northeastern states were developed based on updated hurricane models, and the locations of the contours along the hurricane coastline were adjusted to reflect new research into the decay rate of hurricanes over land.

New maps for the State of Hawaii were included to eliminate the state as a "special wind region" and to provide guidance on the wind patterns that occur because of the unique topography. There are maps for main wind-force resisting systems and component and cladding design in the ASCE 7-16 standard along with four new serviceability maps.

In the 2018 IBC, changes to the maps will decrease design wind speeds for the majority of the United States. The basic design wind speeds have been lowered in most locations based on the latest data available. Along the hurricane coastline from Virginia to Texas, the wind speeds remain nearly unchanged from the 2015 IBC maps. In some cases, there is

1609 continues

1609 continued

a small increase for Risk Category IV (RC IV) structures from values assigned in the 2012 and 2015 IBC for some parts of the country. This is caused by the new mean recurrence interval with a 1.6% probability of exceedance for RC IV buildings. Because of the general reduction in expected wind speeds, some regions will find that even RC IV buildings do not have an increase in design wind speed while other regions will see a slight increase.

The basic wind speeds for all four Risk Category maps decrease significantly west of the continental divide. Wind speeds in the Northern Great Plains states are similar to previous maps. In the rest of the continental United States south and east of the Great Plains, wind speeds generally decrease.

To look up the design wind speed, V, for a specific location two tools are available online. The Applied Technology Council (ATC) has a wind speed tool at windspeed.atcouncil.org for free. The American Society of Civil Engineers (ASCE) has a hazard tool which will include snow, rain, flood, ice wind, seismic and tsunami design information for a specific location for a nominal fee.

1613 Earthquake Loads

CHANGE TYPE: Modification

CHANGE SUMMARY: The site coefficients contained in the IBC have now been brought into alignment with the newest generation of ground motion attenuation equations.

2018 CODE: ~~1613.3.2~~ **1613.2.2 Site class definitions.** Based on the site soil properties, the site shall be classified as Site Class A, B, C, D, E or F in accordance with Chapter 20 of ASCE 7.

Where the soil properties are not known in sufficient detail to determine the site class, Site Class D, <u>subjected to the requirements of Section 1613.2.3,</u> shall be used unless the building official or geotechnical data determines <u>that</u> Site Class E or F soils are present at the site.

<u>Where site investigations that are performed in accordance with Chapter 20 of ASCE 7 reveal rock conditions consistent with Site Class B, but site-specific velocity measurements are not made, the site coefficients F_a and F_v shall be taken at unity (1.0).</u>

~~1613.3.3~~**1613.2.3 Site coefficients and adjusted maximum considered earthquake spectral response acceleration parameters.** The maximum considered earthquake spectral response acceleration for short periods, S_{MS}, and at 1-second period, S_{M1}, adjusted for *site class* effects shall be determined by Equations 16-~~37~~36 and 16-~~38~~37, respectively:

$$S_{MS} = F_a S_s \qquad \text{(Equation 16-}\cancel{37}36\text{)}$$
$$S_{M1} = F_v S_1 \qquad \text{(Equation 16-}\cancel{38}37\text{)}$$

but <u>S_{MS} shall not be taken less than S_{M1} except when determining the seismic design category in accordance with Section 1613.2.5.</u>

where:
- F_a = Site coefficient defined in Table ~~1613.3.3(1)~~1613.2.3(1).
- F_v = Site coefficient defined in Table ~~1613.3.3(2)~~1613.2.3(2).
- S_S = The mapped spectral accelerations for short periods as determined in Section ~~1613.3.1~~1613.2.1.
- S_1 = The mapped spectral accelerations for a 1-second period as determined in Section ~~1613.3.1~~1613.2.1.

1613 continues

TABLE ~~1613.3.3(1)~~1613.2.3(1) Values of Site Coefficient $F_a{}^a$

Site Class	Mapped <u>Risk Targeted</u> Maximum Considered Earthquake (MCE$_R$) Spectral Response Acceleration <u>Parameter</u> at short period					
	$S_s \leq 0.25$	$S_s = 0.50$	$S_s = 0.75$	$S_s = 1.00$	$S_s \underline{\geq} = 1.25$	$\underline{S_s \geq 1.5}$
A	0.8	0.8	0.8	0.8	0.8	0.8
B	0.9~~1.0~~	0.9~~1.0~~	0.9~~1.0~~	0.9~~1.0~~	0.9~~1.0~~	0.9
C	1.3~~1.2~~	1.3~~1.2~~	1.2~~1.1~~	1.2~~1.0~~	1.2~~1.0~~	1.2
D	1.6	1.4	1.2	1.1	1.0	1.0
E	2.4~~2.5~~	1.7	1.3~~1.2~~	Note b~~0.9~~	Note b~~0.9~~	Note b
F	Note b	Note b	Note b	Note b	Note b	Note b

a. Use straight-line interpolation for intermediate values of mapped spectral response acceleration at short period, S_s.
b. Values shall be determined in accordance with Section ~~11.4.7~~ 11.4.8 of ASCE 7.

1613 continued

TABLE ~~1613.3.3(2)~~ <u>1613.2.3(2)</u> Values of Site Coefficient F_v[a]

Site Class	Mapped <u>Risk Targeted Maximum</u> Considered Earthquake (MCE$_R$) Spectral Response Acceleration Parameter at 1-second period					
	$S_1 \leq 0.1$	$S_1 = 0.2$	$S_1 = 0.3$	$S_1 = 0.4$	$S_1 \underline{\geq} = 0.5$	<u>$S_1 \geq 0.6$</u>
A	0.8	0.8	0.8	0.8	0.8	0.8
B	<u>0.8</u> ~~1.0~~	<u>0.8</u> ~~1.0~~	<u>0.8</u> ~~1.0~~	<u>0.8</u> ~~1.0~~	<u>0.8</u> ~~1.0~~	0.8
C	<u>1.5</u> ~~1.7~~	<u>1.5</u> ~~1.6~~	1.5	<u>1.5</u> ~~1.4~~	<u>1.5</u> ~~1.3~~	1.4
D	2.4	2.2[c] ~~2.0~~	2.0[c] ~~1.8~~	1.9[c] ~~1.6~~	1.8[c] ~~1.5~~	<u>1.7[c]</u>
E	~~4.2~~ <u>3.5</u>	3.3[c] ~~3.2~~	<u>2.8[c]</u>	<u>2.4[c]</u>	2.2[c] ~~2.4~~	<u>2.0[c]</u>
F	Note b	Note b	Note b	Note b	Note b	<u>Note b</u>

a. Use straight-line interpolation for intermediate values of mapped spectral response acceleration at 1-second period, S_1.
b. Values shall be determined in accordance with Section ~~11.4.7~~ <u>11.4.8</u> of ASCE 7.
c. See requirements for site-specific ground motions in Section ~~11.4.7~~ <u>11.4.8</u> of ASCE 7.

(As multiple code changes affected substantial portions of Section 1613, the entire code change text is too extensive to be included here. Refer to code changes S114, S119 and S242 in the Complete Revision History to the 2018 I-Codes for the complete text and history of the change.)

<u>Where Site Class D is selected as the default site class per Section 1613.2.2, the value of F_a shall not be less than 1.2. Where the simplified design procedure of ASCE 7 Section 12.14 is used, the value of F_a shall be determined in accordance with ASCE 7 Section 12.14.8.1, and the values of F_v, S_{MS}, and S_{M1} need not be determined.</u>

CHANGE SIGNIFICANCE: The site coefficients of Section 1613.2.3 contained in the IBC date back to soil studies performed in the early 1990s. These site coefficients, based on soil type, were tied to the ground motion attenuation relationships that were used by the United States Geological Survey (USGS) to develop the seismic zone map used by the legacy codes of that era. The USGS maps contained in the 2010 edition of ASCE 7, and in the 2012 IBC and 2015 IBC, were based on an updated set of attenuation relationships known as the next-generation attenuation or NGA equations. The 1990s-era site class coefficients are not appropriate for use with ground motions derived using the NGA equations.

The Building Seismic Safety Council's (BSSC) Provisions Update Committee (PUC) performed extensive studies of the appropriate site class coefficients to use with the NGA-derived ground motions and proposed the new values in Tables 1613.2.3(1) and 1613.2.3(2). These values come from the 2015 National Earthquake Hazard Reduction Program (NEHRP) guidelines. ASCE 7-16 also contains these updated site class coefficients.

In developing the updated site class values, the BSSC discovered that the standard spectral shape derived using the S_{DS} and S_{D1} parameters is not conservative for the design of long period buildings ($T > 1$ second) located on Site Class D or softer soils, when the seismic hazard is dominated by large-magnitude earthquakes. The addition of footnote c in

Table 1613.2.3(2) references ASCE 7 Section 11.4.8 which now requires use of site-specific spectra to represent ground motions for such buildings.

This code change aligns the *2018 International Building Code* (IBC) requirements for determining seismic design category with changes to the provisions of the 2016 edition of *ASCE 7 Minimum Design Loads and Associated Criteria for Buildings and Other Structures* (ASCE 7-16). ASCE 7 parameters for F_a and F_v site coefficients are mirrored within the IBC.

1613.2.1
Seismic Maps

CHANGE TYPE: Modification

CHANGE SUMMARY: The IBC seismic maps have been updated to match new maps in the 2015 NEHRP Provisions and 2016 ASCE 7 standard.

2018 CODE:

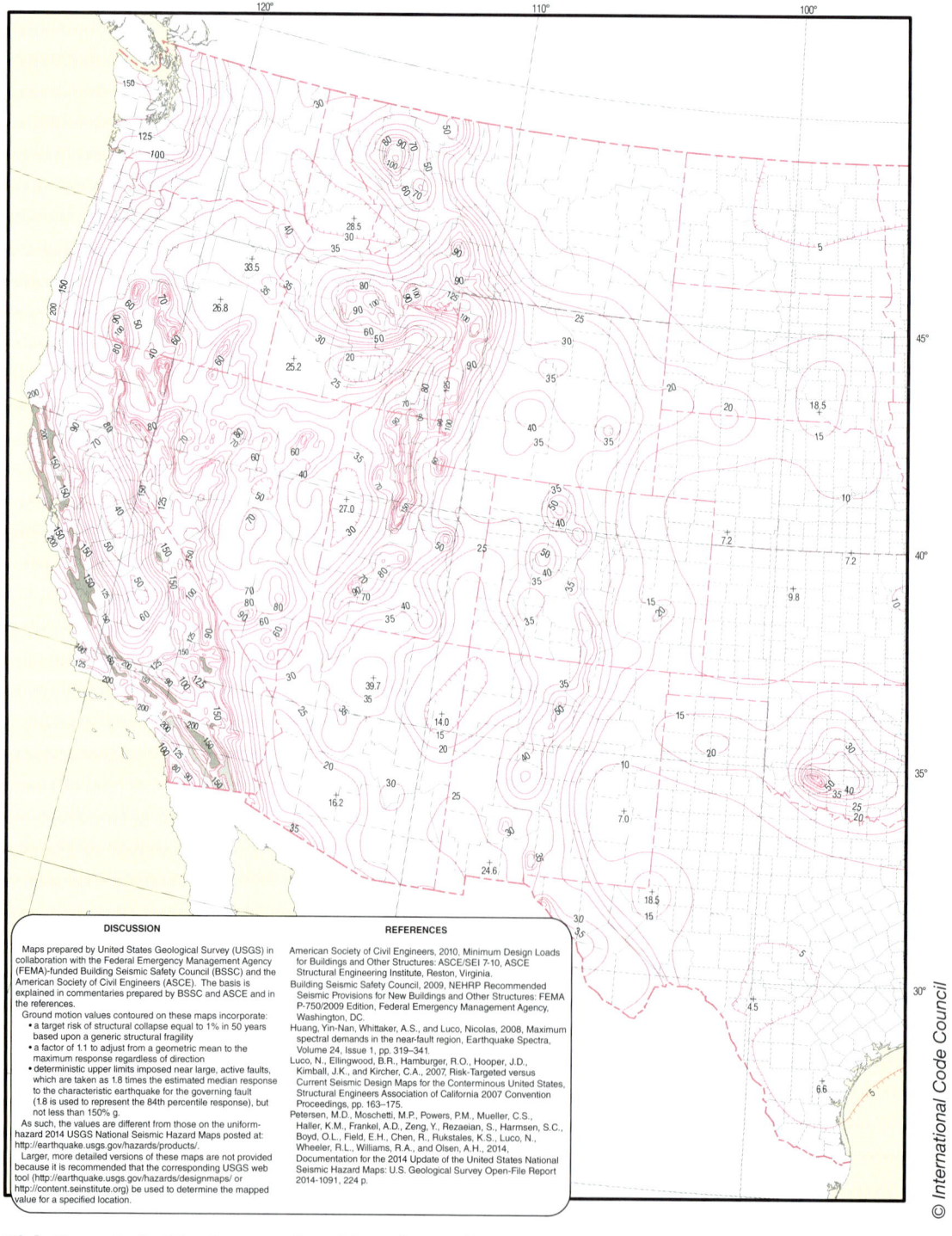

Risk-Targeted Maximum Considered Earthquake (MCE_R) Ground Motion Response Accelerations for the Conterminous United States of 0.2-second Spectral Response Acceleration (5% of Critical Damping)

1613.2.1 ■ Seismic Maps

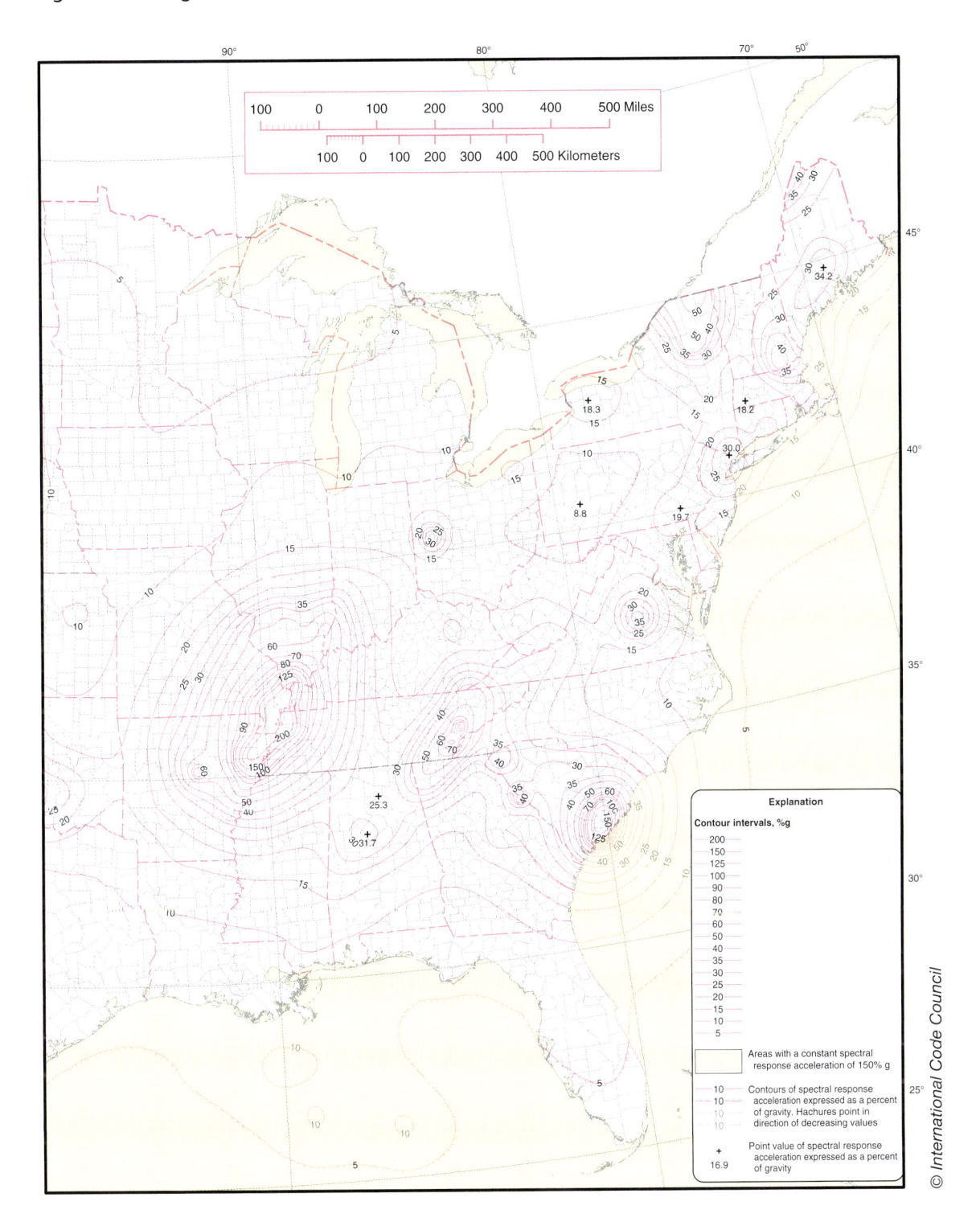

Risk-Targeted Maximum Considered Earthquake (MCE$_R$) Ground Motion Response Accelerations for the Conterminous United States of 0.2-second Spectral Response Acceleration (5% of Critical Damping) continued

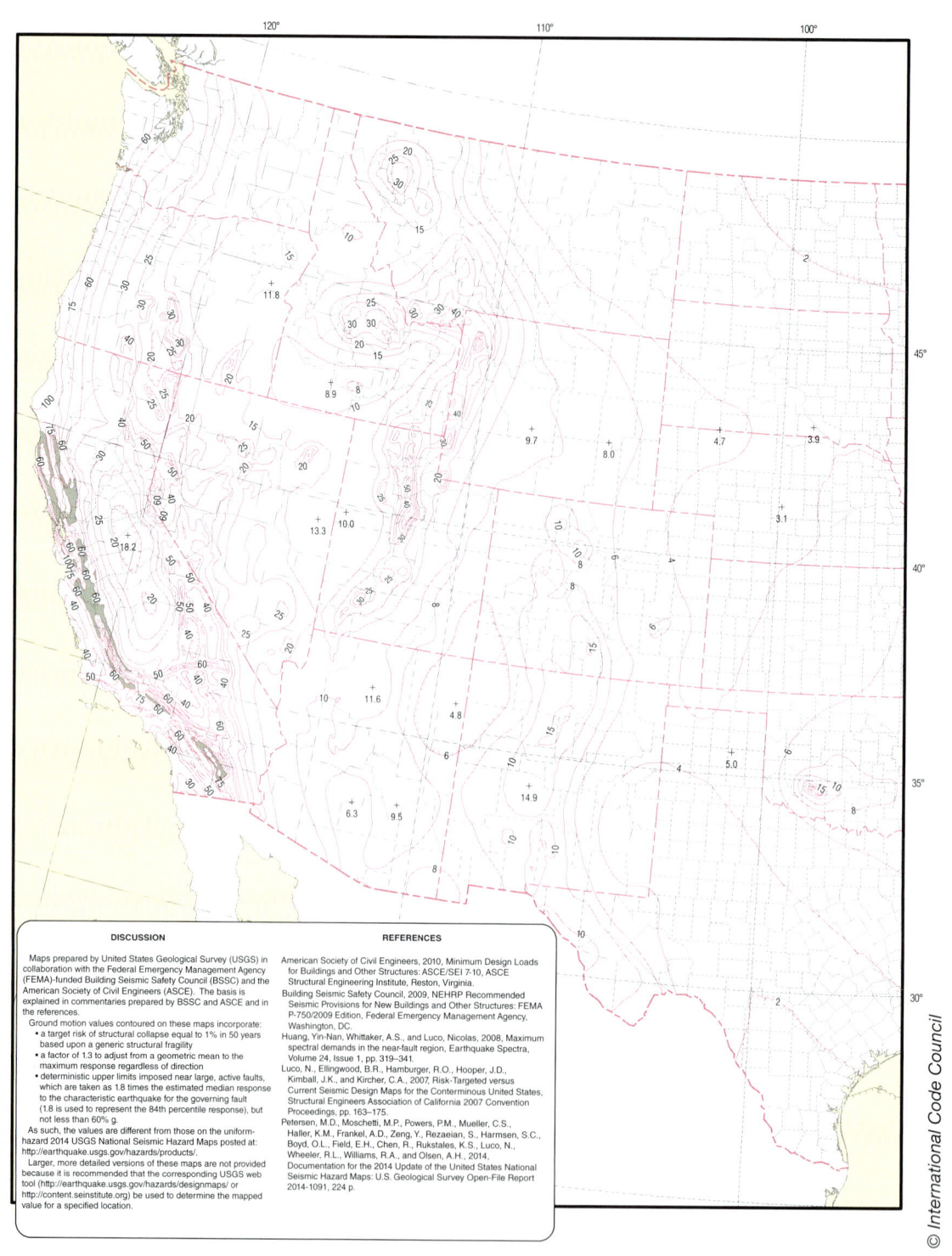

Risk-Targeted Maximum Considered Earthquake (MCE$_R$) Ground Motion Response Accelerations for the Conterminous United States of 1-second Spectral Response Acceleration (5% of Critical Damping)

Significant Changes to the IBC 2018 Edition　　　　　　　　　　1613.2.1 ■ Seismic Maps　**233**

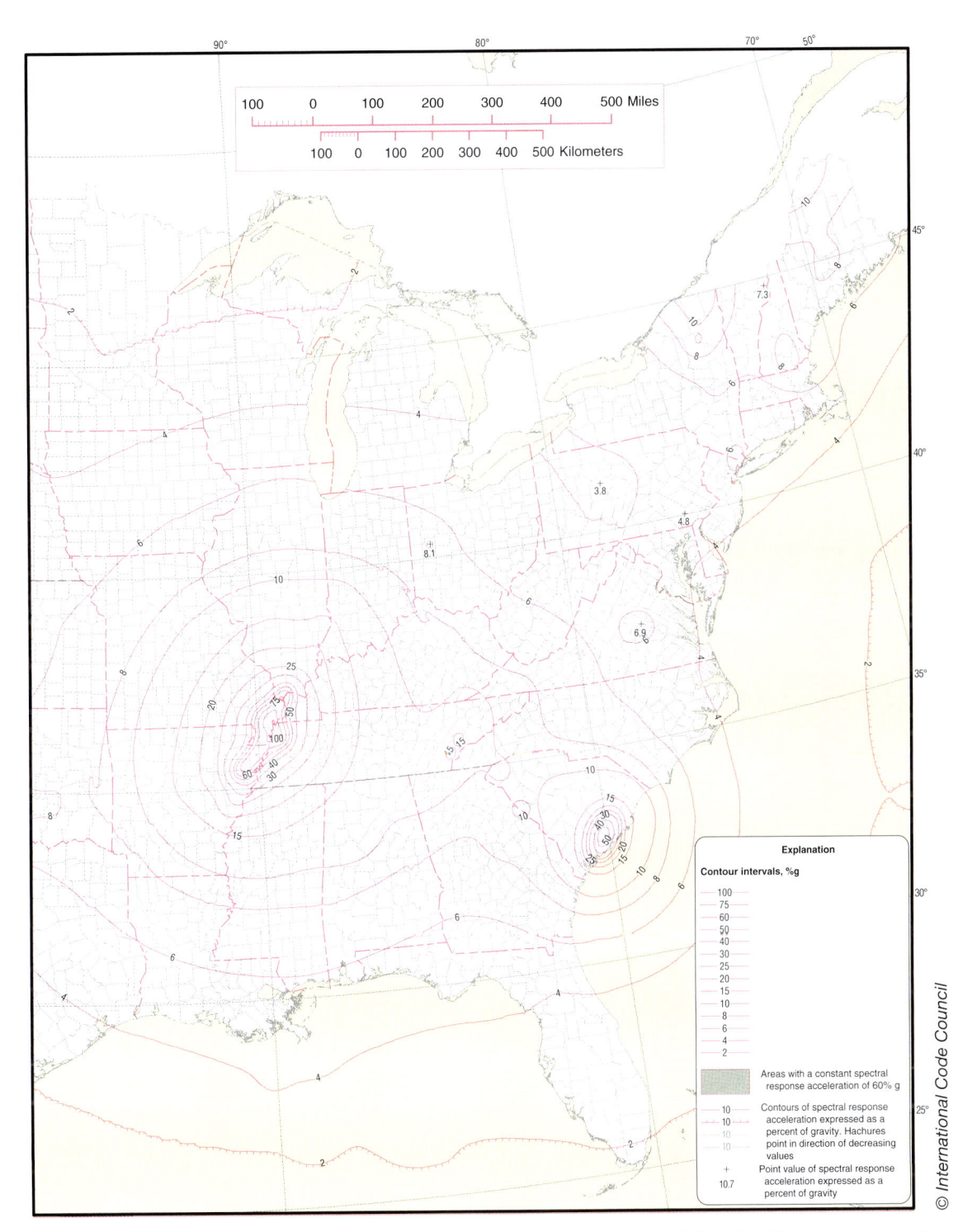

Risk-Targeted Maximum Considered Earthquake (MCE$_R$) Ground Motion Response Accelerations for the Conterminous United States of 1-second Spectral Response Acceleration (5% of Critical Damping) continued

CHANGE SIGNIFICANCE: 2018 IBC Figures 1613.2.1(1) through (8) include updated seismic hazard maps developed by the United States Geological Survey (USGS) for the United States National Seismic Hazard Maps, which include the latest seismic, geologic, and geodetic information on earthquake rates and associated ground shaking.

1613.2.1 continues

1613.2.1 continued

The MCE_R ground motion maps incorporate the latest seismic hazard models based on the national seismic hazard maps prepared by the USGS. The maps are made in collaboration with the Federal Emergency Management Agency (FEMA) and the Building Seismic Safety Council (BSSC). The maps are consistent with those in the *2015 NEHRP Recommended Seismic Provisions for New Buildings and Other Structures* (FEMA P-1050-1) and *ASCE 7 Minimum Design Loads and Associated Criteria for Buildings and Other Structures* (ASCE 7-16). The maps incorporate significant new information on earthquake faults and ground motion attenuation, and are more consistent with the site-specific ground motion procedures of ASCE 7-16 Chapter 21.

The most significant changes for the 2014 USGS model used to make the maps fall into four categories:

1. For Central and Eastern US (CEUS) sources:
   - Development of a moment magnitude-based earthquake catalog through 2012, replacing the 2008 catalog.
   - Updated distribution for maximum magnitude (M_{max}) based on a new analysis of global earthquakes in stable continental regions.
   - Updated New Madrid source model, including fault geometry, recurrence rates of large earthquakes, and alternative magnitudes from M6.6 to M8.0.
   - Updated treatment of earthquakes that are potentially induced by underground fluid injection.

2. For Intermountain West and Pacific Northwest crustal sources:
   - Updated earthquake catalog and treatment of magnitude uncertainty in rate calculations.
   - Updated fault parameters for faults in Utah based on new datasets and models.
   - Introduced new combined geologic and geodetic inversion models for assessing fault slip rates on fault sources.
   - Implemented new models for Cascadia earthquake-rupture geometries and rates based on onshore (paleo-tsunami) and offshore (turbidite) studies.
   - Updated model for deep (intraslab) earthquakes along the coasts of Oregon and Washington, including a new depth distribution for intraslab earthquakes.
   - Allowance for an M_{max} up to M8.0 for crustal and intraslab earthquakes.
   - Addition of the Tacoma fault source and updated the South Whidbey Island fault source in Washington.

3. For California sources:
   - Development of a new seismic source model based on the Uniform California Earthquake Rupture Forecast, Version 3 (UCERF3) and new earthquake forecasts for California, which include many more multi-segment ruptures than in previous editions of the maps.
   - These models were developed over the past several years and involved a major update of the methodology for calculating earthquake recurrence.

4. For ground motion models (or "attenuation relations"):
   - Included new earthquake ground motion models for active shallow crustal earthquakes (NGA-West2) and subduction zone-related interface and intraslab earthquakes.
   - Incorporated new and evaluated older ground motion models: five equations applied for the Western US (WUS), nine for the CEUS, and four for the subduction interface and intraslab earthquakes.
   - Increased the maximum distance from 200 km to 300 km when calculating ground motion from WUS crustal sources.

The new ground motions vary locally depending on complicated changes in the underlying models. In the CEUS, the hazard increases in some places, and the new ground-motion model-weighting scheme generally lowers the ground motions. The resulting maps for the CEUS can differ by ±20% compared to the 2008 maps due to interactions between the various parts of model.

In the Intermountain West region the combined geologic and geodetic inversion models increase the hazard along the Wasatch fault and central Nevada region, but the new NGA-West2 ground motions tend to lower the hazard on the hanging walls of normal faults with respect to the 2008 maps, the type of fault common to Nevada and Utah. These counteracting effects can result in an increase or decrease of hazard.

In the Pacific Northwest, the new Cascadia source model causes the hazard to increase by up to 40% in the southern Cascadia subduction zone (Oregon and Northern California) due to the addition of possible M8 and greater earthquakes, but causes the hazard to decrease slightly along the northern Cascadia subduction zone (Washington and British Columbia) because of reduced earthquake rates relative to the 2008 USGS hazard model. Subduction ground motions from the new models fall off faster with distance than motions in previous models, but they also tend to be higher near fault ruptures.

In California, the new UCERF3 model accounts for a single earthquake that ruptures multiple faults yielding a larger magnitude than applied in the previous model, but with smaller recurrence rates. At a specific site, it is important to examine all model changes to determine why the ground motions may have increased or decreased.

Like previous versions of the USGS national seismic hazard model, the 2014 model purposefully excludes swarms of earthquakes that may be caused by industrial fluid processes such as fracking or wastewater disposal.

Additional details on the technical reasons behind these changes are documented in FEMA P-1050-1, Section C22. A copy of the 2015 NHERP Guidelines (FEMA P-1050-1) is available at fema.gov/media-library/assets/documents/107646.

1615, 1604.5
Tsunami Loads

CHANGE TYPE: Addition

CHANGE SUMMARY: There are many coastal communities in the western United States which need tsunami-resistant design of critical infrastructure and essential facilities. New IBC Section 1615, Tsunami Loads, has been added to address design of these facilities.

2018 CODE:

SECTION 202 DEFINITIONS

TSUNAMI DESIGN GEODATABASE. The ASCE database (version 2016-1.0) of Tsunami Design Zone maps and associated design data for the states of Alaska, California, Hawaii, Oregon, and Washington.

TSUNAMI DESIGN ZONE. An area identified on the Tsunami Design Zone map between the shoreline and the inundation limit, within which certain structures designated in Chapter 16 are designed for or protected from inundation.

1604.5 Risk Category. Each building and structure shall be assigned a risk category in accordance with Table 1604.5. Where a referenced standard specifies an occupancy category, the risk category shall not be taken as lower than the occupancy category specified therein. Where a referenced standard specifies that the assignment of a risk category be in accordance with ASCE 7, Table 1.5-1, Table 1604.5 shall be used in lieu of ASCE 7, Table 1.5-1.

> **Exception:** The assignment of buildings and structures to Tsunami Risk Categories III and IV is permitted to be assigned in accordance with Section 6.4 of ASCE 7.

Tsunami warning sign

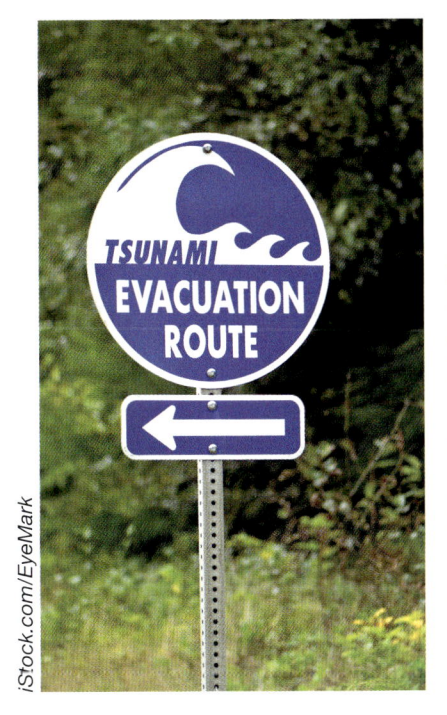

Tsunami route sign

High-rises in locations at low elevation become potential vertical evacuation refuge

SECTION 1615
TSUNAMI LOADS

1615.1 General. The design and construction of Risk Category III and IV buildings and structures located in the Tsunami Design Zones defined in the Tsunami Design Geodatabase shall be in accordance with Chapter 6 of ASCE 7, except as modified by this code.

CHANGE SIGNIFICANCE: Many coastal areas in the western United States are subject to potentially destructive tsunamis. There are many coastal communities in Alaska, Washington, Oregon, California, and Hawaii which need tsunami-resistant design of critical infrastructure and essential facilities to provide vital services necessary for post-disaster response and recovery, and to enable the continued functioning of the community. These communities are at low elevation with long evacuation route distances. The public safety risk has been only partially mitigated through warning and preparedness for evacuation; there are many areas in these five states where complete evacuation prior to tsunami arrival cannot be ensured. Accordingly, these communities need a standard for designated tsunami vertical evacuation refuge structures as an alternative to nonexistent high ground.

A new chapter, Chapter 6 Tsunami Loads and Effects, has been added to the 2016 edition of ASCE 7 *Minimum Design Loads and Associated Criteria for Buildings and Other Structures*. The 2018 *International Building Code* references Chapter 6. When referencing the requirements of ASCE 7 by the IBC, only new construction qualifying as Risk Category III and IV structures must meet requirements to resist tsunami loads. Additionally, there is an exception so that a local jurisdiction may evaluate the tsunami-inundation zone for physical and community risk when assigning these categories to the facilities deemed vital to public health, safety and welfare.

For existing buildings, new structures outside tsunami hazard zones, and new Risk Category I and II buildings within the tsunami hazard zones, there are no mandatory requirements for tsunami protection in the IBC or ASCE 7.

1704.6
Structural Observations

CHANGE TYPE: Modification

CHANGE SUMMARY: Section 1704.6.1 has been added requiring structural observation of buildings that are considered a high-rise or assigned to Risk Category IV.

2018 CODE: 1704.6 Structural observations. Where required by the provisions of Section 1704.6.1, or 1704.6.2 or 1704.6.3, the owner or the owner's authorized agent shall employ a registered design professional to perform structural observations. Structural observation does not include or waive the responsibility for the inspections in Section 110 or the special inspections in Section 1705 or other sections of this code.

Prior to the commencement of observations, the structural observer shall submit to the building official a written statement identifying the frequency and extent of structural observations.

At the conclusion of the work included in the permit, the structural observer shall submit to the building official a written statement that the site visits have been made and identify any reported deficiencies that, to the best of the structural observer's knowledge, have not been resolved.

1704.6.1 Structural observations for structures. Structural observations shall be provided for those structures where one or more of the following conditions exist:

1. The structure is classified as Risk Category IV.
2. The structure is a high-rise building.
3. Such observation is required by the registered design professional responsible for the structural design.
4. Such observation is specifically required by the building official.

Structural observation

~~1704.6.1~~ **1704.6.2 Structural observations for seismic resistance.** Structural observations shall be provided for those structures assigned to Seismic Design Category D, E or F where one or more of the following conditions exist:

1. The structure is classified as Risk Category III or IV.
2. ~~The height of the structure is greater than 75 feet (22 860 mm) above the base as defined in ASCE 7.~~
3. 2. The structure is assigned to Seismic Design Category E, is classified as Risk Category I or II, and is greater than two stories above the grade plane.
4. ~~When so designated by the registered design professional responsible for the structural design.~~
5. ~~When such observation is specifically required by the building official.~~

~~1704.6.2~~ **1704.6.3 Structural observations for wind ~~requirements~~ resistance.** ~~Structural observations shall be provided for those structures sited where~~ Structural observations shall be provided for those structures sited where V_{asd} ~~as determined in accordance with Section 1609.3.1 exceeds 110 mph~~ is 130mph (~~49~~58 m/sec) or greater and the ~~where one or more of the following conditions exist:~~

1. ~~The~~ structure is classified as Risk Category III or IV.
2. ~~The building height is greater than 75 feet (22 860 mm).~~
3. ~~When so designated by the registered design professional responsible for the structural design.~~
4. ~~When such observation is specifically required by the building official.~~

CHANGE SIGNIFICANCE: The 2015 IBC requires structural observation only in the limited situations of high-rise buildings and higher-risk-category structures located in high seismic or high wind regions. A 7-story office building in San Francisco would require structural observation but a 60-story high-rise or a 40,000-seat stadium in New York would not. It was decided that structural observation should be required for all large or important buildings throughout the country.

Quality of construction is increased when an engineering firm which designed a structure verifies that construction is progressing according to the design intent. Structural observation should be required wherever the consequence of structural failure is greater by virtue of complexity of the design, type of occupancy or size of the building, or an increased risk from natural hazards. Structural observation is meant to augment the detailed inspection provided by special inspectors.

In the 2018 IBC, public safety is increased by requiring that all high-rise buildings (having floors more than 75 feet above the lowest level of fire department vehicle access) and those assigned to Risk Category IV are afforded the benefit of structural observation, not just structures at higher risk of an earthquake or hurricane. Section 1704.6.1 has been added to trigger observations for high-rise and high-risk buildings located in any jurisdiction.

1705.5.2

Metal-Plate-Connected Wood Trusses

CHANGE TYPE: Modification

CHANGE SUMMARY: Five-foot-tall wood trusses requiring permanent bracing now require a periodic special inspection to verify that the required bracing has been installed.

2018 CODE: 1705.5.2 Metal-plate-connected wood trusses ~~spanning 60 feet or greater.~~ ~~Where a truss clear span is~~ <u>Special inspections of wood trusses with overall heights of 60 inches (1524 mm) or greater shall be performed to verify that the installation of the permanent individual truss member restraint/bracing has been installed in accordance with the approved truss submittal package. For wood trusses with a clear span</u> of 60 feet (18 288 mm) or greater, the special inspector shall verify <u>during construction,</u> that the temporary installation restraint/bracing is installed in accordance with the approved truss submittal package.

CHANGE SIGNIFICANCE: The BSCI, or *Building Component Safety Information*, is an industry-standard truss-installation document used to determine when and where temporary bracing should be placed. It also gives generic information for permanent bracing requirements. These requirements are often overlooked or misinterpreted by the installer. The installation of the bracing is critical for the safe performance of wood trusses, and if the bracing is not installed, or is installed incorrectly, failure of the trusses is possible during erection or during the life of the building. Failure of trusses in a number of jurisdictions brought this issue to the code hearings. An engineering group became concerned about failures in shorter span trusses. They had inspected existing buildings after failures and discovered that the truss systems often had missing or incorrectly installed bracing.

The engineers believed the inspection requirements of the 2015 IBC were insufficient. A provision has been added to IBC Section 1705.5.2 to require periodic special inspection any time permanent bracing is required and the wood trusses are at least five feet tall. This change now requires most trusses with permanent bracing to have special inspection.

The truss designer, as part a design, determines if and where an individual truss needs to be braced to prevent out-of-plane buckling (roll over) when subjected to the design loads. This permanent truss bracing is often described as "truss restraint."

This code change does not require special inspection when temporary bracing alone is required.

Tall wood trusses

1705.12.1, 1705.13.1
Seismic Force-Resisting Systems

CHANGE TYPE: Clarification

CHANGE SUMMARY: The exceptions for special inspection of seismic force-resisting systems have been clarified for structures in moderate and high seismic regions.

2018 CODE: 1705.12.1.1 Seismic force-resisting systems. Special inspections of structural steel in the seismic force-resisting systems in buildings and structures assigned to Seismic Design Category B, C, D, E or F shall be performed in accordance with the quality assurance requirements of AISC 341.

> **Exception:** ~~Special inspections are not required in the seismic force-resisting systems of buildings and structures assigned to Seismic Design Category B or C that are not specifically detailed for seismic resistance, with a response modification coefficient, R, of 3 or less, excluding cantilever column systems.~~
>
> **Exceptions:**
> 1. In buildings and structures assigned to Seismic Design Category B or C, special inspections are not required for structural steel seismic-force-resisting systems where the response modification coefficient, R, designated for "Steel systems not specifically detailed for seismic resistance, excluding cantilever column systems" in ASCE 7, Table 12.2-1 has been used for design and detailing.
> 2. In structures assigned to Seismic Design Category D, E, or F, special inspections are not required for structural steel seismic-force-resisting systems where design and detailing in accordance with AISC 360 is permitted by ASCE 7, Table 15.4-1.

1705.12.1.2 Structural steel elements. Special inspections of structural steel elements in the seismic force-resisting systems of buildings and structures assigned to Seismic Design Category B, C, D, E or F other than those covered in Section 1705.12.1.1, including struts, collectors, chords and foundation elements shall be performed in accordance with the quality assurance requirements of AISC 341.

> **Exception:** ~~Special inspections of structural steel elements are not required in the seismic force-resisting systems of buildings and structures assigned to Seismic Design Category B or C with a response modification coefficient, R, of 3 or less.~~
>
> **Exceptions:**
> 1. In buildings and structures assigned to Seismic Design Category B or C, special inspections of structural steel elements are not required for seismic force-resisting systems with a response modification coefficient, R, of 3 or less.
> 2. In structures assigned to Seismic Design Category D, E, or F, special inspections of structural steel elements are not

Steel concentric braced frame

1705.12.1, 1705.13.1 continues

1705.12.1, 1705.13.1 continued

required for seismic force-resisting systems where design and detailing other than AISC 341 is permitted by ASCE 7, Table 15.4-1. Special inspection shall be in accordance with the applicable reference standard listed in ASCE 7, Table 15.4-1.

(Changes to the exceptions in Sections 1705.13.1.1 and 1705.13.1.2 mirror changes in Section 1705.12.1.1 and 1705.12.1.2.)

CHANGE SIGNIFICANCE: In the 2015 IBC, Sections 1705.12.1 and 1705.13.1 were in conflict with Sections 2205.2.1 and 2205.2.2 in the steel chapter. Chapter 17 appeared to require special inspections and nondestructive testing in accordance with AISC 341, for structural steel seismic force-resisting (SFRS) systems and structural steel elements in non-steel SFRS that were not required to be designed using AISC 341. The 2018 IBC now clarifies the appropriate standard to use with each exception to special inspection and nondestructive testing for specific structural steel SFRS and structural steel elements in non-steel SFRS in Sections 1705.12.1 and 1705.13.1.

Modifications to the first exception in Sections 1705.12.1.1, and 1705.13.1.1 are editorial so that the sections coordinate with IBC Section 2205.2.1. The new second exception in Sections 1705.12.1.1 and 1705.13.1.1 recognizes that a few structural steel SFRS in high seismic regions – SDC D, E and F -- are designed and detailed using AISC 360 rather than AISC 341, as permitted in ASCE 7-16 Chapter 15, Seismic design requirements for nonbuilding structures. These exceptions specify special inspection and nondestructive testing in accordance with AISC 360.

Modifications to the first exception in Sections 1705.12.1.2 and 1705.13.1.2 align the exceptions with IBC Section 2205.2.2. The new second exception in Sections 1705.12.1.2 and 1705.13.1.2 recognizes that, in high seismic regions, some structural steel elements in concrete or masonry SFRS are designed in accordance with AISC 360 instead of AISC 341, as permitted by ASCE 7-16, Chapter 15. For these structural steel elements, special inspection and nondestructive testing are permitted in accordance with AISC 360.

1705.12.6
Fire Sprinkler Clearance

CHANGE TYPE: Addition

CHANGE SUMMARY: Section 1705.12.6 adds a provision for special inspection of minimum clearance of fire sprinkler components to mechanical, electrical and plumbing systems.

2018 CODE: 1705.12.6 Plumbing, mechanical and electrical components. Periodic special inspection of plumbing, mechanical and electrical components shall be required for the following:

(No changes to Items 1 through 5.)

> 6. Installation of mechanical and electrical equipment, including duct work, piping systems and their structural supports, where automatic fire sprinkler systems are installed in structures assigned to Seismic Design Category C, D, E or F to verify one of the following:
>   6.1. Minimum clearances have been provided as required by Section 13.2.3 ASCE/SEI 7.
>   6.2. A nominal clearance of not less than 3 inches (70 mm) has been be provided between fire protection sprinkler system drops and sprigs and: structural members not used collectively or independently to support the sprinklers; equipment attached to the building structure; and other systems' piping.
>
> Where flexible sprinkler hose fittings are used, special inspection of minimum clearances is not required.

CHANGE SIGNIFICANCE: Experience in recent earthquakes has shown that pounding between sprinkler piping drops and sprigs and adjacent nonstructural components such as pipes and ducts has resulted in pipe

1705.12.6 continues

Fire sprinkler system attachment

1705.12.6 continued

connection failures and accidental activation, which caused flooding and potentially compromised the operability of the system should a fire follow the earthquake. In Section 1705.12.6, sprinkler systems were not directly addressed as designated seismic systems.

ASCE 7 identifies fire protection sprinkler systems as components that are required to function for life-safety purposes after an earthquake, classifying them as a designated seismic system. ASCE 7 Section 13.2.3 requires that interaction, in other words pounding, between designated seismic systems and adjacent components must be avoided. A new minimum clearance requirement is added to the 2016 edition of ASCE 7 for fire sprinklers.

Section 13.2.3.1 requires an installed clearance of three inches between any sprinkler drop or sprig and permanently attached equipment including its support and bracing and as well as other distribution systems. An exception is added for sprinklers with flexible sprinkler hose. This information was previously in Section 13.3.2.

Section 13.3.2 intends that seismic displacements consider both relative displacement between multiple points of support and, for mechanical and electrical components, displacement within the component assemblies. Impact of components must be avoided, unless the components are fabricated of ductile materials that have been shown to be capable of accommodating the expected impact loads. For example, in older suspended ceiling installations, excessive lateral displacement of a ceiling system may fracture sprinkler heads that project through the ceiling. A similar situation may arise if sprinkler heads projecting from a small-diameter branch line pass through a rigid ceiling system. Although the branch line may be properly restrained, it may still displace sufficiently between lateral support points to affect other components or systems.

Maintaining adequate clearances is critical to good seismic performance of fire protection sprinkler systems. Since the clearance requirement is critical, new Section 1705.12.6 Item 6 is added to the 2018 IBC to ensure that the minimum clearance is verified by periodic special inspection in regions of significant seismic risk. The minimum nominal 3-inch clearance from adjacent items is identical to the NFPA 13 clearance requirement from structural members to avoid pounding.

Due to their inherent flexibility, a minimum clearance between listed flexible sprinkler hose fittings and other components, equipment or structural members is not required.

Significant Changes to the IBC 2018 Edition   1804.4 ■ Site Grading   **245**

1804.4
Site Grading

CHANGE TYPE: Modification

CHANGE SUMMARY: As an exception, impervious surfaces are allowed to slope less than 2% near doors to meet the egress requirements in Chapter 10.

2018 CODE: 1804.4 Site grading. The ground immediately adjacent to the foundation shall be sloped away from the building at a slope of not less than one unit vertical in 20 units horizontal (5-percent slope) for a minimum distance of 10 feet (3048 mm) measured perpendicular to the face of the wall. If physical obstructions or lot lines prohibit 10 feet (3048 mm) of horizontal distance, a 5-percent slope shall be provided to an approved alternative method of diverting water away from the foundation. Swales used for this purpose shall be sloped ~~a minimum of~~ <u>not less than</u> 2 percent where located within 10 feet (3048 mm) of the building foundation. Impervious surfaces within 10 feet (3048 mm) of the building foundation shall be sloped a ~~minimum of~~ <u>not less than</u> 2 percent away from the building.

Exceptions:
1. Where climatic or soil conditions warrant, the slope of the ground away from the building foundation shall be permitted to be reduced to not less than one unit vertical in 48 units horizontal (2-percent slope).
2. <u>Impervious surfaces shall be permitted to be sloped less than 2 percent where the surface is a door landing or ramp that is required to comply with Section 1010.1.5, 1012.3 or 1012.6.1.</u>

CHANGE SIGNIFICANCE: While the intent of Section 1804.4 is to require a minimum slope away from the building to allow for proper water drainage, the minimum slope does not take into account walking surfaces, door landings, or ramp landings adjacent to a building that have

1804.4 continues

Front entry slope

iStock.com/shansekala

1804.4 continued

a maximum cross-slope of 2%. This leaves no room for error in construction methods providing not only drainage at a minimum of 2% but also a maximum cross-slope of no more than 2%. Designers often choose a cross-slope of less than 2% in these areas, which in strict accordance with this section, is not compliant for site grading.

The addition of an exception to the 2018 IBC Section 1804.4 provides clarification of the controlling factor in the site grading requirement versus maximum slopes permitted for accessibility. This added exception allows a site slope of less than 2% in areas which need the flat surface for accessibility.

1807.2
Retaining Walls

CHANGE TYPE: Modification

CHANGE SUMMARY: The requirement for consideration of a keyway in the sliding analysis of retaining walls has been deleted from Section 1807.2.

2018 CODE: 1807.2 Retaining walls. Retaining walls shall be designed in accordance with Sections 1807.2.1 through 1807.2.3.

1807.2.1 General. Retaining walls shall be designed to ensure stability against overturning, sliding, excessive foundation pressure and water uplift. ~~Where a keyway is extended below the wall base with the intent to engage passive pressure and enhance sliding stability, lateral soil pressures on both sides of the keyway shall be considered in the sliding analysis.~~

1807.2.2 Design lateral soil loads. Retaining walls shall be designed for the lateral soil loads set forth in Section 1610. <u>For structures assigned to Seismic Design Category D, E, or F, the design of retaining walls supporting more than 6 feet (1829 mm) of backfill height shall incorporate the additional seismic lateral earth pressure in accordance with the geotechnical investigation where required in Section 1803.2.</u>

CHANGE SIGNIFICANCE: The application of soil pressure on both sides of a keyway is a recent addition to the model codes, and has caused concern and opposition from the geotechnical engineering community. The keyway concept is in conflict with accepted engineering practice and the principles of soil mechanics. 2015 IBC language was vague and ambiguous with respect to lateral soil pressures on the keyway.

The application of "lateral earth pressures on both sides of the keyway" is commonly interpreted to require a deepening of the active soil pressure to the bottom of the keyway. Active soil pressure requires movement of the key, which is contrary to the intent of the provision.

As there has been ongoing confusion over the intent of consideration of lateral earth pressure on both sides of the keyway and confusion about the purpose of the keyway, in the 2018 IBC the requirement for a keyway is deleted. A keyway may still be used when designed using the principles of soil mechanics and accepted engineering practice.

New text in Section 1807.2.2 adds a pointer for the structural design of retaining walls to resist lateral loads identified in the geotechnical report. The new requirement provides coordination with the requirements of Section 1803.5.12, geotechnical investigations in Seismic Design Categories D, E, and F, for lateral earth pressure on retaining walls.

Retaining walls

1810.3.8.3
Precast Prestressed Piles

CHANGE TYPE: Modification

CHANGE SUMMARY: Equations in Section 1810.3.8.3 addressing precast prestressed piles have been updated.

2018 CODE: 1810.3.8.3.2 Seismic reinforcement in Seismic Design Category C. For structures assigned to Seismic Design Category C, precast prestressed piles shall have transverse reinforcement in accordance with this section. The volumetric ratio of spiral reinforcement shall not be less than the amount required by the following formula for the upper 20 feet (6096 mm) of the pile.

$$\rho_s = \underline{0.04}(f'_c/f_{yh})[2.8 + \underline{2.34P/f'_cA_g}]$$ (Equation 18-5)

where:
A_g = Pile cross-sectional area square inches (mm²).
f'_c = Specified compressive strength of concrete, psi (MPa).
f_{yh} = Yield strength of spiral reinforcement ≤ 85,000 psi (586 MPa).
P = Axial load on pile, pounds (kN), as determined from Equations 16-5 and 16-7.
ρ_s = Spiral reinforcement index or volumetric ratio (vol. spiral/vol. core).

Not less than ~~At least~~ one-half the volumetric ratio required by Equation 18-5 shall be provided below the upper 20 feet (6096 mm) of the pile.

Exception: The minimum spiral reinforcement index required by Equation 18-5 shall not apply in cases where the design includes full consideration of load combinations specified in ASCE 7, Section 2.3.6 and the applicable overstrength factor, Ω_0. In such cases, minimum spiral reinforcement index shall be as specified in Section 1810.3.8.1.

1810.3.8.3.3 Seismic reinforcement in Seismic Design Categories D through F. For structures assigned to Seismic Design Category D, E or F, precast prestressed piles shall have transverse reinforcement in accordance with the following:

(No changes to items 1 through 4.)

5. Where the transverse reinforcement consists of circular spirals, the volumetric ratio of spiral transverse reinforcement in the ductile region shall comply with the following:

$$\underline{\rho_s} = \underline{0.06}(f'_c/f_{yh})[2.8 \pm \underline{2.34P/f'_cA_g}]$$ (Equation 18-6)

but not exceed:

$$\rho_s = 0.021$$ (Equation 18-7)

where:
A_g = Pile cross-sectional area, square inches (mm²).
f'_c = Specified compressive strength of concrete, psi (MPa).
f_{yh} = Yield strength of spiral reinforcement ≤ 85,000 psi (586 MPa).

P = Axial load on pile, pounds (kN), as determined from Equations 16-5 and 16-7.

ρ_s = Volumetric ratio (vol. spiral/vol. core).

This required amount of spiral reinforcement is permitted to be obtained by providing an inner and outer spiral.

> **Exception:** The minimum spiral reinforcement required by Equation 18-6 shall not apply in cases where the design includes full consideration of load combinations specified in ASCE 7, Section 2.3.6 and the applicable overstrength factor, Ω_0. In such cases, minimum spiral reinforcement shall be as specified in Section 1810.3.8.1.

(No changes to item 6.)

1810.3.8.3.4 Axial load limit in Seismic Design Categories C through F. For structures assigned to Seismic Design Category C, D, E, or F, the maximum factored axial load on precast prestressed piles subjected to a combination of seismic lateral force and axial load shall not exceed the following values:

1. $0.2f'_c A_g$ for square piles
2. $0.4f'_c A_g$ for circular or octagonal piles

(As multiple sections of Section 1810.3.8.3 were affected, the entire code change text is too extensive to be included here. Refer to code change S227 in the Complete Revision History to the 2018 I-Codes for the complete text and history of the change.)

CHANGE SIGNIFICANCE: Recent research considered the relationship between curvature ductility demand on prestressed piles and overall system ductility demand. From results of the research, a new equation was created which results in curvature ductility capacities exceeding 12, established as a minimum limit needed for areas of moderate seismicity—in other words, in Seismic Design Category C (SDC C). Note, below 20 feet of depth, in the lower portion of the pile, required reinforcement is allowed to be reduced by fifty percent.

Stringent code provisions require significant pile ductility in the top 35 feet of the pile for sites in Seismic Design Categories D, E and F. This ductility requirement is driven by concerns over soil-structure interaction model accuracy under seismic loading including the effects of liquefaction and findings from post-earthquake foundation evaluations. Recent research considering the relationship between curvature on prestressed piles and overall system ductility demand influenced derivation of a new equation which results in curvature ductility capacities exceeding 18, which is now a minimum limit for areas of high seismicity. The limit is based on average curvature ductility capacity minus one standard deviation.

For other codes and standards, the highest ductility demand for buildings is in the New Zealand Standard (NZS 3101, 2006) where designs must be based on a minimum curvature ductility capacity of 20. Similarly, ATC-32 (1996) sets the curvature ductility capacity target for vertical compression members at 13 with the expectation that fifty percent more capacity is available.

In the 2018 IBC, the equation for the volumetric ratio of spiral transverse reinforcement in moderate seismic regions has been updated. Additionally, a new exception in Section 1810.3.8.3.2, similar to other

1810.3.8.3 continued

overstrength statements in the IBC, recognizes that the volumetric ratio of spiral reinforcement required may be limited to that required for driving and handling stresses, when the pile foundation system is designed with inclusion of overstrength in the load combinations. The minimum spiral reinforcement required per Section 1810.3.8.1 for driving and handling stresses is sufficient when the design includes the effect of overstrength as the increased resistance to axial forces, shear forces and bending moments in the piling provide a large factor of safety against nonlinear pile behavior.

In Section 1810.3.8.3.3, an updated Equation 18-6 provides a volumetric steel ratio that is 50% higher than that required for SDC C. The updated equations are based on a prescriptive design philosophy that requires spiral confinement in accordance with maximum expected pile curvature ductility demands resulting from the design earthquake. The spiral ratio required is expressed as a function of the curvature ductility capacity of the prestressed pile as follows:

$$\rho_s = 0.06(f'_c/f_{yh})(\mu_\varphi/18)[2.8 + 1.25P/(0.53f'_c A_g)]$$

where μ_φ is the ductility capacity of the prestressed pile.

Precast prestressed concrete piles

1901.2
Seismic Loads for Precast Concrete Diaphragms

CHANGE TYPE: Modification

CHANGE SUMMARY: A new requirement requiring the use of ASCE 7 Section 14.2.4 has been established for the design of precast concrete diaphragms in high seismic regions.

2018 CODE: 1901.2 Plain and reinforced concrete. Structural concrete shall be designed and constructed in accordance with the requirements of this chapter and ACI 318 as amended in Section 1905 of this code. Except for the provisions of Sections 1904 and 1907, the design and construction of slabs on grade shall not be governed by this chapter unless they transmit vertical loads or lateral forces from other parts of the structure to the soil. <u>Precast concrete diaphragms in buildings assigned to Seismic Design Category C, D, E, or F shall be designed in accordance with the requirements of ASCE 7 Section 14.2.4.</u>

CHANGE SIGNIFICANCE: Because the 2015 IBC directly referenced material standards, ASCE 7 Chapter 14, Material specific seismic design and detailing requirements, is exempted from reference by the IBC. This practice of excluding Chapter 14 formed a streamlined process when referencing material standards such as the concrete design standard, ACI 318. The IBC only referenced one document, ACI 318, for general material design.

With the 2016 edition of ASCE 7, one section in Chapter 14 is needed for seismic design of precast concrete assemblies. ASCE 7-16 references Section 14.2.4 for an integral part of the precast diaphragm design procedure. Section 14.2.4 contains a connector qualification methodology that

1901.2 continues

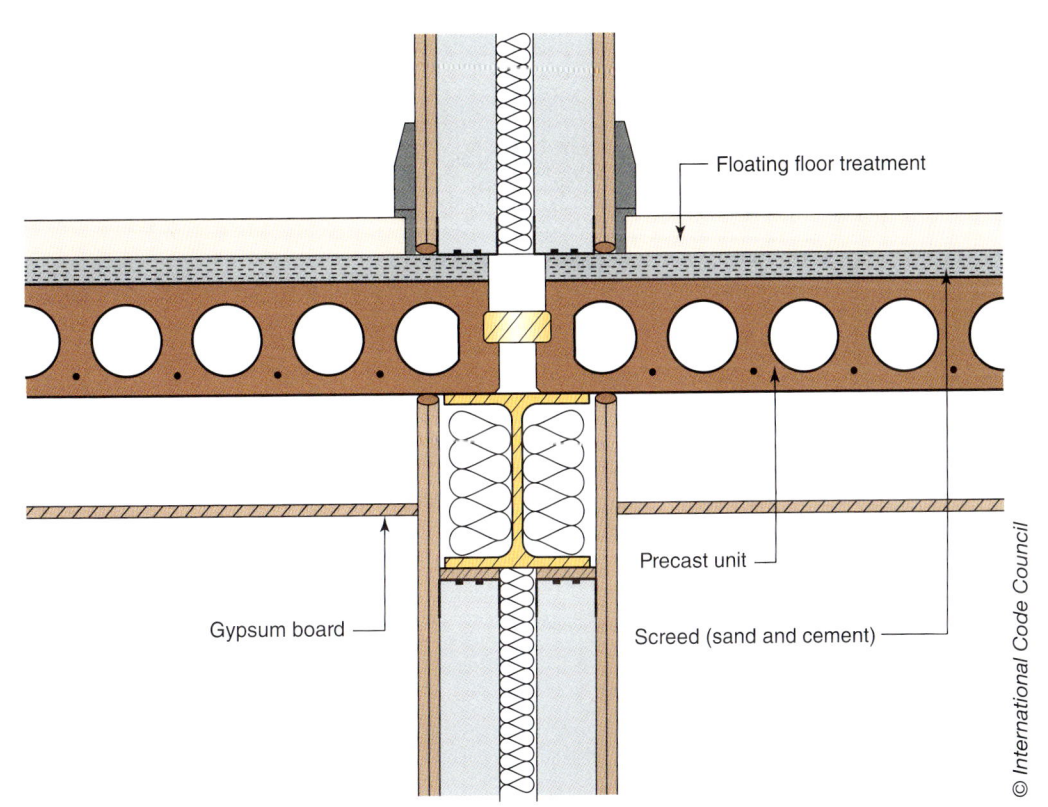

Precast concrete slab

1901.2 continued

was created in the course of development of a seismic design methodology for precast diaphragms. The 2014 edition of ACI 318 does not contain this methodology.

In the 2015 IBC, seismic design of diaphragms referenced ASCE 7-10 Section 12.10. In the 2018 IBC, a new alternative seismic design force for diaphragms has been added in new Section 12.10.3 of ASCE 7-16. The alternative design force is the required design methodology for precast concrete diaphragms in buildings assigned to Seismic Design Category (SDC) C and above. The methodology is also a permitted design for precast concrete diaphragms in lower seismic regions, and for cast-in-place concrete diaphragms and wood diaphragms in any seismic region.

ASCE 7-16 Section 12.10.3 is automatically referenced by the 2018 IBC. However, Section 12.10.3 requires use of Section 14.2.4 which has always explicitly not been adopted by the IBC. To clarify the reference process, Section 1613.1, scoping for seismic design in the 2018 IBC, has been updated to explicitly state that provisions from Chapter 14 of ASCE 7 may be used if referenced in the IBC while Section 1901.2 is updated to require use of ASCE 7 Section 14.2.4 when designing precast concrete diaphragms.

2207.1
SJI Standard

CHANGE TYPE: Modification

CHANGE SUMMARY: The 2015 edition of the combined SJI-100, *Standard Specification for K-Series, LH-Series, and DLH-Series Open Web Steel Joists and Joist Girders*, is the new referenced standard for steel joists.

2018 CODE: 2207.1 General. The design, manufacture and use of open-web steel joists and joist girders shall be in accordance with ~~one of the following Steel Joist Institute (SJI) specifications:~~ either SJI CJ or SJI 100, as applicable.

   1. ~~SJI CJ~~
   2. ~~SJI K~~
   3. ~~SJI LH/DLH~~
   4. ~~SJI JG~~

Chapter 35.

SJI Steel Joist Institute

SJI 100-15, 44th Edition Standard Specification Load Tables and Weight Tables for Steel Joists and Joist Girders K-Series, LH-Series, DHL-Series, Joist Girders

SJI 200—15: Standard Specification for Composite Steel Joists, CJ-Series

   ~~CJ—10 Standard Specification for Composite Steel Joists, CJ-series.~~
   ~~JG—10, Standard Specification for Joist Girders,~~
   ~~K—10, Standard Specification for Open Web Steel Joists, K-series,~~
   ~~LH/DLH—10, Standard Specification for Longspan Steel Joists, LH-series and Deep Longspan Steel Joists, DLH-series~~

CHANGE SIGNIFICANCE: The 2015 edition of the combined SJI-100, *Standard Specification for K-Series, LH-Series, and DLH-Series Open Web Steel Joists and Joist Girders* (44th Edition), is the new referenced standard for structural steel joists. The publication of SJI 100-15 represents a significant change in the presentation of the SJI specifications. Previously, there were three separate specifications (all found in the 43rd Edition), covering K-Series, LH/DLH-Series, and Joist Girders, each one an independent ANSI standard. The newly combined ANSI standard is intended to simplify work for designers.

Beyond combination of the standards, additional changes are found in the new SJI standard, including:

1. Concentrated loads:
   - For concentrated loads, the 100-pound allowance is now included in the specification, provided that certain conditions are met. (Section 4.1.2)
   - For known concentrated load locations, a joist must be designed so no field-applied web members are required. All bracing must be applied at the fab shop. (Section 4.1.2)

2207.1 continues

2207.1 continued

2. For built-up web members comprised of two interconnected shapes, a modified slenderness ratio has been introduced. (Section 4.3.5)

3. SJI welding provisions now match AWS D1.1 and D1.3 requirements with a modified acceptance criteria as permitted by AWS D1.1 Clause 6.8. (Section 4.5.1)

4. Changes have been made to the k factors for web and chord slenderness. (Section 4.3)

5. The K-Series (including KCS) bending exemption for interior panels of less than 24 inches has been removed.

6. The criteria for joint eccentricity have been merged to create criteria based upon the number of web components, but independent of the joist series. (Section 4.5.4)

7. Criteria for bearing seat and bearing plate width, which had previously been only in the SJI Code of Standard Practice, have been added to the specification. (Section 5.4)

8. The criteria for bearing seat depth, to achieve the end reaction farther over the support, has been rewritten for greater clarity. (Section 5.4.3)

9. Connection welds have been added to applicable bridging tables. (Section 5.5)

10. The existing "Minkoff" equation for determination of erection bridging requirements has now been added to the specification. (Section 5.5.2.1)

11. Guidance on seismic loads has been added. (Section 5.13)

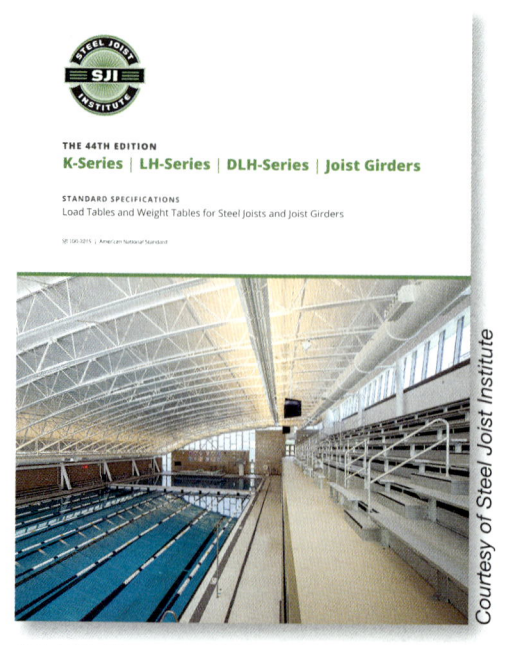

SJI 100-15 standard

2209.2 Cantilevered Steel Storage Racks

CHANGE TYPE: Addition

CHANGE SUMMARY: Reference to the cantilevered storage rack standard, RMI ANSI/MH 16.3, has been added to clarify the characteristics, essential differences, and requirements for cantilevered storage racks.

2018 CODE: 2209.2 Cantilevered steel storage racks. The design, testing, and utilization of cantilevered storage racks made of cold-formed or hot-rolled steel structural members shall be in accordance with RMI ANSI/MH 16.3. Where required by ASCE 7, the seismic design of cantilevered steel storage racks shall be in accordance with Section 15.5.3 of ASCE 7.

CHANGE SIGNIFICANCE: The Rack Manufacturers Institute's (RMI) standard ANSI/MH 16.1 *Specification for the Design, Testing and Utilization of Industrial Steel Storage Racks*, referenced in the 2015 IBC, applies to industrial pallet racks, movable shelf racks and stacker racks made of cold-formed or hot-rolled steel structural members. The standard MH 16.1 specifically does not apply to cantilever racks, portable racks, or drive-in or drive through racks, nor to racks made of materials other than structural steel.

This code change coordinates the definition of cantilevered storage racks with the 2015 IBC definition of steel storage racks. A direct reference is made to the cantilevered storage rack standard, Rack Manufacturers Institute's (RMI) standard ANSI/MH 16.3, *Specification for the Design, Testing and Utilization of Industrial Steel Cantilevered Storage Racks*. Having a separate standard for cantilevered storage racks will help clarify, for designers and users of industrial steel storage racks, the characteristics, essential differences, and requirements in the design, construction, use, and behavior of cantilevered storage racks as distinguished from the more conventional systems commonly known as "pallet racks" or "selective racks."

Cantilevered storage rack

2211
Cold-Formed Steel Light-Frame Construction

CHANGE TYPE: Modification

CHANGE SUMMARY: The 2015 editions of the AISI standards for cold-formed steel, including AISI S240, AISI S400, and AISI S202, have been referenced in the 2018 IBC.

2018 CODE: 2211.1 ~~General~~ Structural framing. ~~The~~ For cold-formed steel light-frame construction, the design and installation of the following structural framing systems, including their members and ~~nonstructural members utilized in cold-formed steel light-frame construction where the specified minimum base steel thickness is not greater than 0.1180 inches (2.997 mm)~~ connections, shall be in accordance with AISI ~~S200~~ S240, and Sections ~~2211.2~~ 2211.1.1 through ~~2211.7, or AISI S220~~ 2211.1.3, as applicable~~.~~:

1. Floor and roof systems.
2. Structural walls.
3. Shear walls, strap-braced walls and diaphragms to resist in-plane lateral loads.
4. Trusses.

2211.1.1 Seismic requirements for cold-formed steel structural systems. The design of cold-formed steel light-frame construction to resist seismic forces shall be in accordance with the provisions of Section 2211.1.1.1 or 2211.1.1.2, as applicable.

2211.1.1.1 Seismic Design Categories B and C. Where a response modification coefficient, R, in accordance with ASCE 7, Table 12.2-1 is

Cold-formed steel framing

used for the design of cold-formed steel light-frame construction assigned to Seismic Design Category B or C, the seismic force-resisting system shall be designed and detailed in accordance with the requirements of AISI S400.

> **Exception:** The response modification coefficient, R, designated for "Steel systems not specifically detailed for seismic resistance, excluding cantilever column systems" in ASCE 7 Table 12.2-1 shall be permitted for systems designed and detailed in accordance with AISI S240 and need not be designed and detailed in accordance with AISI S400.

2211.1.1.2 Seismic Design Categories D through F. In cold-formed steel light-frame construction assigned to Seismic Design Category D, E or F, the seismic force-resisting system shall be designed and detailed in accordance with AISI S400.

~~2211.7~~ **2211.1.2 Prescriptive framing.** *No change to text.*

~~2211.3~~ **2211.1.3 Truss design.** Cold-formed steel trusses shall ~~be designed in accordance~~ comply with ~~AISI S214,~~ the additional provisions of Sections ~~2211.3.1~~ 2211.1.3.1 through ~~2211.3.4 and accepted engineering practice~~ 2211.1.3.3.

~~2211.3.1~~ **2211.1.3.1 Truss design drawings.** The truss design drawings shall conform to the requirements of Section ~~B2.3~~ I1 of AISI ~~S214~~ S202 and shall be provided with the shipment of trusses delivered to the job site. The truss design drawings shall include the details of permanent individual truss member restraint/bracing in accordance with Section ~~B6(a) or B 6(c)~~ I1.6 of AISI ~~S214~~ S202 where these methods are utilized to provide restraint/bracing.

~~2211.3.3~~ **2211.1.3.2 Trusses spanning 60 feet or greater.** *No change to text.*

~~2211.3.4~~ **2211.1.3.3 Truss quality assurance.** Trusses not part of a manufacturing process that provides requirements for quality control done under the supervision of a third-party quality control agency in accordance with AISI S240 Chapter D shall be ~~manufactured~~ fabricated in compliance with Sections 1704.2.5 and 1705.2, as applicable.

2211.2 Nonstructural Members. For cold-formed steel light-frame construction, the design and installation of nonstructural members and connections shall be in accordance with AISI S220.

CHANGE SIGNIFICANCE: The 2015 editions of AISI standards for cold-formed steel have been adopted in the 2018 IBC. These new standards include AISI S240, *North American Standard for Cold-Formed Steel Structural Framing*; AISI S400, *North American Standard for Seismic Design of Cold-Formed Steel Structural Systems*; and AISI S202, *Code of Standard Practice for Cold-Formed Steel Structural Framing*.

2211 continues

2211 continued

AISI S240, the *North American Standard for Cold-Formed Steel Structural Framing*, addresses requirements for construction with cold-formed steel structural framing that are common to prescriptive and engineered light-frame construction. This comprehensive standard was formed by merging the following AISI standards:

- AISI S200—General Provisions
- AISI S210—Floor and Roof System Design
- AISI S211—Wall Stud Design
- AISI S212—Header Design
- AISI S213—Lateral Design
- AISI S214—Truss Design

AISI S240 supersedes all previous editions of these individual AISI standards. Additionally, the standard builds upon this foundation by adding the first comprehensive chapter on quality control and quality assurance (special inspection) for cold-formed steel light-frame construction.

AISI S400-15, *North American Standard for Seismic Design of Cold-Formed Steel Structural Systems*, addresses the design and construction of cold-formed steel structural members and connections used in the seismic force-resisting systems in buildings and other structures. AISI S400 supersedes AISI S110 and the seismic design provisions of AISI S213 and is intended to be applied in conjunction with both AISI S100 and AISI S240, as applicable. This first edition primarily represents a merging of the requirements from the following standards:

- AISI S110—Special Bolted Moment Frame, 2007 edition with Supplement No. 1-2009 edition
- AISI S213—Lateral Design, 2007 edition with Supplement No. 1-2009 edition seismic requirements
- ANSI/AISC 341-10, *Seismic Provisions for Structural Steel Buildings*, which is developed by the American Institute of Steel Construction (AISC) for layout of systems and source of many of the seismic design requirements

Reference to AISI S400 is made for the design of cold-formed steel seismic force-resisting systems. Since the relationship between AISI S240 and AISI S400 is similar to that between AISC 360 and AISC 341, the charging language in IBC Section 2211.1.1 has been modified to parallel the language in Section 2205.2 for structural steel. The IBC references AISI S400 and exempts seismic force-resisting systems only where the seismic design category is B or C and the seismic response modification coefficient, R, equals 3. This is done to recognize that ASCE 7, Table 12.2-1, Section H exempts steel systems from seismic detailing requirements in SDC B and C when R is less than or equal to 3. These systems are designed in accordance with AISI S240. Requirements for the cold-formed steel special-bolted moment frames are now located in AISI S400.

AISI S202, *Code of Standard Practice for Cold-Formed Steel Structural Framing*, is intended to serve as a state-of-the-art mandatory document for establishing contractual relationships between various parties in a construction project where cold-formed steel structural materials, components, and assemblies are used. While the entire standard is not directly referenced in the IBC, portions of AISI S202 establishing minimum requirements for cold-formed steel truss design drawings are directly referenced.

All AISI standards are available for free download at: www.aisistandards.org.

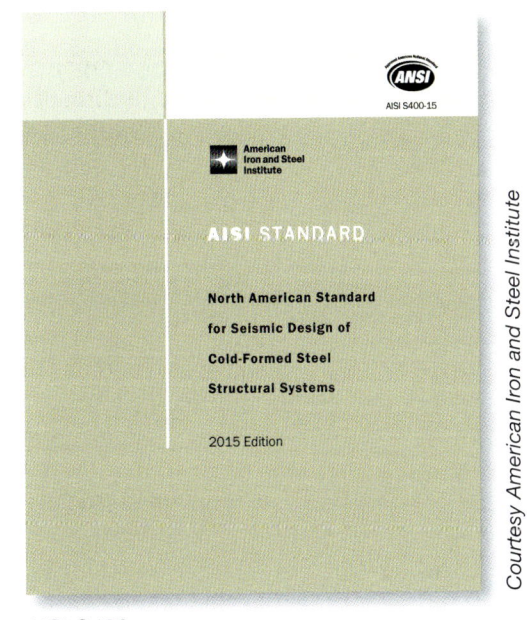

AISI S240

AISI S400

2303.2.2
Fire-Retardant-Treated Wood

CHANGE TYPE: Modification

CHANGE SUMMARY: The types of chemical treatment allowed for fire-retardant-treated lumber have been clarified.

2018 CODE: 2303.2.2 Other means during manufacture. For wood products ~~produced~~ <u>impregnated with chemicals</u> by other means during manufacture, the treatment shall be an integral part of the manufacturing process of the wood product. The treatment shall provide permanent protection to all surfaces of the wood product.<u> The use of paints, coating, stains or other surface treatments is not an approved method of protection as required in this section.</u>

CHANGE SIGNIFICANCE: In the 2015 IBC, requirements for treatment of lumber broke into two paths: (1) pressure impregnation under Section 2303.2.1 and (2) other means during manufacture under Section 2303.2.2. Fire-retardant-treated wood is typically solid sawn lumber or plywood and pressure-treated. For engineered lumber the phrase "other means during manufacture" describes a process where fire-retardant chemicals are applied to wood veneers, chips, or adhesive during manufacture. This process is considered to achieve impregnation with chemicals under the definition of fire-retardant-treated wood and does not require a vacuum chamber or pressure-treatment process. In some cases, the phrase "other means during manufacture" has been broadly interpreted to omit need for the wood product to be impregnated with chemicals in accordance with the definition of fire-retardant-treated wood.

In the 2018 IBC, the minimum alternative process for engineered lumber is clarified in Section 2303.2.2. Wood products produced by other means during manufacture must be impregnated with chemicals.

Additionally, language is clarified to clearly state that wood products protected by paints, stains and other surface treatments are not fire-retardant-treated wood. The suitability of surface treatments as a method of protection must be evaluated as an alternate method under Section 104.11, Alternative materials, design and methods of construction and

Treated lumber

equipment. Surface treatments are evaluated using acceptance criteria which create a path to demonstrate equivalent protection for sheathing and lumber when compared to fire-retardant-treated lumber. Tested attributes of factory-applied surface coatings in acceptance criteria include:

- Durability of the coating
- Potential degradation from exposure to rain during installation
- Flaking or peeling due to shrinkage and expansion
- Effects on strength and stiffness of the wood substrate

Treatment using this alternative process will have an evaluation report or research report showing equivalence with fire-retardant treatment by impregnation of lumber.

2303.6
Nails and Staples

CHANGE TYPE: Modification

CHANGE SUMMARY: Nails and staples are required to conform to the standard ASTM F 1667 including Supplement 1. In addition, minimum average bending moment values have been added for staples.

2018 CODE: 2303.6 Nails and staples. Nails and staples shall conform to requirements of ASTM F 1667, including Supplement 1. Nails used for framing and sheathing connections shall have minimum average bending yield strengths as follows: 80 kips per square inch (ksi) (551 MPa) for shank diameters larger than 0.177 inch (4.50 mm) but not larger than 0.254 inch (6.45 mm), 90 ksi (620 MPa) for shank diameters larger than 0.142 inch (3.61 mm) but not larger than 0.177 inch (4.50 mm) and 100 ksi (689 MPa) for shank diameters of ~~at least~~ not less than 0.099 inch (2.51 mm) but not larger than 0.142 inch (3.61 mm). Staples used for framing and sheathing connections shall have minimum average bending moment as follows: 3.6 in.-lbs (0.41 N-m) for No. 16 gage staples, 4.0 in.-lbs (0.45 N-m) for No. 15 gage staples, and 4.3 in.-lbs (0.49 N-m) for No. 14 gage staples.

CHANGE SIGNIFICANCE: Section 2303.6 of the 2015 IBC contains strength requirements for nails and lacks similar strength requirements for staples. To further describe nail and staple minimum quality; ASTM F 1667, *Standard Specification for Driven Fasteners: Nails, Spikes, and Staples*, contains minimum size requirements for nails and staples. Additionally, the standard addresses bending strength requirements in a supplementary section where it lists minimum requirements for nails. ASTM F 1667 Supplement S1 is a set of supplementary requirements, not a mandatory enforcement unless specifically referenced by the local jurisdiction or state.

The American Wood Council's *National Design Specification for Wood Construction* (AWC NDS) also contains minimum requirements for fastener yield strengths in Appendix I which is non-mandatory. Lastly, AWC NDS and ASTM F 1667 Supplement S1, do not include staple strength requirements.

To determine staple strength, values were taken from testing results listed in the International Staple, Nail, and Tool Association (ISANTA) ICC-ES Report 1539. If the ISANTA report was not used, testing methodology for bending strength in staples was left to the discretion of the building official.

Why is this an issue? During the last two code cycles, staples have become fully integrated into the building code, and are recognized directly in the IBC as an alternative for nails in structural applications. As staples no longer required an alternative material procedure, it became possible for a staple manufacturer to produce code referenced staples which didn't meet minimum strength requirements.

As the two primary references for staples do not contain bending strength requirements, the 2018 IBC has added language for minimum bending moment values for staples. This gives the IBC bending strength values for both nails and staples.

The current sole source of staple performance testing is found in ICC-ES Acceptance Criteria AC201. AC201 uses the ASTM F 1575 nail test method as the basis for its testing procedure and modifies testing as needed for the unique conditions of staples.

The ASTM F 1667 Supplement S1, Section S1.3 requires use of the testing procedure of ASTM F 1575 to determine yield strength, so use of AC 201 mirrors requirements for nails. Additionally, the 2018 IBC now specifically references Supplement S1 of ASTM F 1667 for nails.

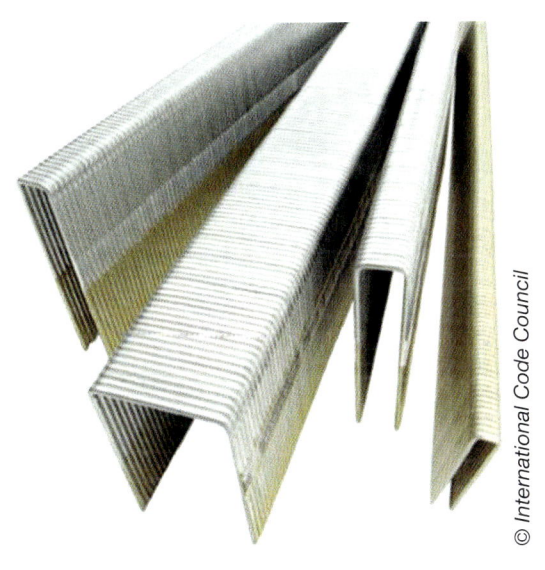

16 gage staples

Table 2304.9.3.2
Mechanically Laminated Decking

CHANGE TYPE: Addition

CHANGE SUMMARY: A new alternative fastener schedule for construction of mechanically laminated decking has been added to the 2018 IBC giving equivalent power-driven fasteners for the 20 penny nail.

2018 CODE:

TABLE 2304.9.3.2 Fastening Schedule for Mechanically Laminated Decking Using Laminations of 2-inch Nominal Thickness

Minimum Nail Size (Length × Diameter) (inches)	Maximum Spacing Between Face Nails[a,b] (inches)		Number of Toenails into Supports[c]
	Decking Supports ≤ 48 inches o.c.	Decking Supports > 48 inches o.c.	
4 × 0.192	30	18	1
4 × 0.162	24	14	2
4 × 0.148	22	13	2
3½ × 0.162	20	12	2
3½ × 0.148	19	11	2
3½ × 0.135	17	10	2
3 × 0.148	11	7	2
3 × 0.128	9	5	2
2¾ × 0.148	10	6	2
2¾ × 0.131	9	6	3
2¾ × 0.120	8	5	3

For SI: 1 inch = 25.4 mm

a. Nails shall be driven perpendicular to the lamination face, alternating between top and bottom edges.
b. Where nails penetrate through two laminations and into the third, they shall be staggered one-third of the spacing in adjacent laminations. Otherwise, nails shall be staggered one-half of the spacing in adjacent laminations.
c. Where supports are 48 inches (1219 mm) on center or less, alternate laminations shall be toenailed to alternate supports; where supports are spaced more than 48 inches (1219 mm) on center, alternate laminations shall be toenailed to every support.

CHANGE SIGNIFICANCE: Building inspectors, plan reviewers, contractors, and designers have all questioned whether proprietary power-driven nails commonly used by contractors onsite replace the code-required 20 penny 4-inch long nail. In the 2015 IBC, an expanded fastener table, Table 2304.10.1, for hand-driven nails attempted to expand the "equivalent sizes" available for power-driven nails. Additionally, the International Staple, Nail and Tool Association (ISANTA) ICC-ES Evaluation Report ESR-1539, Power-Driven Staples and Nails, listed equivalent sizes for the nails in IBC Table 2304.10.1. But there continue to be questions about whether a specific manufacturer's nail is equivalent to or better than the minimum nail called out in the fastening schedule, Table 2304.10.1.

A new alternative fastener schedule for construction of mechanically laminated decking has been added to the 2018 IBC giving equivalent power-driven fasteners for the 20 penny nail. The table provides specific guidance for use of today's typical mechanically driven nails using smaller diameter nails than the 20d common nail. The alternative fastening schedule is based

Mechanically laminated deck

on equivalency to the referenced 20d common nail required in the 2015 IBC Section 2304.9.3.2 for laminations with a 2-inch nominal thickness. Nails listed in the table provide equivalent lateral strength, shear stiffness and withdrawal capacity, as calculated in accordance with the American Wood Council (AWC)'s standard, the *National Design Specification for Wood Construction* (NDS). These nails, while smaller, are spaced more closely than the 20d nail providing equivalent or better strength. Additional proprietary nails can be used by referencing the ISANTA ICC-ES Evaluation Report ESR 1539, Power-Driven Staples and Nails or other applicable evaluation reports.

Table 2304.10.1
Ring Shank Nails

CHANGE TYPE: Modification

CHANGE SUMMARY: The 2018 IBC and IRC are now aligned by requiring 8-penny common or ring shank nails when nailing 6 inches and 12 inches on center for roof sheathing.

2018 CODE:

TABLE 2304.10.1 Fastening Schedule, roof requirements

	Wood structural panels (WSP), subfloor, roof and interior wall sheathing to framing and particleboard wall sheathing to framing		
		Spacing and Location	
Building Element	Number and Type of Fastener	Edges (inches)	Intermediate supports (inches)
~~31.~~ 30. 3⁄8" - 1⁄2"	8d ~~box~~ common or deformed (2½" × ~~0.113~~ 0.131") (roof), or RSRS-01 (2⅜" × 0.113") nail (roof)ᵈ	6	12
	2⅜" × 0.113" nail (roof)	4	8
	1-¾" 16 gage staple, 7⁄16" crown (roof)	3	6
~~32.~~ 31. 19⁄32" - ¾"	8d common or deformed (2½" × 0.131") (roof), or RSRS-01 (2⅜" × 0.113") nail (roof)ᵈ	6	12
	2⅜" × 0.113" nail; or 2" 16 gage staple, 7⁄16" crown	4	8
~~33.~~ 32. 7⁄8" - 1¼"	10d common (3" × 0.148"); or 8d deformed (2½" × 0.131")	6	12

For SI: 1 inch = 25.4 mm.
d. RSRS-01 is a Roof Sheathing Ring Shank nail meeting the specifications in ASTM F 1667.

(No changes to footnotes a–c.)

CHANGE SIGNIFICANCE: In the 2015 *International Residential Code* (IRC), ring shank nails are an option for attaching roof sheathing. Testing has shown them to be equivalent to or better than conventional nail options including 8d common nails. The 2015 *International Building Code* (IBC) did not address ring shank nails and only required an 8d box nail. Testing has also shown that the 8d box nail is insufficient for lower-density roof sheathing in higher wind regions, for example regions with basic, or ultimate, wind speeds of 130 and 140 miles per hour.

The 2018 IBC and IRC are now aligned by requiring an 8d common or ring shank nail for roof sheathing with wider nail spacing. Two additional fasteners, the smaller 8d box nail and a 7⁄16" crown, 16 gage staple, remain additional options but require tighter spacing of the fasteners. This change brings consistency for the two codes when determining minimum nail size for roof sheathing attachment with 6:12 nail spacing.

The deformed nail option (2½" × 0.131") is based on the assumption that the deformed nail, which has non-standard deformations, has at least the same withdrawal capacity and head pull-through performance as the equivalent diameter 8d common (2½" × 0.131") smooth shank nail.

The Roof Sheathing Ring Shank (RSRS) nail is standardized in ASTM F 1667 and added as equivalent to the 8d common nail to resist uplift of roof sheathing. This standardized ring shank nail provides improved withdrawal resistance. A head size of 0.281-inch diameter is specified for the RSRS-01 nail in ASTM F 1667 which is equivalent to the head diameter of the 8d common nail.

Deformed nails

2304.10.5
Fasteners in Treated Wood

CHANGE TYPE: Modification

CHANGE SUMMARY: Staples in preservative-treated wood and fire-retardant-treated wood are now required to be made of stainless steel.

2018 CODE: 2304.10.5 Fasteners and connectors in contact with preservative-treated and fire-retardant-treated wood. Fasteners, including nuts and washers, and connectors in contact with preservative-treated and fire-retardant-treated wood shall be in accordance with Sections 2304.10.5.1 through 2304.10.5.4. The coating weights for zinc-coated fasteners shall be in accordance with ASTM A 153. <u>Stainless steel driven fasteners shall be in accordance with the material requirements of ASTM F1667.</u>

2304.10.5.1 Fasteners and connectors for preservative-treated wood. Fasteners, including nuts and washers, in contact with preservative-treated wood shall be of hot-dipped zinc-coated galvanized steel, stainless steel, silicon bronze or copper. <u>Staples shall be of stainless steel.</u> Fasteners other than nails<u>, staples</u>, timber rivets, wood screws and lag screws shall be permitted to be of mechanically deposited zinc-coated steel with coating weights in accordance with ASTM B 695, Class 55 minimum. Connectors that are used in exterior applications and in contact with preservative-treated wood shall have coating types and weights in accordance with the treated wood or connector manufacturer's recommendations. In the absence of manufacturer's recommendations, ~~a minimum of~~ <u>not less than</u> ASTM A653, Type G185 zinc-coated galvanized steel, or equivalent, shall be used.

> **Exception:** Plain carbon steel fasteners, including nuts and washers, in SBX/DOT and zinc borate preservative-treated wood in an interior, dry environment shall be permitted.

2304.10.5.3 Fasteners for fire-retardant-treated wood used in exterior applications or wet or damp locations. Fasteners, including nuts and washers, for fire-retardant-treated wood used in exterior applications or wet or damp locations shall be of hot-dipped zinc-coated galvanized steel, stainless steel, silicon bronze or copper. <u>Staples shall be of stainless steel.</u> Fasteners other than nails<u>, staples</u>, timber rivets, wood screws and lag screws shall be permitted to be of mechanically deposited zinc-coated steel with coating weights in accordance with ASTM B 695, Class 55 minimum.

CHANGE SIGNIFICANCE: During the last two code cycles, staples have been added as an alternative fastener for use in various types of wood-to-wood connections. The phrase "other than nails and timber rivets" has now been rewritten to include staples as a code-accepted solution. Staples are also now specifically limited to stainless steel where exposed to corrosive environments. The thin wire gages used in staple fasteners (16ga–14ga) are much thinner than those used in nails, and are consequentially more susceptible to corrosion. Due to the thin gage, stainless steel staples are currently the only option in installations requiring increased corrosion resistance.

Stainless steel staples

2304.11
Heavy Timber Construction

CHANGE TYPE: Modification

CHANGE SUMMARY: The heavy timber provisions of Chapter 23 have been reorganized and the 2015 IBC table on engineered lumber dimensional equivalencies previously located in Section 602.4 has been moved into Section 2304.11.

2018 CODE: 2304.11 Heavy timber construction. Where a structure ~~or~~, portion thereof ~~is~~, or individual structural elements are required ~~to be of Type IV construction~~ by ~~other~~ provisions of this code to be of heavy timber, the building elements therein shall comply with the applicable provisions of Sections 2304.11.1 through ~~2304.11.5~~ 2304.11.4. Minimum dimensions of heavy timber shall comply with the applicable requirements in Table 2304.11 based on roofs or floors supported and the configuration of each structural element, or in Sections 2304.11.2 through 2304.11.4. Lumber decking shall also be in accordance with Section 2304.9.

2304.11.1 ~~Columns~~ Details of heavy timber structural members. ~~Columns~~ Heavy timber structural members shall be detailed and constructed in accordance with Sections 2304.11.1.1 through 2304.11.1.3. ~~continuous or superimposed throughout all stories by means of reinforced concrete or metal caps with brackets, or shall be connected by properly designed steel or iron caps, with pintles and base plates, or by timber splice plates affixed to the columns by metal connectors housed within the contact faces, or by other approved methods.~~

2304.11.1.1 ~~Column connections~~ Columns. Minimum dimensions of columns shall be in accordance with Table 2304.11. Columns shall be continuous or superimposed throughout all stories and connected in an approved manner. Girders and beams at column connections shall be closely fitted around columns and adjoining ends shall be cross tied to each other, or intertied by caps or ties, to transfer horizontal loads across joints. Wood

2304.11 continues

Heavy timber construction

2304.11 continued

bolsters shall not be placed on tops of columns unless the columns support roof loads only. <u>Where traditional heavy timber detailing is used, connections shall be by means of reinforced concrete or metal caps with brackets, by properly designed steel or iron caps, with pintles and base plates, by timber splice plates affixed to the columns by metal connectors housed within the contact faces, or by other approved methods.</u>

~~2304.11.2~~<u>2304.11.1.2</u> Floor framing. <u>Minimum dimensions of floor framing shall be in accordance with Table 2304.11.</u> Approved wall plate boxes or hangers shall be provided where wood beams, girders or trusses rest on masonry or concrete walls. Where intermediate beams are used to support a floor, they shall rest on top of girders, or shall be supported by ~~ledgers or blocks securely fastened to the sides of the girders, or they shall be supported by~~ an approved metal hanger into which the ends of the beams shall be closely fitted. <u>Where traditional heavy timber detailing is used, these connections shall be permitted to be supported by ledgers or blocks securely fastened to the sides of the girders.</u>

~~2304.11.3~~<u>2304.11.1.3</u> Roof framing. <u>Minimum dimensions of roof framing shall be in accordance with Table 2304.11.</u> Every roof girder and ~~at least~~ <u>not less than</u> every alternate roof beam shall be anchored to its supporting member~~; and every monitor and every sawtooth construction shall be anchored to the main roof construction. Such anchors shall consist of steel or iron bolts of sufficient strength to resist vertical uplift of the roof.~~ <u>to resist forces as required in Chapter 16.</u>

~~602.4.8~~<u>2304.11.2</u> Partitions and walls. Partitions and walls shall comply with Section ~~602.4.8.1~~<u>2304.11.2.1</u> or ~~602.4.8.2~~<u>2304.11.2.2</u>.

~~602.4.8.2~~<u>2304.11.2.1</u> Exterior walls. Exterior walls shall be <u>permitted to</u> be ~~of one of the following:~~

1. ~~Noncombustible materials.~~

1. ~~Not less than 6 inches (152 mm) in thickness and constructed of one of the following:~~
   1.1. ~~Fire-retardant-treated wood in accordance with Section 2303.2 and complying with Section 602.4.1.~~
   ~~1.1.~~ cross-laminated timber ~~complying with~~ <u>meeting the requirements of</u> Section ~~602.4.2~~<u>2303.1.4</u>.

~~602.4.8.1~~<u>2304.11.2.2</u> Interior walls and partitions. *No change to text.*

~~602.4.6~~<u>2304.11.3</u> Floors. Floors shall be without concealed spaces. Wood floors shall be constructed in accordance with Section ~~602.4.6.1~~ <u>2304.11.3.1</u> or ~~602.4.6.2~~<u>2304.11.3.2</u>.

~~602.4.6.2~~<u>2304.11.3.1</u> Cross-laminated timber floors. Cross-laminated timber shall be not less than 4 inches (102 mm) in <u>actual</u> thickness. Cross-laminated timber shall be continuous from support to support and mechanically fastened to one another. Cross-laminated timber shall be permitted to be connected to walls without a shrinkage gap providing swelling or shrinking is considered in the design. Corbelling of masonry walls under the floor shall be permitted to be used.

~~602.4.6.1~~2304.11.3.2 Sawn or glued-laminated plank floors. *No change to text.*

2304.11.4 Floor decks. ~~Floor decks and covering shall not extend closer than ½ inch (12.7 mm) to walls. Such ½-inch (12.7 mm) spaces shall be covered by a molding fastened to the wall either above or below the floor and arranged such that the molding will not obstruct the expansion or contraction movements of the floor. Corbeling of masonry walls under floors is permitted in place of such molding.~~

~~2304.11.5~~ 2304.11.4 Roof decks. Roofs shall be without concealed spaces and roof decks shall be constructed in accordance with Section 2304.11.4.1 or 2304.11.4.2. Other types of decking shall be an alternative that provides equivalent fire resistance and structural properties. Where supported by a wall, roof decks shall be anchored to walls to resist ~~uplift~~ forces determined in accordance with Chapter 16. Such anchors shall consist of steel bolts, lags, screws or ~~iron bolts~~approved hardware of sufficient strength to resist ~~vertical uplift of the roof.~~prescribed forces.

~~602.4.7~~2304.11.4.1 ~~Roof~~ Cross-laminated timber roofs. ~~Roofs shall be without concealed spaces and wood roof decks shall be sawn or glued laminated, splined or tongue-and-groove plank, not less than 2 inches (51 mm) nominal in thickness; 1⅛-inch-thick (32 mm) wood structural panel (exterior glue); planks not less than 3 inches (76 mm) nominal in width, set on edge close together and laid as required for floors; or of cross- laminated timber. Other types of decking shall be permitted to be used if providing equivalent fire resistance and structural properties.~~ Cross-laminated timber roofs shall be not less than 3 inches (76 mm) nominal in thickness and shall be continuous from support to support and mechanically fastened to one another.

2304.11.4.2 Sawn, wood structural panel, or glued-laminated plank roofs. Sawn, wood structural panel, or glued-laminated plank roofs shall be one of the following:

1. Sawn or glued laminated, splined or tongue-and-groove plank, not less than 2 inches (51 mm) nominal in thickness.
2. 1⅛-inch-thick (32 mm) wood structural panel (exterior glue).
3. Planks not less than 3 inches (76 mm) nominal in width, set on edge close together and laid as required for floors.

2304.11 continues

2304.11 continued

~~TABLE 602.4~~ TABLE 2304.11 ~~Wood Member Size Equivalencies~~ Minimum Dimensions of Heavy Timber Structural Members

Supporting	Heavy Timber Structural Elements	Minimum Nominal Solid Sawn Size		Minimum Glued-laminated Net Size		Minimum Structural Composite Lumber Net Size	
		Width, inch	Depth, inch	Width, inch	Depth, inch	Width, inch	Depth, inch
Floor loads only or combined floor and roof loads	• Columns; • Framed sawn or glue-laminated timber arches which spring from the floor line • Framed timber trusses	8	8	6¾	8¼	7	7½
	• Wood beams and girders	6	10	5	10½	5¼	9½
Roof loads only	• Columns (roof and ceiling loads) • Lower half of wood-frame or glue-laminated arches which spring from the floor line or from grade	6	8	5	8¼	5¼	7½
	• Upper half of wood-frame or glue-laminated arches which spring from the floor line or from grade • Framed timber trusses and other roof framing[a]	6	6	5	6	5¼	5½
	• Framed or glue-laminated arches that spring from the top of walls or wall abutments	4[b]	6	3[b]	6⅞	3½[b]	5½

For SI: 1 inch = 25.4 mm.

a. Spaced members shall be permitted to be composed of two or more pieces not less than 3 inches (76 mm) nominal in thickness where blocked solidly throughout their intervening spaces or where spaces are tightly closed by a continuous wood cover plate of not less than 2 inches (51 mm) nominal in thickness secured to the underside of the members. Splice plates shall be not less than 3 inches (76 mm) nominal in thickness.
b. Where protected by approved automatic sprinklers under the roof deck, framing members shall be not less than 3 inches (76 mm) nominal in width.

CHANGE SIGNIFICANCE: Heavy timber structural elements have long been referenced throughout other parts of the code where a specific heavy timber structural element is detailed for use in another type of construction, in other words in non-heavy timber construction. Requirements for heavy timber construction, including Cross Laminated Timber (CLT), have been moved into one location.

The CLT product standard was added as a reference in the 2015 IBC. A second code change allowed CLT to be utilized for the construction of 2-hour exterior walls in Type IV—heavy timber construction.

In the 2018 IBC, these provisions are moved into one location within Chapter 23. Section 602.4 retains the provisions specific to Type IV construction including sections on fire-retardant-treated wood and cross-laminated timber in exterior walls, use of heavy timber columns and arches in exterior walls, and a reference to the requirements for heavy timber materials located in Section 2304.11.

Section 2304.11 can best be described as the section for all things heavy timber. Heavy timber requirements removed from Section 602.4 are combined and organized with the existing content of Section 2304.11. Table 602.4 is renamed, moved and updated with a description of the components for timber elements based on whether the element supports roof loads and floor loads or only roof loads. Specific footnotes about the size and protection of spaced truss elements and the reduction of roof beam width for sprinklers are noted where applicable. The non-size-related detailing provisions for framing members and connections (columns, floor framing, and roof framing) are placed in Sections 2304.11.1.1, 2304.11.1.2, and 2304.11.1.3.

Sections 2304.11.2 through 2304.11.4 contain pertinent thickness and detailing requirements for walls, roof, and floor deck construction.

TABLE 23-1 New Locations for Heavy Timber Requirements

2018 IBC	2015 IBC	Provision
602.4	602.4	Type IV construction
602.4.1, 602.4.2	602.4.1	Wall assembly thickness
602.4.3	602.4.3	Exterior structural members
2304.11	2304.11	Heavy timber construction
Table 2304.11	Table 602.4	Minimum dimensions
2304.11.1	New	Details of heavy timber structural members
2304.11.1.1	602.4.3, 2304.11.1	Columns
2304.11.1.2	602.4.4, 2304.11.2	Floor framing
2304.11.1.3	602.4.5, 2304.11.3	Roof framing
2304.11.2.1	602.4.8.2	Exterior walls
2304.11.2.2	602.4.8.1	Partitions and interior walls
2304.11.3	602.4.6	Floors
2304.11.3.1	602.4.6.2	CLT floors
2304.11.3.2	602.4.0.1, 2304.11.4	Sawn or glued-laminated plank floors
2304.11.4	2304.11.5	Roof decks

2304.12.2.5, 2304.12.2.6
Supporting Members for Permeable Floors and Roofs

CHANGE TYPE: Modification

CHANGE SUMMARY: The provisions for permeable floors and roofs have been modified to require positive drainage of water and ventilation below the floor or roof to protect supporting wood construction.

2018 CODE: 2304.12.2.5 Supporting members for permeable floors and roofs. Wood structural members that support moisture permeable floors or roofs that are exposed to the weather, such as concrete or masonry slabs, shall be of naturally durable or preservative-treated wood unless separated from such floors or roofs by an impervious moisture barrier. <u>The impervious moisture barrier system protecting the structure supporting floors shall provide positive drainage of water that infiltrates the moisture-permeable floor topping.</u>

<u>**2304.12.2.6 Ventilation beneath balcony or elevated walking surfaces.** Enclosed framing in exterior balconies and elevated walking surfaces that are exposed to rain, snow, or drainage from irrigation shall be provided with openings that provide a net free cross-ventilation area not less than 1/150 of the area of each separate space.</u>

CHANGE SIGNIFICANCE: A key requirement of impervious moisture barrier systems installed under permeable floor systems exposed to water are elements that provide for drainage of water passed through the permeable floor system. Without a properly functioning method to transport this water out, the floor assembly can stay saturated for long periods of time, potentially contributing to failure of the supporting wood structure.

2015 IBC Section 2304.12.2.5 requires an impervious moisture barrier when wood that is not preservative-treated or naturally durable supports moisture-permeable floors or roofs exposed to weather such as concrete and masonry slabs. When such assemblies are a roof, and there is a leak in the impervious barrier, the occupants typically know about it and repairs are made. When the assembly supports a walking surface such as a balcony, there may be no early warning of a leak or decay because leaks can be located over unoccupied areas outside of the structure's building envelope.

The 2015 IBC requirement called for separation of the floor and supporting walls by an impervious moisture barrier when the supporting wood is not preservative-treated or naturally durable. The 2018 IBC

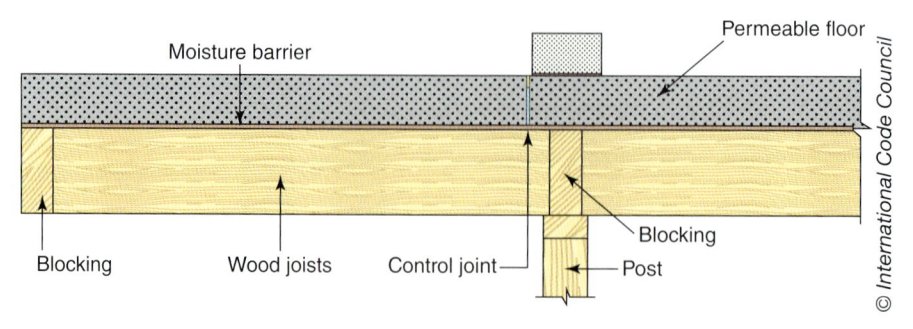

Permeable floor

further requires that the impervious moisture barrier system protect the substructure supporting a floor by providing a positive drainage mechanism for water.

Section 1203 of the 2015 IBC is generally applied by many to require ventilation where wood supports a balcony and is enclosed. The key word is *enclosed*. Whenever the wood framing supporting such structures is enclosed it is more difficult for water in the assembly to depart regardless of the source of the water. It is critical to provide ventilation to enclosed areas, especially to the wood substructure supporting an elevated balcony exposed to the weather.

For the 2018 IBC, the concept in Chapter 12 is duplicated Chapter 23 to emphasize that the requirement for ventilation applies to wood construction and specifically to enclosed balconies. Additionally, the provision clarifies that when a balcony or elevated walking surface serves as a weather-resistant barrier and the joist spaces below are enclosed, cross ventilation is required similar to enclosed rafter spaces in roofs.

Table 2308.4.1.1(1)
Header and Girder Spans—Exterior Walls

CHANGE TYPE: Modification

CHANGE SUMMARY: The header and girder spans for the exterior bearing wall table have been updated to allow No. 2 Southern Pine rather than a minimum No. 1 Southern Pine lumber.

2018 CODE:

TABLE 2308.4.1.1(1) Header and Girder Spans[a,b] for Exterior Bearing Walls

		Ground Snow Load (psf)[e]							
		30						50	
		Building Width[c] (feet)							
		12		24		36		12	
Headers and Girders Supporting	Size	Span[f]	NJ[d]	Span[f]	NJ[d]	Span[f]	NJ[d]	Span[f]	NJ[d]
Roof and Ceiling	1-2 × 6	4 - 0	1	3 - 1	2	2 - 7	2	3 - 5	1
	1-2 × 8	5 - 1	2	3 - 11	2	3 - 3	2	4 - 4	2
	1-2 × 10	6 - 0	2	4 - 8	2	3 - 11	2	5 - 2	2
	1-2 × 12	7 - 1	2	5 - 5	2	4 - 7	3	6 - 1	2
	2-2 × 4	4 - 0	1	3 - 1	1	2 - 7	1	3 - 5	1
	2-2 × 6	6 - 0	1	4 - 7	1	3 - 10	1	5 - 1	1
	2-2 × 8	7 - 7	1	5 - 9	1	4 - 10	2	6 - 5	1
	2-2 × 10	9 - 0	1	6 - 10	2	5 - 9	2	7 - 8	2
	2-2 × 12	10 - 7	2	8 - 1	2	6 - 10	2	9 - 0	2
	3-2 × 8	9 - 5	1	7 - 3	1	6 - 1	1	8 - 1	1
	3-2 × 10	11 - 3	1	8 - 7	1	7 - 3	2	9 - 7	1
	3-2 × 12	13 - 2	1	10 - 1	2	8 - 6	2	11 - 3	2
	4-2 × 8	10 - 11	1	8 - 4	1	7 - 0	1	9 - 4	1
	4-2 × 10	12 - 11	1	9 - 11	1	8 - 4	1	11 - 1	1
	4-2 × 12	15 - 3	1	11 - 8	1	9 - 10	2	13 - 0	1
Roof, ceiling and one center-bearing floor	1-2 × 6	3 - 3	1	2 - 7	2	2 - 2	2	3 - 0	2
	1-2 × 8	4 - 1	2	3 - 3	2	2 - 9	2	3 - 9	2
	1-2 × 10	4 - 11	2	3 - 10	2	3 - 3	3	4 - 6	2

For SI: 1 inch = 25.4 mm, 1 pound per square foot = 0.0479 kPa.

a. Spans are given in feet and inches.
b. Spans are based on minimum design properties for No. 2 grade lumber of Douglas fir-larch, hem-fir, Southern pine and spruce-pine-fir. No. 1 or better grade lumber shall be used for Southern Pine.
c. Building width is measured perpendicular to the ridge. For widths between those shown, spans are permitted to be interpolated.
d. NJ - Number of jack studs required to support each end. Where the number of required jack studs equals one, the header is permitted to be supported by an approved framing anchor attached to the full-height wall stud and to the header.
e. Use 30 psf ground snow load for cases in which ground snow load is less than 30 psf and the roof live load is equal to or less than 20 psf.
f. Spans are calculated assuming the top of the header or girder is laterally braced by perpendicular framing. Where the top of the header or girder is not laterally braced (for example, cripple studs bearing on the header), tabulated spans for headers consisting of 2x8, 2x10, or 2x12 sizes shall be multiplied by 0.70 or the header or girder shall be designed.

Dropped header in exterior wall

Raised header in exterior wall

CHANGE SIGNIFICANCE: The 2015 IBC Table 2308.4.1.1(1), header and girder spans for exterior bearing walls, was updated to deal with changes in Southern Pine (SP) design values. Due to those changes, No. 1 SP lumber was required for use of the table. Locally, No. 1 SP has not been consistently available, and designers, contractors, and building departments requested that the American Wood Council (AWC), which created the original and updated tables, change spans to reflect a No. 2 SP maximum span rather than a No. 1 SP span.

In the 2018 IBC, Table 2308.4.1.1(1) uses No. 2 SP as one of the lumber options. In calculating spans using No. 2 SP as well as other common lumber species, AWC reviewed their original assumptions that were included in the initial table in 1997. Generally, No. 2 SP bending values or No. 2 Hem-fir (HF) deflection controls the allowable span. Loading assumptions changed slightly as well causing an overall decrease in span by about ten percent.

Footnote f has been added to clarify that header spans assume a header is raised into a floor or roof container. For dropped headers, headers not braced on the backside by perpendicular joists, a factor of 0.7 must be applied, reducing the span length by 30% to determine the maximum span. Alternatively, a header or girder calculation including an adjustment for potential buckling can be done to determine the maximum span. If the dropped header is of 2×8, 2×10, or 2×12 framing, the 0.7 reduction factor must be applied. For 2×4 and 2×6 lumber, the reduction is not required but may be applied. An example of a dropped header is a header with cripple or pony wall studs above.

2308.4.1.1(2)

Header and Girder Spans—Interior Walls

CHANGE TYPE: Modification

CHANGE SUMMARY: The header and girder spans for the interior bearing walls table have been updated to allow No. 2 Southern Pine for spans rather than No. 1 Southern Pine lumber. Building width is updated in the table as well, supplying span lengths for narrower building areas.

2018 CODE:

TABLE 2308.4.1.1(2) Header and Girder Spans[a,b] for Interior Bearing Walls

Headers and Girders Supporting	Size	Building Width[c] (feet)					
		12		24		36	
		Span[e]	NJ[d]	Span[e]	NJ[d]	Span[e]	NJ[d]
One floor only	2-2 × 4	4 - 1	1	2 - 10	1	2 - 4	1
	2-2 × 6	6 - 1	1	4 - 4	1	3 - 6	1
	2-2 × 8	7 - 9	1	5 - 5	1	4 - 5	2
	2-2 × 10	9 - 2	1	6 - 6	2	5 - 3	2
	2-2 × 12	10 - 9	1	7 - 7	2	6 - 3	2
	3-2 × 8	9 - 8	1	6 - 10	1	5 - 7	1
	3-2 × 10	11 - 5	1	8 - 1	1	6 - 7	2
	3-2 × 12	13 - 6	1	9 - 6	2	7 - 9	2
	4-2 × 8	11 - 2	1	7 - 11	1	6 - 5	1
	4-2 × 10	13 - 3	1	9 - 4	1	7 - 8	1
	4-2 × 12	15 - 7	1	11 - 0	1	9 - 0	2
Two floors	2-2 × 4	2 - 7	1	1 - 11	1	1 - 7	1
	2-2 × 6	3 - 11	1	2 - 11	2	2 - 5	2
	2-2 × 8	5 - 0	1	3 - 8	2	3 - 1	2
	2-2 × 10	5 - 11	2	4 - 4	2	3 - 7	2
	2-2 × 12	6 - 11	2	5 - 2	2	4 - 3	3
	3-2 × 8	6 - 3	1	4 - 7	2	3 - 10	2
	3-2 × 10	7 - 5	1	5 - 6	2	4 - 6	2
	3-2 × 12	8 - 8	2	6 - 5	2	5 - 4	2
	4-2 × 8	7 - 2	1	5 - 4	1	4 - 5	2
	4-2 × 10	8 - 6	1	6 - 4	2	5 - 3	2
	4-2 × 12	10 - 1	1	7 - 5	2	6 - 2	2

For SI: 1 inch = 25.4 mm, 1 pound per square foot = 0.0479 kPa.

a. Spans are given in feet and inches.
b. Spans are based on minimum design properties for No. 2 grade lumber of Douglas fir-larch, hem-fir, Southern pine and spruce-pine-fir. No. 1 or better grade lumber shall be used for Southern Pine.
c. Building width is measured perpendicular to the ridge. For widths between those shown, spans are permitted to be interpolated.
d. NJ - Number of jack studs required to support each end. Where the number of required jack studs equals one, the header is permitted to be supported by an approved framing anchor attached to the full-height wall stud and to the header.
e. Spans are calculated assuming the top of the header or girder is laterally braced by perpendicular framing. Where the top of the header or girder is not laterally braced (for example, cripple studs bearing on the header), tabulated spans for headers consisting of 2x8, 2x10, or 2x12 sizes shall be multiplied by 0.70 or the header or girder shall be designed.

Dropped header in interior wall

Raised girder in interior wall

CHANGE SIGNIFICANCE: The 2015 IBC Table 2308.4.1.1(2), header and girder spans for interior bearing walls, was updated to deal with changes in SP design values. Due to those changes, No. 1 SP lumber was required in the table. Locally, No. 1 SP has not been consistently available and designers, contractors and building departments requested that the American Wood Council (AWC), which created the original tables and the updated tables, change spans to reflect a No. 2 SP maximum span rather than a No. 1 SP maximum span.

Table 2308.4.1.1(2) in the 2018 IBC is updated with maximum spans based on No. 2 SP lumber. For a building width of 36 feet, spans have decreased or remained the same depending upon whether the span for No. 2 SP was shorter than the maximum span for the other lumber varieties. In calculating spans using No. 2 SP as well as other common lumber species, AWC reviewed their original assumptions that were included in the initial table in 1997. Generally, No. 2 SP bending values or deflection of No. 2 Hem-fir (HF) controls the allowable span. Loading assumptions changed slightly as well causing an overall decrease in span by about ten percent.

The previous spans of 20 and 28 feet have been updated to a span of 12 and 24 feet for building width. This allows a longer span when looking at building widths less than 20 feet. Often an irregular-shaped structure will have a segment that is narrower than 20 feet. If the former spans of 20 or 28 feet are desired, the tables may be interpolated to determine a maximum span length rather than using the next larger building width to determine the span length.

Footnote e has been added to clarify that header spans assume a header is raised into a floor or roof container. For dropped headers, headers not braced on the backside by perpendicular joists, a factor of 0.7 must be applied, decreasing the span length by 30% to determine the maximum span. Alternatively, a header or girder calculation including an adjustment for potential buckling can be done to determine the maximum span. When the dropped header is 2×8, 2×10, or 2×12 framing, the 0.7 reduction factor must be applied. For 2×4 and 2×6 lumber, the reduction is not required but may be applied. An example of a dropped header is a header with cripple or pony wall studs above.

2308.5.5.1
Openings in Exterior Bearing Walls

CHANGE TYPE: Modification

CHANGE SUMMARY: Single-member lumber headers are now permitted in prescriptive wood framing.

2018 CODE: 2308.5.5.1 Openings in exterior bearing walls. Headers shall be provided over each opening in exterior bearing walls. The size and spans in Table 2308.4.1.1(1) are permitted to be used for one- and two-family dwellings. Headers for other buildings shall be designed in accordance with Section 2301.2, Item 1 or 2. Headers shall be of two or more pieces of nominal 2-inch (51 mm) framing lumber set on edge as shall be permitted by in accordance with Table 2308.4.1.1(1) and nailed together in accordance with Table 2304.10.1 or of solid lumber of equivalent size.

Single member headers of nominal 2-inch (51 mm) thickness shall be framed with a single flat 2-inch-nominal (51 mm) member or wall plate not less in width than the wall studs on the top and bottom of the header in accordance with Figures 2308.5.5.1(1) and 2308.5.5.1(2) and face nailed to the top and bottom of the header with 10d box nails (3 inches × 0.128 inches [76 mm × 3.3 mm]) spaced 12 inches (305 mm) on center.

Wall studs shall support the ends of the header in accordance with Table 2308.4.1.1(1). Each end of a lintel or header shall have a bearing length of not less than 1½ inches (38 mm) for the full width of the lintel.

CHANGE SIGNIFICANCE: The 2018 IBC allows single headers under limited loading conditions to increase energy efficiency of the building. Installation of a single header in an exterior wall results in a greater thickness of cavity insulation to reduce heat loss through the header. The single-header configuration may be used for small openings only.

Single member header in an exterior wall

Components of a single member header

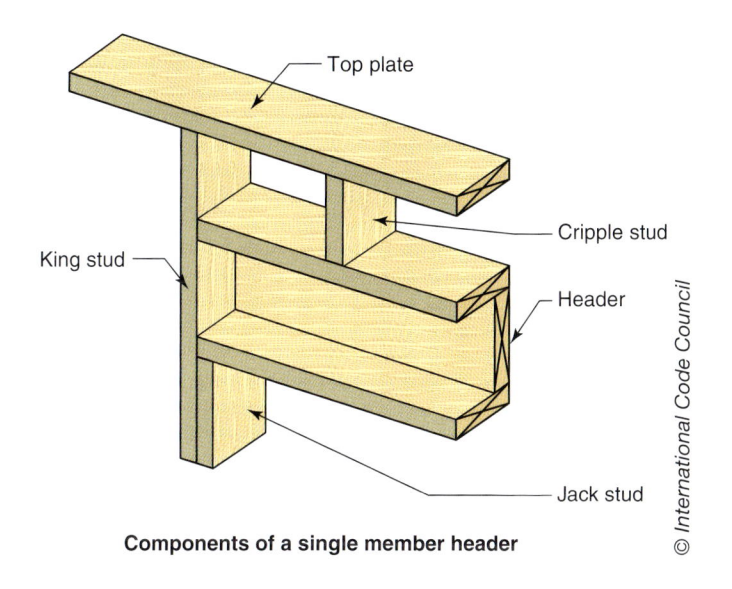

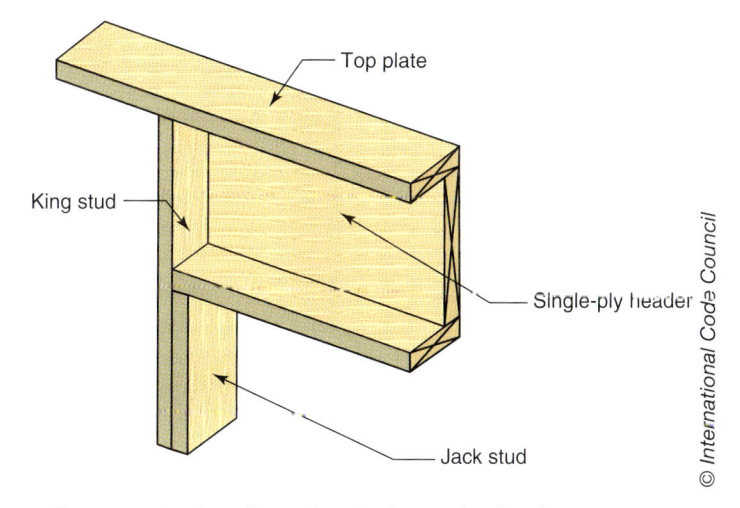

Components of an alternative single member header

Table 2308.4.1.1(1) allows use of the single header on any story, but the maximum span is typically only two to four feet.

Section 2305.5.1 adds minimum connection requirements for single headers. These new single-member (single-ply) header requirements are consistent with the *International Residential Code* and *Wood Frame Construction Manual (WFCM)*. Additionally, provisions of Section 2308.5.5.1 are revised to coordinate with the tabulated header sizes of Table 2308.4.1.1(1).

Two new figures clarify the installation details for single headers with two different top plate conditions. In the first figure, a cripple wall, or pony wall, may be built above the header to finish out the space between the top plate and header. Or the header may fill the entire space between the top plate above the header and the plate at the top of the opening, as shown in the second figure.

2407.1
Structural Glass Baluster Panels

CHANGE TYPE: Modification

CHANGE SUMMARY: Requirements for glass panels that are used as a structural component in a guard have been clarified.

2018 CODE: 2407.1.1 Loads. The panels and their support system shall be designed to withstand the loads specified in Section 1607.8. ~~A design~~ Glass guard elements shall be designed using a factor of safety of four ~~shall be used for safety~~.

2407.1.2 ~~Support~~ Structural glass baluster panels. ~~Each handrail~~ Guards with structural glass baluster panels shall be installed with an attached top rail or ~~guard section~~ handrail. The top rail or handrail shall be supported by ~~a minimum of~~ not fewer than three glass ~~balusters~~ baluster panels, or shall be otherwise supported to remain in place should one glass baluster panel fail. ~~Glass balusters shall not be installed without an attached handrail or guard.~~

> **Exception:** ~~A~~ An attached top rail ~~shall~~ or handrail is not ~~be~~ required where the glass ~~balusters~~ baluster panels are laminated glass with two or more glass plies of equal thickness and of the same glass type ~~when approved by the building official~~. The panels shall be ~~designed to withstand the loads specified in Section 1607.8~~ tested to remain in place as a barrier following impact or glass breakage in accordance with ASTM E 2353.

Structural glass baluster panels
iStock.com/Andry5

CHANGE SIGNIFICANCE: Code requirements for glass panels used as a structural component in a guard have been clarified. Imperfections in glass can cause the panel to fail at loads that are well below its nominal resistance value. A top rail or a handrail at stairs provides additional fall protection for a person leaning on the guard, should a glass panel fail. Having a handrail attached to at least three panels also provides backup when grabbing the handrail to prevent a fall. As an alternative, an exception allows glass-only guards (without an attached top rail or handrail) if the balusters are laminated glass. The 2015 IBC exempts glass balusters from having a top rail, if approved by the building official. While laminated glass provides backup against total panel failure, the glass baluster system must be designed to support the full loads for guards, as specified in Section 2407.1.1.

In the 2018 IBC, requirement for approval by the building official has been deleted as there was no criterion to base approval upon. A new provision requiring testing to ASTM E 2353-14 has been added. ASTM E 2353, *Standard Test Methods for Performance of Glazing in Permanent Railing Systems, Guards and Balustrades*, was developed to test the ability of glazing materials in guard assemblies to remain in place as a barrier after impact or glass breakage. Testing glass baluster systems that have no top rails in accordance with this standard helps to ensure that they remain in place as a barrier after impact or glass breakage.

2510.6
Water-Resistive Barrier

CHANGE TYPE: Modification

CHANGE SUMMARY: An exception has been added for Climate Zones 1A, 2A, and 3A in which a ventilated space is required between a water-resistive barrier and stucco finish.

2018 CODE: 2510.6 Water-resistive barriers. Water-resistive barriers shall be installed as required in Section 1403.2 and, where applied over wood-based sheathing, shall include a water-resistive vapor-permeable barrier with a performance at least equivalent to two layers of water-resistive barrier complying with ASTM E 2556, Type I. The individual layers shall be installed independently such that each layer provides a separate continuous plane and any flashing (installed in accordance with Section 1404.4) intended to drain to the water-resistive barrier is directed between the layers.

Exceptions:

1. Where the water-resistive barrier that is applied over wood-based sheathing has a water resistance equal to or greater than that of a water-resistive barrier complying with ASTM E 2556, Type II and is separated from the stucco by an intervening, substantially nonwater-absorbing layer or drainage space.

2. Where the water-resistive barrier is applied over wood-based sheathing in Climate Zone 1A, 2A, or 3A, a ventilated air space shall be provided between the stucco and water-resistive barrier.

CHANGE SIGNIFICANCE: In many climates, having a vapor-permeable water-resistive barrier that is too vapor-permeable can result in inward moisture movement through the exterior sheathing into the wall assembly, causing an increased risk of moisture damage and mold. Rain falls, stucco finishes get wet, and then the sun comes out, heating the exterior of the stucco, and moisture moves inward behind the water-resistant barrier. Significant moisture can remain behind stucco in humid climates.

2510.6 continues

Two-Layer System
- Each layer of water-resistive barrier is individually installed in a ship lapped fashion
- Interior layer forms continuous drainage plane and is integrated with flashing

Water-resistive barrier

2510.6 continued Water-resistive barrier materials with too high a vapor permeance allow excessive moisture to move into walls. Moist climates 1A through 3A typically have the largest mold issues; for example, in areas in the southeastern United States.

In the 2018 IBC, a new exception addresses the problem of moisture infiltration. A ventilated airspace, now required between stucco and the water-resistive barrier, is intended provide a means to mitigate the potential for moisture migration into the wall assembly.

2603.13 Cladding Attachment over Foam Sheathing to Wood Framing

CHANGE TYPE: Addition

CHANGE SUMMARY: Requirements for cladding over foam sheathing and wood framing have been added to the *International Building Code* consistent with the *International Residential Code* and cold-formed steel stud requirements.

2018 CODE: 2603.13 Cladding attachment over foam sheathing to wood framing. Cladding shall be specified and installed in accordance with Chapter 14 and the cladding manufacturer's installation instructions. Where used, furring and furring attachments shall be designed to resist design loads determined in accordance with Chapter 16. In addition, the cladding or furring attachments through foam sheathing to framing shall meet or exceed the minimum fastening requirements of Section 2603.13.1 or Section 2603.13.2, or an approved design for support of cladding weight.

Exceptions:

1. Where the cladding manufacturer has provided approved installation instructions for application over foam sheathing, those requirements shall apply.
2. For exterior insulation and finish systems, refer to Section 1407.
3. For anchored masonry or stone veneer installed over foam sheathing, refer to Section 1404.

2603.13 continues

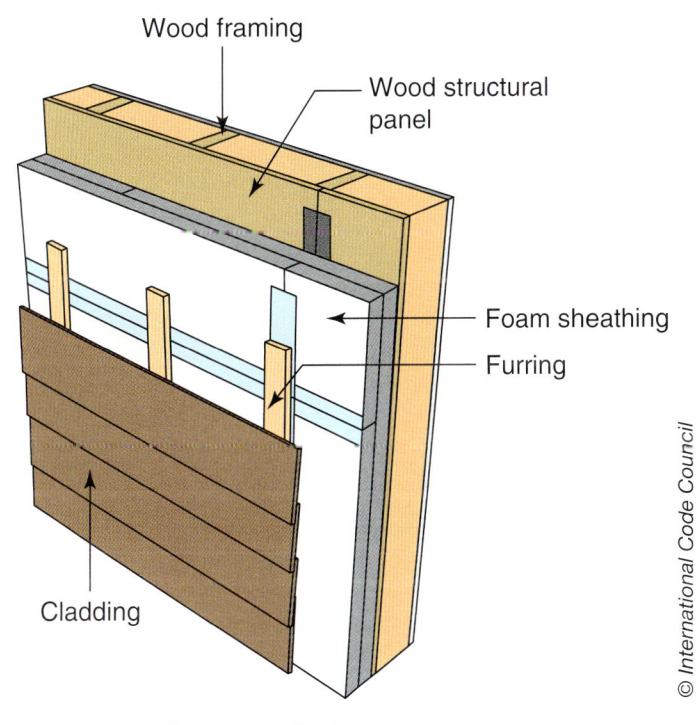

Cladding attachment over furring

2603.13 continued

2603.13.1 Direct attachment. Where cladding is installed directly over foam sheathing without the use of furring, minimum fastening requirements to support the cladding weight shall be as specified in Table 2603.13.1.

2603.13.2 Furred cladding attachment. Where wood furring is used to attach cladding over foam sheathing, furring minimum fastening requirements to support the cladding weight shall be as specified in Table 2603.13.2. Where placed horizontally, wood furring shall be preservative-treated wood in accordance with Section 2303.1.9 or naturally durable wood and fasteners shall be corrosion resistant in accordance with Section 2304.10.5.

TABLE 2603.13.1 Cladding Minimum Fastening Requirements for Direct Attachment over Foam Plastic Sheathing to Support Cladding Weight[a]

Cladding Fastener Through Foam Sheathing into:	Cladding Fastener - Type and Minimum Size[b]	Cladding Fastener Vertical Spacing (inches)	Maximum Thickness of Foam Sheathing[c] (inches)							
			16" o.c. Fastener Horizontal Spacing				24" o.c. Fastener Horizontal Spacing			
			Cladding Weight:				Cladding Weight:			
			3 psf	11 psf	18 psf	25 psf	3 psf	11 psf	18 psf	25 psf
Wood Framing (minimum 1¼-inch penetration)	0.113" diameter nail	6	2.00	1.45	0.75	DR	2.00	0.85	DR	DR
		8	2.00	1.00	DR	DR	2.00	0.55	DR	DR
		12	2.00	0.55	DR	DR	1.85	DR	DR	DR
	0.120" diameter nail	6	3.00	1.70	0.90	0.55	3.00	1.05	0.50	DR
		8	3.00	1.20	0.60	DR	3.00	0.70	DR	DR
		12	3.00	0.70	DR	DR	2.15	DR	DR	DR
	0.131" diameter nail	6	4.00	2.15	1.20	0.75	4.00	1.35	0.70	DR
		8	4.00	1.55	0.80	DR	4.00	0.90	DR	DR
		12	4.00	0.90	DR	DR	2.70	0.50	DR	DR
	0.162" diameter nail	6	4.00	3.55	2.05	1.40	4.00	2.25	1.25	0.80
		8	4.00	2.55	1.45	0.95	4.00	1.60	0.85	0.50
		12	4.00	1.60	0.85	0.50	4.00	0.95	DR	DR

For SI: 1 inch = 25.4 mm; 1 pound per square foot (psf) = 0.0479 kPa
DR = design required
o.c. = on center

a. Wood framing shall be spruce-pine-fir or any wood species with a specific gravity of 0.42 or greater in accordance with ANSI/AWC NDS.
b. Nail fasteners shall comply with ASTM F 1667, except nail length shall be permitted to exceed ASTM F 1667 standard lengths.
c. Foam sheathing shall have a minimum compressive strength of 15 psi in accordance with ASTM C 578 or ASTM C 1289.

TABLE 2603.13.2 Furring Minimum Fastening Requirements for Application over Foam Plastic Sheathing to Support Cladding Weight[a,b]

					Maximum Thickness of Foam Sheathing[d] (inches)							
					16" o.c. Furring[e]				24" o.c. Furring[e]			
		Fastener Type and Minimum Size	Minimum Penetration into Wall Framing (inches)	Fastener Spacing in Furring (inches)	Siding Weight:				Siding Weight:			
Furring Material	Framing Member				3 psf	11 psf	18 psf	25 psf	3 psf	11 psf	18 psf	25 psf
Minimum 1x Wood Furring[c]	Minimum 2x Wood Stud	0.131" diameter nail	1¼	8	4.00	2.45	1.45	0.95	4.00	1.60	0.85	DR
				12	4.00	1.60	0.85	DR	4.00	0.95	DR	DR
				16	4.00	1.10	DR	DR	3.05	0.60	DR	DR
		0.162" diameter nail	1¼	8	4.00	4.00	2.45	1.60	4.00	2.75	1.45	0.85
				12	4.00	2.75	1.45	0.85	4.00	1.65	0.75	DR
				16	4.00	1.90	0.95	DR	4.00	1.05	DR	DR
		No. 10 wood screw	1	12	4.00	2.30	1.20	0.70	4.00	1.40	0.60	DR
				16	4.00	1.65	0.75	DR	4.00	0.90	DR	DR
				24	4.00	0.90	DR	DR	2.85	DR	DR	DR
		¼" lag screw	1½	12	4.00	2.65	1.50	0.90	4.00	1.65	0.80	DR
				16	4.00	1.95	0.95	0.50	4.00	1.10	DR	DR
				24	4.00	1.10	DR	DR	3.25	0.50	DR	DR

For SI: 1 inch = 25.4 mm; 1 pound per square foot (psf) = 0.0479 kPa
DR = design required
o.c. = on center

a. Wood framing and furring shall be spruce-pine-fir or any wood species with a specific gravity of 0.42 or greater in accordance with ANSI/AWC NDS.
b. Nail fasteners shall comply with ASTM F 1667, except nail length shall be permitted to exceed ASTM F 1667 standard lengths.
c. Where the required cladding fastener penetration into wood material exceeds ¾ inch (19 mm) and is not more than 1½ inches (38 mm), a minimum 2x wood furring or an approved design shall be used.
d. Foam sheathing shall have a minimum compressive strength of 15 psi in accordance with ASTM C 578 or ASTM C 1289.
e. Furring shall be spaced not greater than 24 inches (610 mm) on center in a vertical or horizontal orientation. In a vertical orientation, furring shall be located over wall studs and attached with the required fastener spacing. In a horizontal orientation, the indicated 8-inch (203 mm) and 12-inch (305 mm) fastener spacing in furring shall be achieved by use of two fasteners into studs at 16 inches (406 mm) and 24 inches (610 mm) on center, respectively.

CHANGE SIGNIFICANCE: Requirements for cladding attachment over foam sheathing to wood framing were approved in the 2015 *International Residential Code* (IRC). Similar requirements for steel framing with cladding over foam sheathing were approved in the 2015 IBC and IRC. This code change in the 2018 IBC creates congruent information within the IBC and IRC for exterior wall covering assemblies on wood-frame walls that include foam sheathing.

At the request of the brick industry, this provision includes an 18 psf cladding weight category intended for use with brick veneer. The foam thickness values are rounded to the nearest 0.05-inch thickness to align with actual thicknesses of foam sheathing products rather than nominal thicknesses. Similar changes occur for steel framing in the IBC and wood and steel framing in the IRC.

The prescriptive fastening requirements for cladding materials installed over foam sheathing are based on a project sponsored by the New York State Energy Research and Development Agency (NYSERDA). The project goal was to ensure adequate performance of cladding over rigid

2603.13 continues

2603.13 continued foam sheathing. The project included testing of cladding attachments through various thicknesses of foam sheathing using various fastener types on steel frame wall assemblies, including supplemental test data to address attachments to wood framing. The cladding attachment requirements and foam sheathing thickness limits are based on the *National Design Specification for Wood Construction* (NDS) yield equations verified by test data to limit cladding connection movement to no more than 0.015-inch slip under cladding weight or dead load. This deflection-controlled approach generally resulted in safety factors in the range of 5 to 8 for the average shear capacity and demonstrated adequate long-term deflection control.

PART 7

Building Services, Special Devices, and Special Conditions

Chapters 27 through 33

- **Chapter 27** Electrical
  No changes addressed
- **Chapter 28** Mechanical Systems
  No changes addressed
- **Chapter 29** Plumbing Systems
  No changes addressed
- **Chapter 30** Elevators and Conveying Systems
- **Chapter 31** Special Construction
- **Chapter 32** Encroachments into the Public Right-of-Way
  No changes addressed
- **Chapter 33** Safeguards during Construction

Although building services such as electrical systems (Chapter 27), mechanical systems (Chapter 28) and plumbing systems (Chapter 29) are regulated primarily through separate and distinct codes, limited provisions are set forth in the *International Building Code*. Chapter 30 regulates elevators and similar conveying systems to a limited degree, as most requirements are found in American Society of Mechanical Engineers (ASME) standards. The special construction provisions of Chapter 31 include those types of elements or structures that are not conveniently addressed in other portions of the code. By "special construction," the code is referring to membrane structures, pedestrian walkways, tunnels, awnings, canopies, marquees, and similar building features that are unregulated elsewhere. Chapter 32 governs the encroachment of structures into the public right-of-way, and Chapter 33 addresses safety during construction and the protection of adjacent public and private properties. ■

3001.2
Emergency Elevator Communication Systems

3006.2.1
Corridors Adjacent to Elevator Hoistway Openings

3007.1
Extent of Fire Service Access Elevator Travel

3008.1.1
Required Number of Occupant Evacuation Elevators

3113
Relocatable Buildings

3310.1
Stairways in Buildings under Construction

3314
Fire Watch during Construction

3001.2
Emergency Elevator Communication Systems

CHANGE TYPE: Addition

CHANGE SUMMARY: Additional communication capabilities are now required in accessible elevators to enhance the usability of the two-way communication system by individuals with varying degrees of hearing or speech impairments.

2018 CODE: <u>**3001.2 Emergency elevator communication systems for the deaf, hard of hearing and speech impaired.** An emergency two-way communication system shall be provided that:</u>

<u>1. Is a visual and text-based and a video-based 24/7 live interactive system.</u>

<u>2. Is fully accessible by the deaf, hard of hearing and speech impaired, and shall include voice-only options for hearing individuals.</u>

<u>3. Has the ability to communicate with emergency personnel utilizing existing video conferencing technology, chat/text software or other approved technology.</u>

CHANGE SIGNIFICANCE: In multistory buildings, passenger elevators are typically utilized as vertical accessible routes. As such, the elevators and associated elevator landings must have accessible features complying with ICC A117.1, the IBC-referenced technical standard for accessibility. Applicable to accessible elevators, the A117.1 standard requires an emergency two-way communication system between the elevator car and a point outside the hoistway. The system must fully comply with the provisions in ASME A17.1/CSA B44, the *Safety Code for Elevators and Escalators* as published by the American Society of Mechanical Engineers International. Additional communication capabilities are now required to enhance usability of the system by individuals with varying degrees of hearing or speech impairments.

Elevator control panel

In order to provide totally accessible communication between individuals in elevators who are deaf, hard of hearing or speech impaired and local governmental emergency services, additional communication features must be provided within the elevator. Key features of the provisions include:

1. The system is to be visual text-based as well as video-based and provided with a live interactive system.
2. Full accessibility must be available to individuals who are deaf, hard of hearing and speech impaired.
3. The two-way communication is to occur between the elevator cab and the appropriate local emergency personnel.

Although these types of technical requirements are typically found in the A117.1 standard rather than in the code itself, currently the accessibility standard does not contain such criteria. Therefore, its inclusion in the IBC recognizes the high degree of importance assigned to the new requirements.

It is expected that current technologies will be able to be readily adapted to provide the level of accessibility set forth in the specific provisions. The three conditions are intended to provide a high degree of flexibility and options that will make compliance readily achievable.

3006.2.1
Corridors Adjacent to Elevator Hoistway Openings

CHANGE TYPE: Clarification

CHANGE SUMMARY: Where an elevator hoistway door opens into a fire-resistance-rated corridor, the door opening must be protected in accordance with one of the three general methods established in Section 3006.3 which recognizes the use of elevator lobbies, additional doors, and pressurization of the hoistway.

2018 CODE: <u>**3006.2.1 Rated corridors.** Where corridors are required to be fire-resistance rated in accordance with Section 1020.1, elevator hoistway openings shall be protected in accordance with Section 3006.3.</u>

CHANGE SIGNIFICANCE: Where a corridor is identified within a building, it may be necessary to introduce fire-resistance-rated construction and protected openings in order to isolate the means of egress path as defined by the corridor from surrounding spaces. Fire-resistance-rated corridors are intended to provide a protected path of travel within the exit access that will allow occupants to evacuate or relocate during a fire condition that occurs in spaces adjacent to the corridor. Such a corridor must also provide a significant degree of resistance to the movement of smoke into the egress environment. Construction of a fire-resistance-rated corridor is provided through the use of fire partitions, as well as fire-protection-rated openings having smoke- and draft-control opening protection.

Where an elevator opens into such a corridor, it has been previously questioned as to whether or not the hoistway opening requires the same degree of smoke- and draft-control protection as other openings into the fire-resistance-rated corridor. Although the elevator hoistway door can typically achieve the required degree of opening fire protection, it cannot comply with the air leakage limits mandated for a smoke- and draft-control assembly. Reference is often made to Section 701.2 which indicates that fire assemblies that serve multiple purposes in a building must comply with all of the requirements that are applicable for each of the individual fire assemblies. Provisions specific to the condition of an elevator hoistway door opening into a fire-resistance-rated corridor now indicate that such an opening must be protected in accordance with one of the methods established in Section 3006.3.

Hoistway opening protection can be achieved through compliance with any one of three different approaches. One method is to provide an enclosed elevator lobby at each floor to separate the elevator hoistway shaft enclosure doors from the corridor, using fire partitions or smoke

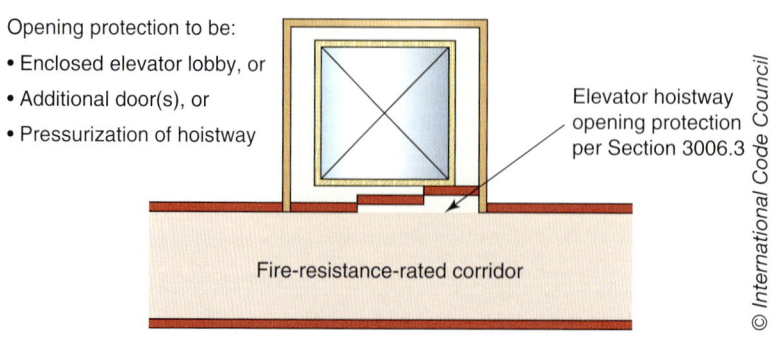

Corridor opening at elevator hoistway

partitions for lobby construction as applicable. A second option is the installation of an additional door at each elevator hoistway door opening that complies with the smoke- and draft-control door assembly requirements of Section 716.2.2.1.1. In this case, the door assembly must be tested in accordance with UL 1784 without an artificial bottom seal. The third approach is to pressurize the elevator hoistway in accordance with Section 909.21. Any of these options is considered a permissible approach to address the separation required between a fire-resistance-rated corridor and an elevator hoistway shaft enclosure.

The protection provided at the elevator hoistway doors restricts the movement of smoke from the fire floor to other floors in the building by way of the elevator shaft enclosure. Hoistway opening protection is already required in some multistory buildings as set forth in Section 3006.2. The new provisions expand the protection to additional buildings, both sprinklered and nonsprinklered, as identified below:

- Sprinklered buildings that are not considered as high-rise buildings where the corridor serves:
  - Group R occupancies with an occupant load greater than 10
  - Group H-1, H-2, and H-3 occupancies
  - Group H-4 and H-5 occupancies with an occupant load greater than 30
- Nonsprinklered buildings where the hoistway connects only three stories and the corridor serves:
  - Group A, B, E, F, M, S, and U occupancies with an occupant load greater than 30

3007.1

Extent of Fire Service Access Elevator Travel

CHANGE TYPE: Modification

CHANGE SUMMARY: Fire service access elevators, where required, now only need to provide access to those floor levels at and above the lowest level of fire department access. In addition, elevators that only connect a parking garage to a building's lobby need not serve as fire service access elevators.

2018 CODE: 3007.1 General. Where required by Section 403.6.1, every floor <u>above and including the lowest level of fire department vehicle access</u> of the building shall be served by fire service access elevators complying with Sections 3007.1 through 3007.9. Except as modified in this section, fire service access elevators shall be installed in accordance with this chapter and ASME A17.1/CSA B44.

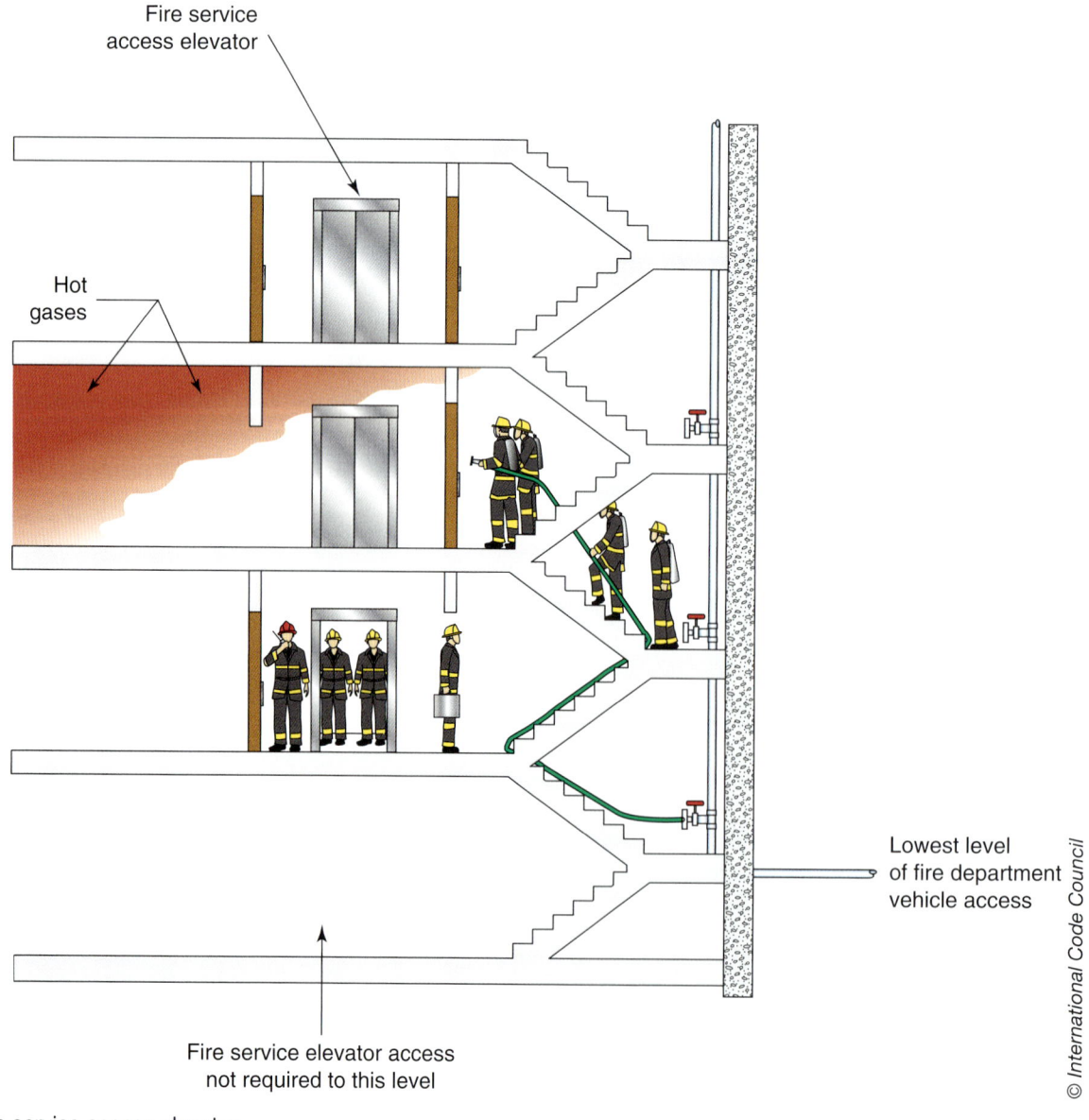

Fire service access elevator

Exception: Elevators that only service an open or enclosed parking garage and the lobby of the building shall not be required to serve as fire service access elevators.

CHANGE SIGNIFICANCE: To facilitate the rapid deployment of firefighters, the IBC has mandated fire service access elevators in tall buildings since the 2009 edition. In high-rise buildings with an occupied floor more than 120 feet above the lowest level of fire department vehicle access, at least two elevators complying with Section 3007 must be provided for use by fire service personnel and other emergency responders. The scoping limits of where such elevators are required has been modified to limit their required access to only those floor levels at and above the lowest level of fire department access. In addition, elevators that only connect a parking garage to a building's lobby need not serve as fire service access elevators.

The original provisions addressing fire service access elevators required access to all floor levels of the building, including any levels that were situated below grade. Typically, access for the fire service in regulated high-rise buildings is focused on travel to stories located very high in the building. Fire departments will very seldom ever take an elevator provided the stairways are manageable.

A second modification addresses the belief that fire service access elevators are not necessary in parking garages. Applicable to both open and enclosed parking garages, the new exception reflects the expectation that firefighters will not choose to take an elevator past the fire floor, preferring instead to use the stairways.

3008.1.1
Required Number of Occupant Evacuation Elevators

CHANGE TYPE: Modification

CHANGE SUMMARY: A reduction in the minimum number of elevators that must be considered as occupant evacuation elevators now reflects a more reasonable performance-based approach while still retaining the capacity to evacuate a high-rise building more quickly than stairs alone.

2018 CODE: 3008.1 General. ~~Where elevators are to be~~ Elevators used for occupant self-evacuation during fires~~, all passenger elevators for general public use~~ shall comply with Sections 3008.1 through 3008.10. ~~Where other elevators are used for occupant self-evacuation, those elevators shall comply with these sections.~~

3008.1.1 Number of occupant evacuation elevators. The number of elevators available for occupant evacuation shall be determined based on an egress analysis that addresses one of the following scenarios.

1. Full-building evacuation where the analysis demonstrates that the number of elevators provided for evacuation results in an evacuation time less than 1 hour.
2. Evacuation of the five consecutive floors with the highest cumulative occupant load where the analysis demonstrates that the number of elevators provided for evacuation results in an evacuation time less than 15 minutes.

Not less than one elevator in each bank shall be designated for occupant evacuation. Not less than two shall be provided in each occupant evacuation elevator lobby where more than one elevator opens into the lobby. Signage shall be provided to denote which elevators are available for occupant evacuation.

3008.8.1 Determination of standby power load. Standby power loads shall be based upon the determination of the number of occupant evacuation elevators in Section 3008.1.1.

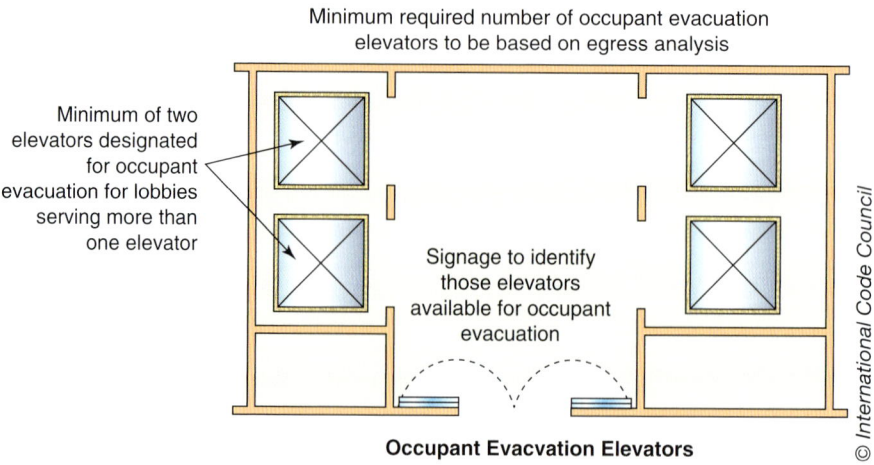

Occupant evacuation elevators

CHANGE SIGNIFICANCE: Provisions relating to elevators that are intended to be used for occupant evacuation during emergency conditions were part of a package of high-rise building regulations introduced in the 2009 IBC. Although not required by the code, such self-evacuation elevators are permitted to be installed in high-rise buildings where in compliance with the requirements set forth in Section 3008. The use of occupant elevators does, however, provide one significant alternative approach to the design of the means of egress through their permissible substitution for the additional interior exit stairway required by Section 403.5.2 in buildings more than 420 feet in height. Previously, all passenger elevators intended for general public use were required to comply with the occupant evacuation elevator requirements. A reduction in the minimum number of elevators that must be considered as occupant evacuation elevators now reflects a more reasonable performance-based approach while still retaining the capacity to evacuate a high-rise building more quickly than stairs alone.

Although the use of elevators for occupant evacuation in super high-rise buildings is a viable and more efficient option than stairways, it can require an excessive amount of standby power. Previously, where all passenger elevators were required to be considered as occupant evacuation elevators, the required two hours of standby power for every elevator simultaneously often created a condition viewed as excessive, even more so where the occupant loads are low.

Two options have now been established for the determination of the minimum number of elevators required to meet the performance intent. Scenario #1 focuses on the full evacuation of the building's occupants within a one-hour time period. This approach does not mandate full building evacuation, but rather is to be considered as a benchmark for the analysis. Scenario #2 is generally based on a more typical phased evacuation process. The 15-minute limit criterion addresses the removal of occupants from the area to which the fire department will respond. The use of five consecutive floors in the analysis provides for a safety factor for the minimum required number of occupant evacuation elevators.

Additional criteria mandate that at least one elevator in every elevator bank be provided for occupant evacuation. Where multiple passenger elevators open into a lobby, a minimum of two occupant evacuation elevators are required. These conditions ensure that as individuals reach an elevator lobby location there will always be one or more elevators available to them for evacuation purposes. In order to better identify the specific elevators that are intended for evacuation activities, signage must be provided indicating their availability.

3113
Relocatable Buildings

CHANGE TYPE: Addition

CHANGE SUMMARY: A process of acceptance for relocatable modular buildings has been established in order to provide clear and consistent direction in the relocation, reuse, and/or repurposing of such buildings.

2018 CODE:

**SECTION 202
DEFINITIONS**

RELOCATABLE BUILDING. A partially or completely assembled building constructed and designed to be reused multiple times and transported to different building sites.

**SECTION 3113
RELOCATABLE BUILDINGS**

3113.1 General. The provisions of this section shall apply to relocatable buildings. Relocatable buildings manufactured after the effective date of this code shall comply with the applicable provisions of this code.

Exception: This section shall not apply to manufactured housing used as dwellings.

3113.1.1 Compliance. A newly constructed relocatable building shall comply with the requirements of this code for new construction. An existing relocatable building that is undergoing alteration, addition, change of occupancy or relocation shall comply with Chapter 14 of the *International Existing Building Code.*

Relocatable building

3113.2 Supplemental information. Supplemental information specific to a relocatable building shall be submitted to the authority having jurisdiction. It shall, as a minimum, include the following in addition to the information required by Section 105.

1. Manufacturer's name and address.
2. Date of manufacture.
3. Serial number of module.
4. Manufacturer's design drawings.
5. Type of construction in accordance with Section 602.
6. Design loads including: roof live load, roof snow load, floor live load, wind load and seismic site class, use group and design category.
7. Additional building planning and structural design data.
8. Site-built structure or appurtenance attached to the relocatable building.

3113.3 Manufacturer's data plate. Each relocatable module shall have a data plate that is permanently attached on or adjacent to the electrical panel, and shall include the following information:

1. Occupancy group.
2. Manufacturer's name and address.
3. Date of manufacture.
4. Serial number of module.
5. Design roof live load, design floor live load, snow load, wind and seismic design.
6. Approved quality assurance agency or approved inspection agency.
7. Codes and standards of construction.
8. Envelope thermal resistance values.
9. Electrical service size.
10. Fuel burning equipment and size.
11. Special limitations if any.

3113.4 Inspection agencies. The building official is authorized to accept reports of inspections conducted by approved inspection agencies during off-site construction of the relocatable building, and to satisfy the applicable requirements of Sections 110.3 through 110.3.11.1.

CHANGE SIGNIFICANCE: Site-built buildings are intended to remain on their original lot for the life of the building. However, there are those partially or completely assembled structures that are intended to be reused multiple times, typically on different sites. It is important that such "relocatable buildings" be addressed in the IBC in a manner specific to the unique conditions of their use. New provisions have been established addressing the process of acceptance of a relocatable building in order to provide clear and consistent direction in the relocation, reuse, and/or repurposing of such buildings.

3113 continues

3113 continued

The regulation of relocatable buildings is often done at the state level through statutes that are specifically established for modular buildings intended to be relocated from site to site. These new provisions set forth in Section 3113 are now available for use by those states without mandated requirements, intended to address the confusion and inconsistency that has previously occurred.

There are a significant number of relocatable modular buildings in use today. According to the Modular Building Institute, there are estimated to be over 600,000 code-compliant relocated buildings currently in use in North America. Of that number, public school districts collectively own and operate about 30% of the total number. Of the remainder, many buildings are used for business purposes such as construction site offices and sales offices.

It is important to note that many provisions of the IBC currently apply to both site-built and relocatable modular buildings, particularly for new construction. However, there previously have been several conflicting code provisions that could not be applied to both types of buildings, including those related to submittal documents and inspections. The new section specific to relocatable buildings introduces requirements addressing code compliance, required supplemental information, the manufacturer's data plate, and the use of approved third-party inspection agencies.

3310.1
Stairways in Buildings under Construction

CHANGE TYPE: Modification

CHANGE SUMMARY: At least one temporary or permanent stairway must now be provided in a building under construction once the building has reached a height of 40 feet as measured from the lowest level of fire department vehicle access.

2018 CODE: 3310.1 Stairways required. Where ~~a~~ building ~~has been constructed to a building height of 50~~ construction exceeds 40 feet ~~(15 240 mm) or four stories, or where an existing building exceeding 50 feet (15 240 mm)~~ (12 192 mm) in ~~building height is altered~~ height above the lowest level of fire department vehicle access, ~~no fewer than one~~ a temporary ~~lighted~~ or permanent stairway shall be provided ~~unless~~. As construction progresses, such stairway shall be extended to within one ~~or more~~ floor of the ~~permanent stairways are erected as the~~ highest point of construction ~~progresses~~ having secured decking or flooring.

CHANGE SIGNIFICANCE: During the construction of a building, there are a variety of hazards, both known and unanticipated, that could affect the safety of those individuals on the job site. The potential for a fire or other emergency that would require immediate evacuation of the structure often exists due to the equipment, materials, and activities that are present. Where the height of a building under construction results in extended evacuation time for the individuals, it is critical that the vertical travel be as efficient as possible. Therefore, the IBC has historically regulated the means of egress in such buildings. Provisions have been revised to now require that at least one stairway be provided where the building has been constructed to a height of 40 feet.

During construction work, it has previously been required that at least one temporary or permanent stairway be provided where the building height reached a point 50 feet or four stories. The threshold at which the stairway is required has been reduced to 40 feet above the lowest level

3310.1 continues

Temporary stairway during construction

3310.1 continued

of fire department vehicle access to be consistent with the current provisions addressing the installation of a standpipe. In addition, as construction progresses the required extension of the stairway is now based on the highest point of construction where there is secured flooring or decking. This point of application is also consistent with current standpipe provisions.

3314 Fire Watch During Construction

CHANGE TYPE: Addition

CHANGE SUMMARY: In order to protect adjacent properties from fire in a building of considerable height when under construction, new provisions have been established to give authority to the fire code official to require a fire watch during those hours where no construction work is being done.

2018 CODE:

SECTION 3314
FIRE WATCH DURING CONSTRUCTION

3314.1 Fire watch during construction. Where required by the fire code official, a fire watch shall be provided during nonworking hours for construction that exceeds 40 feet in height above the lowest adjacent grade.

CHANGE SIGNIFICANCE: Some of the most hazardous conditions related to buildings often are present during the construction process. Recent fires that have occurred at construction sites during times of no activity have demonstrated the need for early notification that can only be provided by fire watch personnel. The lack of fire alarm and detection devices during the construction process requires an alternative approach to identifying and communicating the presence of a fire event. In order to protect adjacent properties from fire in a building of considerable height when under construction, new provisions have been established to give authority to the fire code official to require a fire watch during those hours where no construction work is being done.

Fires in sizable buildings under construction have the potential for significant heat release due to the fire loading created by building components and other materials used in the building's construction. For this

3314 continues

Building under construction

3314 continued

reason, when required by the fire code official, a fire watch is to be provided where the height of construction exceeds 40 feet above the lowest adjacent grade. The 40-foot threshold is consistent with other fire and life safety requirements for buildings under construction, such as the provisions for standpipes and means of egress stairways.

It is expected that the new requirement will apply only to new construction. It is not intended for the provisions to be applied to alterations and other types of minor construction activity. Existing buildings would be regulated by a comprehensive fire safety plan. Although the potential for a sizable fire load requiring implementation of a fire watch program would be more probable for a building of combustible construction, there are no conditions based on the building's construction type. All new buildings, regardless of occupancy or type of construction, that exceed the 40-foot threshold are subject to the fire watch requirement if mandated by the fire code official.

Although the primary benefit of identifying a fire early in its development will typically be the protection of adjoining properties and neighboring buildings, the reduction in property loss and protection of fire personnel are also important aspects of a fire watch activity.

PART 8

Appendices

Appendices A through N

- **Appendix A through F** — No changes addressed
- **Appendix G** — Flood-Resistant Construction
- **Appendix H through M** — No changes addressed
- **Appendix N** — Replicable Buildings

G103.6
Watercourse Alteration

APPENDIX N
Guidelines for Replicable Buildings

As stated in Chapter 1 of the IBC, provisions in the appendices do not apply unless specifically referenced in the adopting ordinance. The appendices are developed in much the same manner as the main body of the model code. However, the appendix information is judged to be outside the scope and purpose of the code at the time of code publication. Many times an appendix offers supplemental information, alternative methods, or recommended procedures. The information may also be specialized and applicable or of interest to only a limited number of jurisdictions. Although an appendix may provide some guidelines or examples of recommended practices or assist in the determination of alternative materials or methods, it will have no legal status and cannot be enforced until it is specifically recognized in the adopting legislation. Appendix chapters or portions of such chapters that gain general acceptance over time can move into the main body of the model code through the code-development process.

G103.6
Watercourse Alteration

CHANGE TYPE: Modification

CHANGE SUMMARY: Notification of a watercourse alteration should be given to all adjacent jurisdictions, not just those jurisdictions that will be affected.

2018 CODE: G103.6 Watercourse alteration. Prior to issuing a permit for any alteration or relocation of any watercourse, the building official shall require the applicant to provide notification of the proposal to the appropriate authorities of all affected adjacent government jurisdictions, as well as appropriate state agencies. A copy of the notification shall be maintained in the permit records and submitted to FEMA.

CHANGE SIGNIFICANCE: The National Flood Insurance Program (NFIP) regulations specify that communities notify adjacent communities when a proposal to alter or relocate a watercourse is received. When a local jurisdiction uses IBC Appendix G, the current phrasing in Section G103.6 requires judgment to determine whether an adjacent jurisdiction is affected by a proposed watercourse alteration. Only with engineering analysis it is feasible to determine whether adjacent communities are affected.

Rather than require the analysis, the 2018 IBC provision is modified to align with the NFIP regulations to require that all adjacent communities are notified. This allows each community to check on whether it is likely to be affected by the change in the watercourse.

Flood control along a river

Flood control in mountainous regions

Appendix N
Guidelines for Replicable Buildings

CHANGE TYPE: Addition

CHANGE SUMMARY: Guidelines for replicable buildings have been added to the appendix in order to give jurisdictions a tool they can adopt to help streamline the plan review process in regard to code compliance.

2018 CODE:

<u>**APPENDIX N**
REPLICABLE BUILDINGS

SECTION N101
ADMINISTRATION

N101.1 Purpose. The purpose of this appendix is to provide a format and direction regarding the implementation of a replicable building program.

N101.2 Objectives. Such programs allow a jurisdiction to recover from a natural disaster faster and allow for consistent application of the codes for replicable building projects. It will result in faster turnaround for the end user, and a quicker turnaround through the plan review process.

SECTION N102
DEFINITIONS

N102.1 Definitions. The following words and terms shall, for the purposes of this appendix, have the meanings shown herein.</u>

Appendix N continues

Replicable sales building

Appendix N continued

REPLICABLE BUILDING. A building or structure utilizing a replicable design.

REPLICABLE DESIGN. A prototypical design developed for application in multiple locations with minimal variation or modification.

SECTION N103
REPLICABLE DESIGN REQUIREMENTS

N103.1 Prototypical construction documents. A replicable design shall establish prototypical construction documents for application at multiple locations. The construction documents shall include details appropriate to each wind region, seismic design category, and climate zone for locations in which the replicable design is intended for application. Application of replicable design shall not vary with regard to the following, except for allowable variations in accordance with Section N106.

1. Use and occupancy classification.
2. Building heights and area limitations.
3. Type of construction classification.
4. Fire-resistance ratings.
5. Interior finishes.
6. Fire protection system.
7. Means of egress.
8. Accessibility.
9. Structural design criteria.
10. Energy efficiency.
11. Type of mechanical and electrical systems.
12. Type of plumbing system and number of fixtures.

SECTION N104
REPLICABLE DESIGN SUBMITTAL REQUIREMENTS

N104.1 General. A summary description of the replicable design and related construction documents shall be submitted to an approved agency. Where approval is requested for elements of the replicable design that is not within the scope of the *International Building Code*, the construction documents shall specifically designate the codes for which review is sought. Construction documents shall be signed, sealed and dated by a registered design professional.

N104.1.1 Architectural plans and specifications. Where approval of the architectural requirements of the replicable design is sought, the submittal documents shall include architectural plans and specifications as follows:

1. Description of uses and the proposed occupancy groups for all portions of the building.
2. Proposed type of construction of the building.

3. Fully dimensioned drawings to determine building areas and height.
4. Adequate details and dimensions to evaluate means of egress, including occupant loads for each floor, exit arrangement and sizes, corridors, doors, and stairs.
5. Exit signs and means of egress lighting, including power supply.
6. Accessibility scoping provisions.
7. Description and details of proposed special occupancies such as a covered mall, high-rise, mezzanine, atrium, and public garage.
8. Adequate details to evaluate fire-resistive-rated construction requirements, including data substantiating required ratings.
9. Details for plastics, insulation and safety glazing installation.
10. Details of required fire protection systems.
11. Material specifications demonstrating fire-resistance criteria.

N104.1.2 Structural plans, specifications, and engineering details. Where approval of the structural requirements of the replicable design is sought, the submittal documents shall include details for each wind region, seismic design category and climate zone for which approval is sought; and shall include the following:

1. Signed and sealed structural design calculations that support the member sizes on the drawings.
2. Design load criteria, including; frost depth, live loads, snow loads, wind loads, earthquake design date, and other special loads
3. Details of foundations and superstructure.
4. Provisions for special inspections.

N104.1.3 Energy conservation details. Where approval of the energy conservation requirements of the replicable design is sought, the submittal documents shall include details for each climate zone for which approval is sought; and shall include the following:

1. Climate zones for which approval is sought.
2. Building envelope details.
3. Building mechanical system details.
4. Details of electrical power and lighting systems.
5. Provisions for system commissioning.

SECTION N105
REVIEW AND APPROVAL OF REPLICABLE DESIGN

N105.1 General. Proposed replicable designs shall be reviewed by an approved agency. The review shall be applicable only to the replicable design features submitted in accordance with Section N104. The review shall determine compliance with this code and additional codes specified in Section N104.1.

Appendix N continues

Appendix N continued

N105.2 Documentation. The results of the review shall be documented indicating compliance with the code requirements.

N105.3 Deficiencies. Where the review of the submitted construction documents identifies elements where the design is deficient and will not comply with the applicable code requirements, the approved agency shall notify the proponent of the replicable design, in writing, of the specific areas of noncompliance and request correction.

N105.4 Approval. Where the review of the submitted construction documents determines that the design is in compliance with the codes designated in Section N104.1, and where deficiencies identified in Section N105.3 have been corrected the approved agency shall issue a summary report of Approved Replicable Design. The summary report shall include any limitations on the approved replicable design including, but not limited to climate zones, wind regions and seismic design categories.

SECTION N106
SITE-SPECIFIC APPLICATION OF
APPROVED REPLICABLE DESIGN

N106.1 General. Where site-specific application of a replicable design that has been approved under the provisions of Section N105 is sought, the construction documents submitted to the building official shall comply with this section.

N106.2 Submittal documents. A summary description of the replicable design and related construction document shall be submitted. Construction documents shall be signed, sealed, and dated by the registered design professional. A statement, signed sealed and dated by the registered design professional, that the replicable design submitted for local review is the same as the replicable design reviewed by the approved agency, shall be submitted.

N106.2.1 Architectural plans and specifications. Architectural plans and specifications shall include the following:

1. Construction documents for variations from the replicable design.
2. Construction for portions that are not part of the replicable design.
3. Documents for local requirements as identified by the building official.
4. Construction documents detailing the foundation system.

SECTION N107
SITE-SPECIFIC REVIEW AND
APPROVAL OF REPLICABLE DESIGN

N107.1 General. Proposed site-specific application of replicable design shall be submitted to the building official in accordance with the provisions of Chapter 1 and Appendix N.

N107.2 Site-specific review and approval of replicable design.
The building official shall verify that the replicable design submitted for site-specific application is the same as the approved replicable design reviewed by the approved agency. In addition, the building official shall review the following for code compliance.

1. Construction documents for variations from the replicable design.
2. Construction for portions of the building that are not part of the replicable design.
3. Documents for local requirements as identified by the building official.

CHANGE SIGNIFICANCE: The concept of replicable building design applies where a given prototype is to be built in a variety of locations while maintaining consistent overall design parameters. The resulting replicable building is reviewed and deemed code compliant by a designated expert, allowing local jurisdictions to accept the structure as substantially code compliant. Such buildings could be approved for construction without a complete plan review, with the construction documents examined by the local authority for compliance with local amendments and conditions only. An appendix chapter has been added to the IBC to establish guidelines for replicable buildings that can give jurisdictions a tool they can adopt to help streamline the plan review process in regard to code compliance.

The ICC *Guideline for Replicable Buildings*, ICC G1-2010, was published in 2010 to provide jurisdictions with guidance regarding the process of expedited plan review for replicable buildings. The new appendix provisions, based on the G1-2010 guideline, are intended to benefit both owners and local jurisdictions by:

- Enhancing public safety through a more uniform review process,
- Conserving local resources through the elimination of repetitive reviews of transportable plans, and
- Reducing the time between permit submittal and construction mobilization.

One of the major benefits of the *International Codes* (I-Codes) is the establishment of consistent code provisions for states and local jurisdictions to enforce. Consistent enforcement is very important to national corporations and franchise chains that wish to construct similar or identical buildings across the country. The review process by an approved agency will greatly enhance the consistency of code application for replicable buildings and support consistent enforcement of the I-Codes while maintaining local jurisdictional authority.

Implementing a building document review process to examine and verify the replicable construction documents comply with the current applicable I-Codes could save considerable state and local resources and time by eliminating repetitive code-compliance reviews. Local jurisdictions can then utilize their resources to focus on reviews of complex and high-risk projects. By coupling a global review of replicable documents with a local review of unique jurisdictional requirements that differ from

Appendix N continues

Appendix N continued

the applicable I-Code, replicable buildings that utilize this optional regulatory review process will be constructed more cost effectively and with greater consistency. Allowing owners, architects, builders, and engineers to submit a design, with its various facades and options, will also allow for more thorough plan examinations, reduced construction times, fewer in-field change orders, more meaningful inspections, and quicker occupancy of the finished building.

The application of a replicable design shall be consistent in all of the fundamental code areas, such as fire- and life-safety, accessibility, structural stability, and energy efficiency, as well as plumbing, electrical, and mechanical systems. The proposed design criteria as submitted must be reviewed by an approved individual or agency for every code discipline reviewed. Site-specific issues must be identified and approved, including variations from the replicable design, local requirements, and foundation design.

PART 9

2018 International Existing Building Code (IEBC)

Chapters 1 through 16 and Appendices A through C

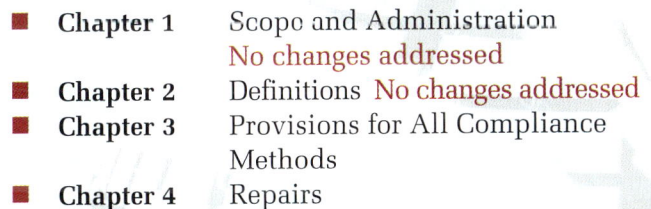

- **Chapter 1**   Scope and Administration
  No changes addressed
- **Chapter 2**   Definitions No changes addressed
- **Chapter 3**   Provisions for All Compliance Methods
- **Chapter 4**   Repairs
- **Chapter 5**   Prescriptive Compliance Method
- **Chapter 6**   Classification of Work
  No changes addressed
- **Chapter 7**   Alterations—Level 1
- **Chapter 8**   Alterations—Level 2
- **Chapter 9**   Alterations—Level 3
- **Chapter 10**   Change of Occupancy
- **Chapter 11**   Additions
- **Chapter 12**   Historic Buildings
  No changes addressed
- **Chapter 13**   Performance Compliance Methods
- **Chapter 14**   Relocated or Moved Buildings
- **Chapter 15**   Construction Safeguards
  No changes addressed
- **Chapter 16**   Reference Standards
  No changes addressed
- **Appendix A**   Guidelines for the Seismic Retrofit of Existing Buildings
- **Appendix B**   Supplementary Accessibility Requirements for Existing Buildings and Facilities
  No changes addressed
- **Appendix C**   Gable End Retrofit for High-Wind Areas
  No changes addressed

Applicable to all existing buildings, the IEBC is intended to provide flexibility to permit the use of alternative approaches to achieve compliance with minimum requirements to safeguard the public health, safety, and welfare. Both structural and life safety changes are addressed. The IEBC covers existing buildings by creating three paths for compliance. Chapter 3 gives an overview of the paths.

The compliance paths include prescriptive compliance, classification of work, and performance compliance. Prescriptive compliance requirements are found in Chapter 5 while performance compliance is covered in Chapter 13. Classification of work is found in Chapters 6 through 11, while more focused topics on repairs, historic buildings, moved buildings, and construction safeguards are found in Chapters 4, 12, 14, and 15 respectively. The appendices supply additional requirements and information on seismic retrofit, high wind retrofit, and accessibility. ■

IEBC 303.1
Live Loads

IEBC 303.3.2, IEBC APPENDIX A5
Earthquake Hazard Reduction in Existing Concrete Buildings

IEBC 305; CHAPTERS 4, 5, 6, 13, 14
Reorganization

IEBC 405.2.1.1
Snow Damage

IEBC 502.4
Loading of Existing Structural Elements

IEBC 502.7, 503.15, 804, 1105
Carbon Monoxide Detectors in Group I-1, I-2, I-4, and R Occupancies

IEBC 502.8, 1106, 1301.2.3.1
Storm Shelters in Group E Additions

IEBC 503.7
Anchorage for Concrete and Reinforced Masonry Walls

IEBC 503.10
Anchorage of Unreinforced Masonry Partitions

IEBC 505.4, 701.4
Emergency Escape Opening Operation

IEBC 506.4
Structural Loads

IEBC 507.4
Structural Loads in Historic Buildings

IEBC 805.3.1.1
Single-Exit Buildings

IEBC 904.1.4
Automatic Sprinkler System at Floor of Alteration

IEBC 906.7
Anchorage of Unreinforced Masonry Partitions

IEBC 1006
Seismic Loads and Access to Risk Category IV Structures

IEBC 1103
Changes to Gravity and Lateral Loads with an Addition

IEBC 303.1
Live Loads

CHANGE TYPE: Modification

CHANGE SUMMARY: Requirements for live loads from Chapters 4 and 8 have been combined and placed in Chapter 3 as they apply for all compliance methods.

2018 CODE:

SECTION 303
Structural Design Loads and Evaluation and Design Procedures

303.1 Live Loads <u>Where an addition or alteration does not result in increased design live load, existing gravity load-carrying structural elements shall be permitted to be evaluated and designed for live loads approved prior to the addition or alteration. If the approved live load is less than that required by Section 1607 of the *International Building Code*, the area designated for the nonconforming live load shall be posted with placards of approved design indicating the approved live load. Where the addition or alteration results in increased design live load, the live load required by Section 1607 of the *International Building Code* shall be used.</u>

~~**402.3.1 Design live load.** Where the addition does not result in increased design live load, existing gravity load-carrying structural elements shall be permitted to be evaluated and designed for live loads approved prior to the addition. If the approved live load is less than that required by Section 1607 of the *International Building Code*, the area designed for the nonconforming live load shall be posted with placards of approved design indicating the approved live load. Where the addition does result in increased design live load, the live load required by Section 1607 of the *International Building Code* shall be used.~~

~~**403.3.1 Design live load.** Where the alteration does not result in increased design live load, existing gravity load-carrying structural elements shall be permitted to be evaluated and designed for live loads approved prior to the alteration. If the approved live load is less than that required by Section 1607 of the *International Building Code*, the area designed for the nonconforming live load shall be posted with placards of approved design indicating the approved live load. Where the alteration does result in increased design live load, the live load required by Section 1607 of the *International Building Code* shall be used.~~

~~**807.3 Minimum design loads.** The minimum design loads on existing elements of a structure that do not support additional loads as a result of an alteration shall be the loads applicable at the time the building was constructed.~~

CHANGE SIGNIFICANCE: The *2015 International Existing Building Code* (IEBC) maintained three separate provisions for minimum live loads for additions and alterations using the prescriptive and work area methods. The provisions allowed live loads smaller than those stated in the 2015 IBC Table 1607 listing minimum live loads. The 2015 IEBC Section 807.3 title also appeared to cover all load types but only addressed live loads. Snow, wind, and earthquake loads were all addressed more specifically by other provisions, especially in the work area method to which Section 807.3 applied.

IEBC 303.1 continues

IEBC 303.1 continued

These provisions are now combined and have been moved into the 2018 IEBC Chapter 3, which has become the location for provisions which apply to all three compliance methods. Contradictory text found in multiple sections and contradictory references to multiple sections have been deleted.

The new provision replaces Sections 402.3.1 and 403.3.1. The concept of "previously approved design live load" from Sections 402.3.1 and 403.3.1 is retained in Section 303.1. This ensures that a comparison is made between the new intended live load and the original live load which may be less than is required in the 2018 IBC Table 1607.1 live loads for new buildings. Additionally, the placard requirement has been retained.

In making these changes, the methodology for use of the IEBC has been revised. The work flow for structural design now includes the following steps:

1. Determine project type—alteration, addition, repair, change of occupancy.
2. Go to a specific project type topic or pick your preferred method—prescriptive, performance or work area.
3. If multiple building changes exist in one project (like an alteration being done together with a repair), use the sections focused on each change as appropriate for the project.

All design methods now use the same provisions for repairs, for relocations, and for accessibility independent of method or project type. For additions, alterations, and change of occupancy, a preferred compliance method must be chosen.

Live loads

IEBC 303.3.2, IEBC Appendix A5
Earthquake Hazard Reduction in Existing Concrete Buildings

CHANGE TYPE: Deletion

CHANGE SUMMARY: In order to clarify and simplify concrete evaluation and retrofit, ASCE 41 continues to be directly referenced while Appendix Chapter A5, Earthquake Hazard Reduction in Existing Concrete Buildings, is deleted from the 2018 IEBC.

2018 CODE: ~~301.1.4.2~~ **303.3.2 Compliance with reduced** ~~International Building Code~~-**level seismic forces.** Where seismic evaluation and design is permitted to ~~meet~~ use reduced seismic forces, the criteria used shall be in accordance with one of the following:

1. The *International Building Code* using 75 percent of the prescribed forces. Values of R, Ω_0 and C_d used for analysis shall be as specified in Section ~~301.1.4.1~~ 303.3.1 of this code.

2. Structures or portions of structures that comply with the requirements of the applicable chapter in Appendix A as specified in Items 2.1 through ~~2.5~~ 2.4 and subject to the limitations of the respective Appendix A chapters shall be deemed to comply with this section.
   2.1. The seismic evaluation and design of unreinforced masonry bearing wall buildings in Risk Category I or II are permitted to be based on the procedures specified in Appendix Chapter A1.
   2.2. Seismic evaluation and design of the wall anchorage system in reinforced concrete and reinforced masonry wall buildings with flexible diaphragms in Risk Category I or II are permitted to be based on the procedures specified in Chapter A2.
   2.3. Seismic evaluation and design of cripple walls and sill plate anchorage in residential buildings of light-frame wood construction in Risk Category I or II are permitted to be based on the procedures specified in Chapter A3.
   2.4. Seismic evaluation and design of soft, weak, or open-front wall conditions in multiple unit residential buildings of wood construction in Risk Category I or II are permitted to be based on the procedures specified in Chapter A4.
   ~~2.5. Seismic evaluation and design of concrete buildings assigned to Risk Category I, II or III are permitted to be based on the procedures specified in Chapter A5.~~

3. ASCE 41, using the performance objective in Table ~~302.1.4.2~~ 303.3.2 for the applicable risk category.

APPENDIX A Guidelines for the Seismic Retrofit of Existing Buildings

~~**CHAPTER A5** —EARTHQUAKE HAZARD REDUCTION IN EXISTING CONCRETE BUILDINGS~~

(Deleted in its entirety.)

IEBC 303.3.2, IEBC Appendix A5 continues

IEBC 303.3.2, IEBC Appendix A5
continued

CHANGE SIGNIFICANCE: In the *International Existing Building Code* (IEBC), there are multiple optional procedures to reduce seismic loads. These options include Appendix A, Guidelines for the Seismic Retrofit of Existing Buildings. Within Appendix A of the 2015 IEBC are five chapters:

- Chapter A1—Seismic Strengthening Provisions for Unreinforced Masonry Bearing Wall Buildings
- Chapter A2—Earthquake Hazard Reduction in Existing Reinforced Concrete and Reinforced Masonry Wall Buildings with Flexible Diaphragms
- Chapter A3—Prescriptive Provisions for Seismic Strengthening of Cripple Walls and Sill Plate Anchorage of Light, Wood-Frame Residential Buildings
- Chapter A4—Earthquake Risk Reduction in Wood-Frame Residential Buildings with Soft, Weak, or Open Front Walls
- Chapter A5—Earthquake Hazard Reduction in Existing Concrete Buildings

Chapter A5 has been a useful, simplified procedure for concrete evaluation and retrofit. However, a reference to ASCE 41 already exists in Section 301.1.4 of the IEBC. The 2015 IEBC Chapter A5 prescriptive requirements plus a pointer to ASCE 41 created some confusion with contradictory references within the Chapter.

With the update of ASCE 41, *Seismic Evaluation and Retrofit of Existing Buildings*, written to systematically determine when to reduce new construction seismic loads for existing buildings, Chapter A5 has been deleted from the 2018 IEBC. There continues to be a path to ASCE 41 in Section 303.3.2 for concrete buildings with no contradictory provisions.

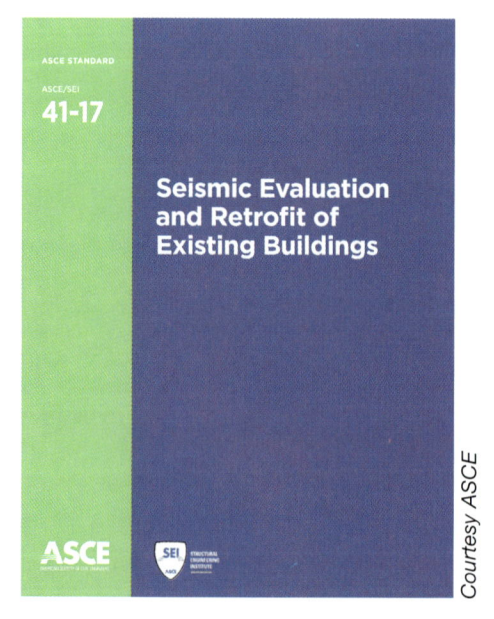

Seismic evaluation and retrofit standard

IEBC 305; Chapters 4, 5, 6, 13, 14
Reorganization

CHANGE TYPE: Clarification

CHANGE SUMMARY: Section 410, Accessibility, has been relocated to a new Section 305. Chapters 4, 5, 6, 13, and 14 have been relocated, resulting in a reorganization of the provisions and new chapter numbering.

2018 CODE:

SECTION ~~410~~ 305
ACCESSIBILITY FOR EXISTING BUILDINGS

CHAPTER ~~6~~ 4
REPAIRS

CHAPTER ~~4~~ 5
PRESCRIPTIVE COMPLIANCE METHOD

CHAPTER ~~5~~ 6
CLASSIFICATION OF WORK

CHAPTER ~~13~~ 14
RELOCATED OR MOVED BUILDINGS

CHAPTER ~~14~~ 13
PERFORMANCE COMPLIANCE METHODS

CHANGE SIGNIFICANCE: A new Section 305 has been created and all accessibility provisions of 2015 IEBC Section 410 have been moved to this new section. All accessibility requirements for existing buildings are now placed in one section, allowing a focused and clear set of requirements

IEBC 305; Chapters 4, 5, 6, 13, 14 continues

TABLE IEBC 3-1

IEBC Chapter 2018	IEBC Chapter 2015	Title
1	1	Scope and Administration
2	2	Definitions
3	3	Provisions for All Compliance Methods
4	6	Repairs
5	4	Prescriptive Compliance Method
6	5	Classification of Work
7	7	Alterations—Level 1
8	8	Alterations—Level 2
9	9	Alterations—Level 3
10	10	Change of Occupancy
11	11	Additions
12	12	Historic Buildings
13	14	Performance Method
14	13	Relocated or Moved Buildings
15	15	Construction Safeguards

IEBC 305; Chapters 4, 5, 6, 13, 14 continued

for users to understand. The intent of this change is a reorganization of accessibility provisions to avoid duplication of the same requirements in multiple code chapters. The 2015 IEBC prescriptive and work area methods have separate but generally consistent accessibility requirements, which now are fully consistent and found in Section 305. Additionally, the "Repair" chapter has been relocated to Chapter 4 and the text on Repairs in the "Classification of Work" chapter has been deleted, making it clear that repair activities apply to all types of construction in existing buildings and are not just part of the work area method. Similarly, the chapter on "Relocated Buildings" has moved to Chapter 14, as the text and reference to it have been deleted from the prescriptive, performance, and work area methods. This relocation and change clarifies that relocated buildings are not necessarily a part of other activities in existing buildings and they stand on their own.

IEBC 405.2.1.1
Snow Damage

CHANGE TYPE: Addition

CHANGE SUMMARY: Structural components damaged by snow events must be repaired assuming snow loads for new buildings using the IBC.

2018 CODE: **405.2.1.1 Snow damage** <u>Structural components whose damage was caused by or related to snow load effects shall be repaired, replaced, or altered to satisfy the requirements of Section 1608 of the *International Building Code.*</u>

CHANGE SIGNIFICANCE: Design snow loads, especially with the effects of climate change, are increasing in many areas. Existing framing carrying dead and live loads generally does not require upgrading, even in those cases where it is nonconforming, because it has a history of adequate service. Design-level snow loads do not have that history. And unlike wind or earthquake loads, snow loads at damaging or design levels are likely to occur frequently. Thus, deficient components that are repaired to existing levels may be damaged again during future winter conditions.

In the 2018 IEBC, Chapter 4 has become a chapter focused on repairs and addresses repairs using any compliance method. Chapter 7, Level 1 Alterations, continues to cover roof replacement and recover using the work area method.

In Section 405.2.1.1 a limited upgrade requirement for structural damage caused by snow has been added. Instead of allowing repair to the pre-damage condition, any repaired or replaced elements are required to be designed following provisions for new construction. Similar elements in the structure that did not sustain damage are not required to be upgraded.

Damage to roof due to large snow loads

IEBC 502.4
Loading of Existing Structural Elements

CHANGE TYPE: Modification

CHANGE SUMMARY: The provisions covering loading of existing structural elements where an addition is built have been modified and a new exception for buildings designed using the *International Residential Code* has been added.

2018 CODE: ~~402.3~~ <u>502.4</u> **Existing structural elements carrying gravity load.** Any existing gravity load-carrying structural element for which an addition and its related alterations cause an increase in design ~~gravity~~ <u>dead, live, or snow</u> load, <u>including snow drift effects,</u> of more than 5 percent shall be ~~strengthened, supplemented,~~ replaced or ~~otherwise~~ altered as needed to carry the ~~increased~~ gravity ~~load~~ <u>loads</u> required by the *International Building Code* for new structures. Any existing gravity load-carrying structural element whose ~~gravity~~ <u>vertical</u> load-carrying capacity is decreased <u>as part of the addition and its related alterations</u> shall be considered an altered element subject to the requirements of Section ~~403.3~~ <u>503.3</u>. Any existing element that will form part of the lateral load path for any part of the addition shall be considered <u>to be</u> an existing lateral load-carrying structural element subject to the requirements of Section ~~402.4~~ <u>502.5</u>.

> **Exception:** <u>Buildings of Group R occupancy with not more than five dwelling or sleeping units used solely for residential purposes where the existing building and the addition together comply with the conventional light-frame construction methods of the *International Building Code* or the provisions of the *International Residential Code.*</u>

Addition of a hospital entrance

~~**402.3.1 Design live load.** Where the addition does not result in increased design live load, existing gravity load-carrying structural elements shall be permitted to be evaluated and designed for live loads approved prior to the addition. If the approved live load is less than that required by Section 1607 of the *International Building Code*, the area designed for the nonconforming live load shall be posted with placards of approved design indicating the approved live load. Where the addition does result in increased design live load, the live load required by Section 1607 of the *International Building Code* shall be used.~~

CHANGE SIGNIFICANCE: The provisions for existing structural elements carrying gravity loads in Chapters 4, 8, and 11 are frequently redundant. The 2015 IEBC Sections 402.3.1 and 403.3.1, Design Live Loads, in the prescriptive method are reconciled with Section 807.3, Minimum Design Loads, in the work area method by deleting all three sections and placing the provision for live loads into Chapter 3. The 2018 IEBC Chapter 3 has become the location for provisions which apply to all three compliance methods. Contradictory text is no longer located in multiple sections; contradictory references are deleted. The prescriptive compliance method requirements are moved from Chapter 4 into Chapter 5, while Chapter 4 is labeled Repairs and addresses repairs using any compliance method.

2018 IEBC Section 502.4 focuses on elements which require an alteration due to the creation of an addition. Language is updated to clarify that existing elements that have decreased load-carrying capacity due to the creation of an addition must be considered altered elements and meet the requirements of Section 503.3. Additionally, the section now explicitly requires snow loads, including consideration of snow drifts, and live and dead loads to be part of the gravity load considered during analysis of an element.

IEBC 502.7, 503.15, 804, 1105

Carbon Monoxide Detectors in Group I-1, I-2, I-4, and R Occupancies

CHANGE TYPE: Addition

CHANGE SUMMARY: Carbon monoxide provisions have been added in the Prescriptive Method Additions, Alterations Level 2 Additions, and Additions for Group I-1, I-2, I-4, and R occupancies.

2018 CODE: <u>**502.7 Carbon monoxide alarms in existing portions of a building.** Where an addition is made to a building or structure of Group I-1, I-2, I-4 or R occupancy, the existing building shall be provided with carbon monoxide alarms in accordance with Section 1103.9 of the *International Fire Code* or Section R315 of the *International Residential Code*, as applicable.</u>

<u>**Exceptions:**</u>

<u>1. Work involving the exterior surfaces of buildings, such as the replacement of roofing or siding, or the addition or replacement of windows or doors, or the addition of porches or decks.</u>

<u>2. Installation, alteration or repairs of plumbing or mechanical systems, other than fuel-burning appliances.</u>

<u>**503.15 Carbon monoxide alarms.** Carbon monoxide alarms shall be provided to protect sleeping units and dwelling units in Group I-1, I-2, I-4 and R occupancies in accordance with Section 1103.9 of the *International Fire Code*.</u>

<u>**Exceptions:**</u>

<u>1. Work involving the exterior surfaces of buildings, such as the replacement of roofing or siding, or the addition or replacement of windows or doors, or the addition of porches or decks.</u>

Carbon monoxide detector

Combined smoke and carbon monoxide detector

2. Installation, alteration or repairs of plumbing or mechanical systems, other than fuel-burning appliances.

SECTION 804
CARBON MONOXIDE DETECTION

804.1 Carbon monoxide alarms. Any work area in Group I-1, I-2, I-4 and R occupancies shall be equipped with carbon monoxide alarms in accordance with Section 1103.9 of the *International Fire Code*.

Exceptions:
1. Work involving the exterior surfaces of buildings, such as the replacement of roofing or siding, or the addition or replacement of windows or doors, or the addition of porches or decks.
2. Installation, alteration or repairs of plumbing or mechanical systems, other than fuel-burning appliances.

SECTION 1105
CARBON MONOXIDE ALARMS IN
GROUPS I-1, I-2, I-4 AND R

1105.1 Carbon monoxide alarms in existing portions of a building. Where an addition is made to a building or structure of a Group I-1, I-2, I-4 or R occupancy, the existing building shall be equipped with carbon monoxide alarms in accordance with Section 1103.9 of the *International Fire Code* or Section R315 of the *International Residential Code*, as applicable.

CHANGE SIGNIFICANCE: The inclusion of provisions for carbon monoxide alarms in the IEBC makes the IEBC coordinated and consistent with the IFC and IRC. Section 1103.9 of the IFC requires carbon monoxide alarms in existing Group I-1, I-2, I-4, and R occupancies in accordance with IFC Section 915. The IRC has provisions for such alarms in Section R315. Reference is made from the IEBC to applicable sections of the IFC or IRC to keep the requirements consistent between the IEBC and the IFC or IRC depending on the use under consideration. IFC Section 1103.9 requires carbon monoxide alarms retroactively in all existing buildings of Group I-1, I-2, I-4, and R occupancy, regardless of any alteration or addition taking place; however, it does allow such alarms to be battery-operated, and the same would be true when applying the IEBC.

IEBC 502.8, 1106, 1301.2.3.1

Storm Shelters in Group E Additions

CHANGE TYPE: Addition

CHANGE SUMMARY: Where storm shelters are required based on IBC and ICC 500 for Group E occupancies, any addition to such existing occupancies where the occupant load of the addition is 50 or more will trigger the construction of a storm shelter.

2018 CODE: <u>**502.8 Additions to Group E facilities.** For additions to Group E occupancies, storm shelters shall be provided in accordance with Section 1106.1.</u>

<u>**SECTION 1106**</u>
<u>**STORM SHELTERS**</u>

<u>**1106.1 Addition to a Group E occupancy.** Where an addition is added to an existing Group E Occupancy located in an area where the shelter design wind speed for tornados is 250 mph in accordance with Figure 304.2(1) of ICC 500 and the occupant load in the addition is 50 or more, the addition shall have a storm shelter constructed in accordance with ICC 500.</u>

<u>**Exceptions:**</u>

1. <u>Group E day care facilities.</u>
2. <u>Group E occupancies accessory to places of religious worship.</u>
3. <u>Additions meeting the requirements for shelter design in ICC 500.</u>

<u>**1106.1.1 Required occupant capacity.** The required occupant capacity of the storm shelter shall include all the buildings on the site, and shall be the greater of the following:</u>

1. <u>The total occupant load of the classrooms, vocational rooms and offices in the Group E occupancy.</u>
2. <u>The occupant load of any indoor assembly space that is associated with the Group E occupancy.</u>

Group E, educational facility destruction due to tornado

Exceptions:
1. Where an addition is being added on an existing Group E site, and where the addition is not of sufficient size to accommodate the required occupant capacity of the storm shelter for all the buildings on the site, the storm shelter shall at a minimum accommodate the required capacity for the addition.
2. Where approved by the code official, the required occupant capacity of the shelter shall be permitted to be reduced by the occupant capacity of any existing storm shelters on the site.

1106.1.2 Location. Storm shelters shall be located within the buildings they serve, or shall be located where the maximum distance of travel from not fewer than one exterior door of each building to a door of the shelter serving that building does not exceed 1000 ft. (305 m).

1301.2.3.1 Additions to Group E facilities. For additions to Group E occupancies, storm shelters shall be provided in accordance with Section 1106.1.

CHANGE SIGNIFICANCE: The IBC requires storm shelters be built in Group E occupancies where there is a high risk of severe tornados. The determination, design, and construction must be in accordance with ICC 500, *ICC/NSSA Standard for the Design and Construction of Storm Shelters*. The IBC, however, does not address storm shelter requirements for additions to existing school buildings. The need for the same level of protection in existing educational facilities for schoolchildren, staff, and teachers resulted in provisions now found in new IEBC Section 1106. This section provides the criteria for storm shelters in existing Group E occupancy additions. Additions that have an occupant load of 50 or more are accordingly required to incorporate a storm shelter in the same geographic areas where new school buildings would require a storm shelter. As always, if the entire addition meets ICC 500, there is not a need for a separate storm shelter.

IEBC 503.7
Anchorage for Concrete and Reinforced Masonry Walls

CHANGE TYPE: Addition

CHANGE SUMMARY: For alterations of buildings located in higher seismic areas, when a work area includes more than one-half of the building's floor area, wall anchors must be installed at the roof line along concrete and reinforced masonry walls.

2018 CODE: <u>**503.7 Anchorage for concrete and reinforced masonry walls.** Where the work area exceeds 50 percent of the building area, the building is assigned to Seismic Design Category C, D, E or F, and the building's structural system includes concrete or reinforced masonry walls with a flexible roof diaphragm, the alteration work shall include installation of wall anchors at the roof line, unless an evaluation demonstrates compliance of existing wall anchorage. Use of reduced seismic forces shall be permitted.</u>

CHANGE SIGNIFICANCE: In the 2015 IEBC, the work area method has a comprehensive provision in Section 907.4.5 that requires roof-to-wall anchors in Level 3 alterations for concrete and reinforced masonry walls as well as unreinforced masonry walls (URMs). The prescriptive method addresses only URM walls in Section 403.6. Anchorage should also be placed in reinforced walls as well; concrete tilt-up and reinforced masonry walls need to have anchorage added during a retrofit, when a building was not designed with anchors at the roof and floors.

In the 2018 IEBC, Section 503 is updated to include anchorage of reinforced concrete and masonry walls in the prescriptive compliance section on alterations when the work area will include more than half the building area. This update to provisions in the 2018 IEBC resolves the inconsistency between the work area and prescriptive compliance methods.

Positive anchorage of roof to wall

IEBC 503.10
Anchorage of Unreinforced Masonry Partitions

CHANGE TYPE: Addition

CHANGE SUMMARY: A mitigation trigger has been added to the 2018 IEBC to address a common nonstructural falling hazard: unreinforced masonry partitions.

2018 CODE: <u>**503.10 Anchorage of unreinforced masonry partitions in major alterations.** Where the work area exceeds 50 percent of the building area, and where the building is assigned to Seismic Design Category C, D, E, or F, unreinforced masonry partitions and nonstructural walls within the work area and adjacent to egress paths from the work area shall be anchored, removed, or altered to resist out-of-plane seismic forces, unless an evaluation demonstrates compliance of such items. Use of reduced seismic forces shall be permitted.</u>

CHANGE SIGNIFICANCE: In the 2015 IEBC, both the prescriptive and work area methods include mitigation requirements for unreinforced masonry (URM) parapets and bearing walls, triggered by major alterations (Level 3). Alteration projects involving 50% of a building's area will typically trigger parapet removal or replacement in their scope. A related, unaddressed hazard involves the failure of interior unreinforced masonry partitions, especially around stairwells and in egress corridors. Mitigation of this well-understood and common hazard is justified in a Level 3 alteration.

The 2018 IEBC adds a mitigation trigger to address this common nonstructural falling hazard. The trigger is only intended to apply to URM nonstructural walls (nonbearing walls), not to gypsum bearing and nonbearing walls within or outside the work area. Additionally, the provision intends to only require mitigation within the work area and along egress paths from the work area to building exits.

Retrofited brick wall

IEBC 505.4, 701.4

Emergency Escape Opening Operation

CHANGE TYPE: Addition

CHANGE SUMMARY: Emergency escape and rescue openings are required to be operational. Related provisions for being operational have been added to the Prescriptive Compliance Method and Alterations Level 1.

2018 CODE: <u>**505.4 Emergency escape and rescue openings.** Emergency escape and rescue openings shall be operational from the inside of the room without the use of keys or tools. Bars, grilles, grates or similar devices are permitted to be placed over emergency escape and rescue openings provided the minimum net clear opening size complies with the code that was in effect at the time of construction and such devices shall be releasable or removable from the inside without the use of a key, tool or force greater than that which is required for normal operation of the escape and rescue opening. Where such bars, grilles, grates or similar devices are installed, they shall not reduce the net clear opening of the emergency escape and rescue openings. Smoke alarms shall be installed in accordance with Section 907.2.10 of the *International Building Code* regardless of the valuation of the alteration.</u>

<u>**701.4 Emergency escape and rescue openings.** Emergency escape and rescue openings shall be operational from the inside of the room without the use of keys or tools. Bars, grilles, grates or similar devices placed over emergency escape and rescue openings shall comply with the minimum net clear opening size required by the code that was in effect at the time of construction. Such devices shall be releasable or removable from the inside without the use of a key, tool or force greater than that which is required for normal operation of the escape and rescue opening. Where such bars, grilles, grates or similar devices are installed, they shall not reduce the net clear opening of the emergency escape and</u>

Emergency escape and rescue opening operation in existing buildings rescue openings. Smoke alarms shall be installed in accordance with Section 907.2.10 of the *International Building Code* regardless of the valuation of the alteration.

CHANGE SIGNIFICANCE: Emergency escape and rescue openings under the IBC and IRC are required to be operational without the use of a key, special knowledge, or effort; however, this very important criterion is not found in the 2015 IEBC prescriptive compliance and work area methods. This change introduces in the IEBC the same requirements found in the IBC and IRC for consistency between the I-Codes as this is a critical life-safety feature in all buildings. The criteria for the operation of such openings are that they should remain operational and that any bars, grilles or grates are releasable or removable from the inside without the use of a key, tool or force greater than the opening would normally require. Sections 505.2, 505.3, 702.4, and 702.5 address other aspects of emergency egress and rescue openings related to the operation of such windows. The specific criteria added are in Prescriptive Compliance Method and Alterations Level 1; however, the same requirements would apply in Alterations Levels 2 and 3 due to the cascading effect that require compliance with the provisions of Alterations Level 1.

IEBC 506.4
Structural Loads

CHANGE TYPE: Modification

CHANGE SUMMARY: Buildings undergoing a change of occupancy shall have live, snow, wind, and seismic loads checked against design loads based on IBC level forces.

2018 CODE: <u>**506.4 Structural.** Any building undergoing a change of occupancy shall satisfy the requirements of this section.</u>

<u>**506.4.1 Live loads.** Structural elements carrying tributary live loads from an area with a change of occupancy shall satisfy the requirements of Section 1607 of the *International Building Code.* Design live loads for areas of new occupancy shall be based on Section 1607 of the *International Building Code.* Design live loads for other areas shall be permitted to use previously approved design live loads.</u>

> <u>**Exception:** Structural elements whose demand-capacity ratio considering the change of occupancy is not more than 5 percent greater than the demand-capacity ratio based on previously approved live loads need not comply with this section.</u>

<u>**506.4.2 Snow and wind loads.** Where a change of occupancy results in a structure being assigned to a higher risk category, the structure shall satisfy the requirements of Sections 1608 and 1609 of the *International Building Code* for the new risk category.</u>

> <u>**Exception:** Where the area of the new occupancy is less than 10 percent of the building area, compliance with this section is not required. The cumulative effect of occupancy changes over time shall be considered.</u>

<u>**506.4.3 Seismic loads (seismic-force-resisting system).**</u> ~~**407.4 Structural. When**~~ <u>Where a change of occupancy results in a building being reclassified to a higher risk category, the building shall satisfy the</u>

Brick warehouse renovated and used for business occupancy

requirements of Section 1613 of the *International Building Code* for the new risk category using full seismic forces. ~~The structure shall conform to the seismic requirements for a new structure of the higher risk category. For purposes of this section, compliance with ASCE 41, using a Tier 3 procedure and the two-level performance objective in Table 301.1.4.1 for the applicable risk category, shall be deemed to meet the requirements of Section 1613 of the *International Building Code*. All work shall comply with the structural provisions of Section 1007.~~

Exceptions:

~~Specific seismic detailing requirements of Section 1613 of the *International Building Code* for a new structure shall not be required to be met where the seismic performance is shown to be equivalent to that of a new structure. A demonstration of equivalence shall consider the regularity, overstrength, redundancy and ductility of the structure.~~

1. Where the area of the new occupancy is less than 10 percent of the building area and the new occupancy is not assigned to Risk Category IV, compliance with this section is not required. The cumulative effect of occupancy changes over time shall be considered.

2. ~~When~~ Where a change of use results in a structure being reclassified from Risk Category I or II to Risk Category III and the structure is located where the seismic coefficient, S_{DS}, is less than 0.33, compliance with this section ~~the seismic requirements of Section 1613 of the *International Building Code*~~ is not required.

3. Unreinforced masonry bearing wall buildings assigned to Risk Category III, and to Seismic Design Category A or B shall be permitted to use Appendix Chapter A1 of this code.

506.4.4 Access to Risk Category IV. Any structure that provides operational access to an adjacent structure assigned to Risk Category IV as the result of a change of occupancy shall itself satisfy the requirements of Sections 1608, 1609, and 1613 of the *International Building Code*. For compliance with Section 1613, *International Building Code*-level seismic forces shall be used. Where operational access to the Risk Category IV structure is less than 10 feet (3048 mm) from either an interior lot line or from another structure, access protection from potential falling debris shall be provided.

CHANGE SIGNIFICANCE: In the 2015 IEBC, the prescriptive method has only one load-specific structural provision related to change of occupancy, Section 407.4, which triggers a seismic upgrade when the risk category increases. Other upgrades are triggered generally by Section 407.1, which requires any building with change of occupancy to meet all requirements for the new occupancy. The 2015 IEBC work area method contains different requirements for change of occupancy.

To reconcile requirements in the prescriptive compliance and work area methods, the 2018 IEBC provisions substantially reduced upgrade requirements for wind and snow loads by loosening the trigger and adding an exception applicable to those cases where the area of the change of

IEBC 506.4 continues

IEBC 506.4 continued

occupancy is less than 10% of the total building area. Provisions in Section 506 now match the updated provisions of Section 1006, structural loads for change of occupancy using the work area method. Additionally, a new exception for unreinforced masonry buildings in low-seismic regions adds a pointer to Appendix A for an alternate design path.

A new provision, mirroring Section 1006.4, requires consideration of access to a Risk Category IV structure which must be maintained during and after a wind or snow event as well as after an earthquake.

IEBC 507.4
Structural Loads in Historic Buildings

CHANGE TYPE: Addition

CHANGE SUMMARY: In Chapter 5, the prescriptive compliance method, structural requirements for historic buildings are added.

2018 CODE: 507.4 Structural. Historic buildings shall comply with the applicable structural provisions in this chapter.

Exceptions:
1. The code official shall be authorized to accept existing floors and existing live loads and to approve operational controls that limit the live load on any floor.
2. Repair of substantial structural damage is not required to comply with Sections 405.2.3 and 405.2.4. Substantial structural damage shall be repaired in accordance with Section 405.2.1.

CHANGE SIGNIFICANCE: In the 2015 IEBC, there is a significant difference between the prescriptive and work area methods for consideration of structural loads for historic buildings. In the prescriptive method, Section 408.1 states that improvements to the existing building need be made only if they are specifically required. The balance of Section 408 has no specific structural checks or upgrade triggers, not even for added dead load or removal of a structural element, so depending upon the interpretation, Section 408.1 states that historic buildings are exempt from any structural work. By contrast, in the work area method, Section 1206.1 indicates specifically that the code's structural provisions *do* apply to historic buildings.

A 2018 IEBC code change reconciles the prescriptive compliance method and the work area method, clarifying that the Chapter 5 structural requirements are safety-related and should be enforced in historic buildings. The wording of new Section 507.4 comes from Section 1205.1,

IEBC 507.4 continues

Historic building: WSU library and clock tower, Pullman, Washington

IEBC 507.4 continued

Historic building: Sacred Heart Church, Dayton, Ohio

where historic buildings are subject to the same structural upgrade triggers as non-historic buildings. Then exceptions are allowed where the upgrade will threaten the historic significance of the building. The use of the exceptions requires an alternate path to maintain the safety of the public.

It is often debated whether historic structures should be exempt from the code's wind and seismic upgrade triggers. As a minimum, checks of dead, live and snow loads must be done to confirm adequacy when the structure is altered. This new provision allows the use of operational controls (in other words, restriction of the public and signage stating loading limits to be used) rather than altering a building and losing its historic significance in the process.

Renovated historic building: warehouse to condos, Portland, Oregon

IEBC 805.3.1.1
Single-Exit Buildings

CHANGE TYPE: Modification

CHANGE SUMMARY: Single-exit buildings and spaces under Alteration Levels 2 and 3 have been modified to be more consistent with the IBC.

2018 CODE: ~~**805.3.1.1 Single-exit buildings.** Only one exit is required from buildings and spaces of the following occupancies:~~

1. ~~In Group A, B, E, F, M, U and S occupancies, a single exit is permitted in the story at the level of exit discharge when the occupant load of the story does not exceed 50 and the exit access travel distance does not exceed 75 feet (22 860 mm).~~

2. ~~Group B, F-2, and S-2 occupancies not more than two stories in height that are not greater than 3,500 square feet per floor (326 m2), when the exit access travel distance does not exceed 75 feet (22 860 mm). The minimum fire-resistance rating of the exit enclosure and of the opening protection shall be 1 hour.~~

3. ~~Open~~ parking ~~structures where vehicles are mechanically parked.~~

4. ~~In Group R-4 occupancies, the maximum occupant load excluding staff is 16.~~

5. ~~Groups R-1 and R-2 not more than two stories in height, when there are not more than four dwelling units per floor and the exit access travel distance does not exceed 50 feet (15 240 mm). The minimum fire-resistance rating of the exit enclosure and of the opening protection shall be 1 hour.~~

6. ~~In multilevel dwelling~~ units in ~~buildings of occupancy Group R-1 or R-2, an exit shall not be required from every level of the dwelling unit provided that one of the following conditions is met:~~
   6.1. ~~The travel distance within the dwelling unit does not exceed 75 feet (22 860 mm); or~~
   6.2. ~~The building is not more than three stories in height and all third-floor space is part of one or more dwelling units located in part on the second floor; and no habitable room within any such dwelling unit shall have a travel distance that exceeds 50 feet (15 240 mm) from the outside of the habitable room entrance door to the inside of the entrance door to the dwelling unit.~~

7. ~~In Group R-2, H-4, H-5 and I occupancies and in rooming houses and child care centers, a single exit is permitted in a one-story building with a maximum occupant load of 10 and the exit access travel distance does not exceed 75 feet (22 860 mm).~~

8. ~~In buildings of Group R-2 occupancy that are equipped throughout with an automatic fire sprinkler system, a single exit shall be permitted from a basement or story below grade if every dwelling unit on that floor is equipped with an approved window providing a clear opening of at least 5 square feet (0.47 m2) in area, a minimum net clear opening of 24 inches (610 mm) in height and 20 inches (508 mm) in width, and a sill height of not more than 44 inches (1118 mm) above the finished floor.~~

3400 S.F. office building, travel distance 70 feet, single exit

IEBC 805.3.1.1 continues

IEBC 805.3.1.1 continued

9. ~~In buildings of Group R-2 occupancy of any height with not more than four dwelling units per floor; with a smokeproof enclosure or outside stairway as an exit; and with such exit located within 20 feet (6096 mm) of travel to the entrance doors to all dwelling units served thereby.~~

10. ~~In buildings of Group R-3 occupancy equipped throughout with an automatic fire sprinkler system, only one exit shall be required from basements or stories below grade.~~

805.3.1.1 Single-exit buildings. A single exit or access to a single exit shall be permitted from spaces, any story or any occupied roof where one of the following conditions exist:

1. The occupant load, number of dwelling units and exit access travel distance do not exceed the values in Table 805.3.1.1(1) or 805.3.1.1(2).

2. In Group R-1 or R-2, nonsprinklered buildings, individual single-story or multistory dwelling or sleeping units shall be permitted to have a single exit or access to a single exit from the dwelling or sleeping unit provided one of the following criteria are met:
   2.1. The occupant load is not greater than 10 and the exit access travel distance within the unit does not exceed 75 feet (22 860 mm).
   2.2. The building is not more than three stories in height; all third story space is part of dwelling with an exit access doorway on the second story; and the portion of the exit access travel distance from the door to any habitable room within any such unit to the unit entrance doors does not exceed 50 feet (15 240 mm).

3. In buildings of Group R-2 occupancy of any number of stories with not more than four dwelling units per floor served by an interior exit stairway; with a smokeproof enclosure in accordance with Sections 909.20 and 1023.11 of the *International Building Code* or an exterior stairway as an exit; where the portion of the exit access travel distance from the dwelling unit entrance door to the exit not greater than of 20 feet (6096 mm).

TABLE 805.3.1.1(1) Stories with One Exit or Access to One Exit for R-2 Occupancies

Story	Occupancy	Maximum Number of Dwelling Units	Maximum Exit Access Travel Distance (feet)
Basement, First or second story above grade plane	R-2[a]	4 dwelling units	50
Third story above grade plane and higher	NP	NA	NA

For SI: 1 foot = 3048
NP = Not Permitted.
NA = Not Applicable.
a. Group R-2, non-sprinklered and provided with emergency escape and rescue openings in accordance with Section 1030 of the International Building Code.

TABLE 805.3.1.1(2) Stories with One Exit or Access to One Exit for Other Occupancies

Story	Occupancy	Maximum Occupant Load Per Story	Maximum Exit Access Travel Distance (feet)
First story above or below grade plane	B, F-2, S-2[a]	35	75
Second story above grade plane	B, F-2, S-2[a]	35	75
Third story above grade plane and higher	NP	NA	NA

For SI: 1 foot = 3048
NP = Not Permitted.
NA = Not Applicable.
a. The length of exit access travel distance in a Group S-2 open parking garage shall be not more than 100 feet.

CHANGE SIGNIFICANCE: Criteria addressing single-exit buildings and spaces under the IEBC should be similar to the IBC or have more flexibility. The justification for this modification was that the IEBC could be interpreted to be more restrictive than the IBC; therefore, the change deletes most of the current language in 2015 IEBC and replaces it with new text and two new tables. This change has been made in Alterations Level 2 only; however, it would also be applicable in Alterations Level 3 due to the cascading criteria of Section 901.2 that requires Level 3 comply with Level 2 and Level 1 Alterations provisions. From inception, the IEBC has intended not to be more restrictive than the code for new construction and some specific provisions have addressed this issue. An example of this is found in Section 305.3 (Section 410.3 of 2015 IEBC). This concept, however, would be a logical approach for all provisions in the IEBC regardless of the subject as, in Section 101.3, the code makes it clear that the intent of the IEBC is to "provide more flexibility."

IEBC 904.1.4
Automatic Sprinkler System at Floor of Alteration

CHANGE TYPE: Modification

CHANGE SUMMARY: The Alterations Level 2 provision requiring that water for automatic fire sprinkler systems be available at the floor of the alteration without the need for a fire pump has been moved to Chapter 9 for Alterations Level 3 and the fire pump criterion has been deleted.

2018 CODE: ~~804.2.4~~ <u>904.1.4</u> **Other required automatic sprinkler systems.** In buildings and areas listed in Table 903.2.11.6 of the *International Building Code*, work areas that have exits or corridors shared by more than one tenant or that have exits or corridors serving an occupant load greater than 30 shall be provided with an automatic sprinkler system under the following conditions:

1. The *work area* is required to be provided with an automatic sprinkler system in accordance with the *International Building Code* applicable to new construction.

2. The building <u>site</u> has sufficient municipal water supply for design <u>and installation</u> of an automatic sprinkler system ~~available to the floor without installation of a new fire pump~~.

CHANGE SIGNIFICANCE: In Alterations Level 3, Section 904.1 references Level 2 fire sprinkler provisions of Section 803.2 (804.2 of 2015 IEBC). The 2015 IEBC Section 804.2.4 text addressing "other required automatic sprinkler systems" has been relocated to Chapter 9 for Alterations Level 3. This means that the general fire sprinkler provisions of 803.2 deal with this topic adequately for Level 2 Alterations and there is no need for any additional provisions. The relocation to Chapter 9, however, means that the additional criteria for fire sprinklers will continue to be a requirement for Level 3 Alterations.

The other change in this section is related to the need for a fire pump to make available adequate water for sprinkler use to the areas under consideration. This change is based on the logic that Level 3 Alterations are large enough that for cases where sprinklers would be required, the need for installation of a fire pump to supplement the water flow and pressure would not be the deciding factor. The needed fire safety will trigger the fire sprinkler requirement as long as the building site has sufficient municipal water supply.

Sprinkler riser room

IEBC 906.7
Anchorage of Unreinforced Masonry Partitions

CHANGE TYPE: Addition

CHANGE SUMMARY: The 2018 IEBC adds a mitigation trigger to address a common nonstructural falling hazard: unreinforced masonry partitions.

2018 CODE: **906.7 Anchorage of unreinforced masonry partitions.** Where the building is assigned to Seismic Design Category C, D, E, or F, unreinforced masonry partitions and nonstructural walls within the work area and adjacent to egress paths from the work area shall be anchored, removed, or altered to resist out-of-plane seismic forces, unless an evaluation demonstrates compliance of such items. Use of reduced seismic forces shall be permitted.

CHANGE SIGNIFICANCE: In the 2015 IEBC, both the prescriptive and work area methods include mitigation requirements for unreinforced masonry (URM) parapets and bearing walls, triggered by major alterations (Level 3). Alteration projects involving at least 50% of a building's area will typically trigger parapet removal or replacement in their scope. A related, unaddressed hazard involves the failure of interior unreinforced masonry partitions, especially around stairwells and in egress corridors. Mitigation of this well-understood and common hazard is justified in a Level 3 Alteration.

The 2018 IEBC adds a mitigation trigger to address this common nonstructural falling hazard. The trigger is only intended to apply to URM nonstructural walls (nonbearing walls), not to gypsum bearing and nonbearing walls within or outside the work area. Additionally, the provision intends to only require mitigation within the work area and along egress paths from the work area to building exits.

Retrofitted brick wall

IEBC 1006

Seismic Loads and Access to Risk Category IV Structures

CHANGE TYPE: Modification

CHANGE SUMMARY: When a change of occupancy occurs placing a building in a higher risk category, the seismic loads on the building must be evaluated using full seismic forces. Access to the building must be maintained when passing through or near other buildings and structures.

2018 CODE: ~~1007.3~~ **1006.3 Seismic loads.** ~~Existing buildings with a change of occupancy shall comply with the seismic provisions of Sections 1007.3.1 and 1007.3.2.~~

~~**1007.3.1 Compliance with International Building Code-level seismic forces.**~~ ~~Where a building or portion thereof is subject to~~ Where a change of occupancy ~~that~~ results in a building being assigned to a higher risk category ~~based on Table 1604.5 of the~~ *International Building Code*, the building shall ~~comply with~~ satisfy the requirements of Section 1613 of the *for* International Building Code ~~-level seismic forces as specified in Section 301.1.4.1~~ for the new risk category using full seismic forces.

Exceptions:

1. Where a change of use results in a building being reclassified from Risk Category I or II to Risk Category III and the seismic coefficient, S_{DS}, is less than 0.33. ~~Where approved by the code official, specific detailing provisions required for a new structure are not required to be met where it can be shown that an equivalent level of performance and seismic safety is obtained for the applicable risk category based on the provision for reduced International Building Code-level seismic forces as specified in Section 301.1.4.2.~~

Fire station

2. Where the area of the new occupancy ~~with a higher hazard category~~ is less than ~~or equal to~~ 10 percent of the ~~total~~ building ~~floor~~ area and the new occupancy is not <u>assigned to</u> ~~classified as~~ Risk Category IV. ~~For the purposes of this exception, buildings occupied by two or more occupancies not included in the same risk category, shall be subject to the provisions of Section 1604.5.1 of the *International Building Code*.~~ The cumulative effect ~~of the area~~ of occupancy changes <u>over time</u> shall be considered ~~for the purposes of this exception~~.

3. Unreinforced masonry bearing wall buildings ~~in~~ <u>assigned to</u> Risk Category III <u>and</u> to Seismic Design Category A or B, shall be ~~allowed to be strengthened to meet the requirements of~~ <u>permitted to use</u> Appendix Chapter A1 of this code ~~[Guidelines for the Seismic Retrofit of Existing Buildings [(GSREB)]~~.

<u>**1006.4**</u> ~~**1007.3.2**~~ **Access to Risk Category IV.** ~~Where a change of occupancy is such that compliance with Section 1007.3.1 is required and the building is~~ <u>Any structure that provides operational access to an adjacent structure</u> assigned to Risk Category IV~~, operational access to the building shall not be through an adjacent structure, unless that structure conforms to~~ ~~for Risk Category IV structures.~~ <u>as the result of a change of occupancy shall itself satisfy</u> the requirements <u>of Sections 1608, 1609, and 1613 of the *International Building Code*. For compliance with Section 1613, the full seismic forces shall be used</u>. Where operational access <u>to a Risk Category IV structure</u> is less than 10 feet (3048 mm) from either an interior lot line or from another structure, access protection from potential falling debris shall be provided ~~by the owner of the Risk Category IV structure~~.

CHANGE SIGNIFICANCE: The provisions for a seismic upgrade triggered by a change of risk category are simplified, with identical wording placed in the prescriptive and work area methods. The provision's Exception 1 was deleted as it was found to be redundant.

A new Section 1006.3 exception for buildings changing from Risk Category I or II to Risk Category III in regions with a design spectral response acceleration at a short period, S_{DS}, less than 0.33g has been added. In low seismic regions, Seismic Design Categories A and B, consideration of changes to the seismic loading on the seismic force resisting system is not required.

Provisions in Section 1006 are now consistent with the updated provisions of Section 506, structural loads for change of occupancy using the prescriptive compliance method. In the 2018 IEBC, Section 1007.3.2, renumbered as Section 1006.4, recognizes that any structure which provides access to a Risk Category IV structure must maintain that access during and after a wind or snow event as well as after an earthquake.

IEBC 1103
Changes to Loads with an Addition

CHANGE TYPE: Modification

CHANGE SUMMARY: Section 1103, Structural is revised to align with the prescriptive compliance method and to have better flow within the section.

2018 CODE:

SECTION 1103
STRUCTURAL

1103.1 ~~Compliance with the International Building Code.~~ ~~Additions to existing buildings or structures are new construction and shall comply with the *International Building Code*.~~

~~1103.2~~ Additional gravity loads. ~~Existing~~ Any existing gravity load-carrying structural ~~elements supporting~~ element for which an addition and its related alterations cause an increase in design dead, live, or snow load, including snow drift effects, of more than 5 percent shall be replaced or altered as needed to carry the gravity loads required by the *International Building Code* for new structures. Any existing gravity load-carrying structural element whose gravity load-carrying capacity is decreased as part of the addition and its related alterations shall be considered to be an altered element subject to the requirements of Section 806.2. Any existing element that will form part of the lateral load path for any ~~additional gravity loads as a result~~ part of ~~additions~~ the addition shall ~~comply with~~ be considered an existing lateral load-carrying structural element subject to the *~~International Building Code~~* requirements of Section 1103.3.

Addition of a hospital entrance

Exceptions:

1. ~~Structural elements whose stress is not increased by more than 5 percent.~~

2. Buildings of Group R occupancy with ~~no~~ not more than five dwelling units or sleeping units used solely for residential purposes where the existing building and the addition together comply with the conventional light-frame construction methods of the *International Building Code* or the provisions of the *International Residential Code*.

~~1103.3~~ **1103.2 Lateral force-resisting system.** ~~The~~ Where the addition is structurally independent of the existing structure, existing lateral ~~force-resisting system~~ load-carrying structural elements shall be permitted to remain unaltered. Where the addition is not structurally independent of ~~existing buildings to which additions are made~~ the existing structure, the existing structure and its addition acting together as a single structure shall ~~comply with~~ meet the requirements of Sections ~~1103.3.1, 1103.3.2~~ 1609 and ~~1103.3.3~~ 1613 of the *International Building Code* using ~~International Building Code-level~~ full seismic forces.

Exceptions:

1. Buildings of Group R occupancy with ~~no~~ not more than five dwelling or sleeping units used solely for residential purposes where the existing building and the addition comply with the conventional light-frame construction methods of the *International Building Code* or the provisions of the *International Residential Code*.

2. Any existing lateral load-carrying structural element whose demand-capacity ratio with the addition considered is not more than 10 percent greater than its demand-capacity ratio with the addition ignored shall be permitted to remain unaltered. For purposes of calculating demand-capacity ratios, the demand shall consider applicable load combinations with design lateral loads or forces in accordance with Sections 1609 and 1613 of the *International Building Code*. For purposes of this exception, comparisons of demand-capacity ratios and calculation of design lateral loads, forces, and capacities shall account for the cumulative effects of additions and alterations since original construction. ~~For purposes of calculating demand-capacity ratios, the demand shall consider applicable load combinations involving International Building Code-level seismic forces in accordance with Section 301.1.4.1.~~

~~**1103.3.1 Vertical addition.** Any element of the lateral force-resisting system of an existing building subjected to an increase in vertical or lateral loads from the vertical addition shall comply with the~~ *International Building Code* ~~wind provisions and the~~ *International Building Code* ~~level seismic forces specified in Section 301.1.4.1 of this code.~~

IEBC 1103 continues

IEBC 1103 continued

~~**1103.3.2 Horizontal addition.** Where horizontal additions are structurally connected to an existing structure, all lateral force-resisting elements of the existing structure affected by such addition shall comply with the~~ *International Building Code* ~~wind provisions and the IBC level seismic forces specified in Section 301.1.4.1 of this code.~~

~~**1103.3.3 Voluntary addition of structural elements to improve the lateral force-resisting system.** Voluntary addition of structural elements to improve the lateral force-resisting system of an existing building shall comply with Section 807.6.~~

~~**1103.4 Snow drift loads.** Any structural element of an existing building subjected to additional loads from the effects of snow drift as a result of an addition shall comply with the~~ *International Building Code.*

> **Exceptions:**
> 1. ~~Structural elements whose stress is not increased by more than 5 percent.~~
> 2. ~~Buildings of Group R occupancy with no more than five dwelling units or sleeping units used solely for residential purposes where the existing building and the addition comply with the~~ conventional light-frame ~~construction methods~~ light-frame construction methods of the *International Building Code* ~~or the provisions of the~~ *International Residential Code.*

CHANGE SIGNIFICANCE: This code change reconciles provisions in the prescriptive compliance and work area methods. There are a number of redundant sections on loads in the 2015 IEBC Chapters 4 and 11. Deleting the redundant sections and placing them into the 2018 IEBC Chapter 3 allows streamlining of the two design methods. Language in Section 1103 has been updated to be consistent with language in 2018 IEBC Section 502, Additions. Similar text for alterations is located in Section 806 for the work area method. Contradictory text has been deleted and provisions are streamlined and revised to have similar wording in each chapter.

Index

Note: The letter 't' followed by numbers represents 'tables'.

A

AAMA. *See* American architectural manufacturers association (AAMA)
Accessibility, 184–191
Accessory storage spaces, 21–22
ADA Standards for Accessible Design, 156, 191
Adhered masonry, 196
Administration
    definition references, removal of, 7–8
    greenhouse, 2–3
    repair garage, 4
    sleeping unit, 5–6
Aerosol fire-extinguishing systems, 129
Agricultural greenhouses, 26–27
AISI S110, 258
AISI S202. *See Code of standard practice for cold-formed steel structural framing*
AISI S213, 258
AISI S230, 222
AISI S240. *See North American standard for cold-formed steel structural framing*
AISI S400-15. *See North American standard for seismic design of cold-formed steel structural systems*
AISI standards, 258
Allowable area factor, 62t
American architectural manufacturers association (AAMA), 212
American wood council (AWC), 265, 277, 279
Anchored masonry, 196
ANSI/AISC 341-10, 258
ANSI/ASSE Z 359.1, 169
Architectural cast stone, 197
ASME A17.1/CSA B44, 290
ASTM D 1970, 205
ASTM D 226, 207
ASTM D 3161, 199
ASTM D 4869, 207
ASTM D 6757, 207
ASTM F 1667 supplement S1, 263
Attic sprinkler protection, 123
Attics, 122–124
Automatic fire extinguishing systems, 125
Automatic sprinkler systems, 125, 166, 340
Automatic water mist systems, 125
AWC. *See* American wood council (AWC)
AWC NDS. *See* National design specification for wood construction (AWC NDS)
AWC WFCM, 222

B

Balconies and decks, 120
BSCI. *See* Building component safety information (BSCI)
BSSC. *See* Building seismic safety council's (BSSC)
Building component safety information (BSCI), 239
Building-integrated photovoltaic panels, 206
    attachment, 207
    deck requirements, 206
    deck slope, 206
    high wind attachment, 206–207
    ice barrier, 207
    material standards, 207
    underlayment, 206
    underlayment application, 206
    wind resistance, 207
Building planning, 11–79
Building seismic safety council's (BSSC), 228
Building-integrated photovoltaic roof panel, 207–208

C

Cantilevered steel storage racks, 255
Carbon dioxide extinguishing systems, 125
Carbon monoxide detectors, 324–325
Children's play structures, 47–48
Cladding attachment over foam sheathing to wood framing, 285
    direct attachment, 286
    furred cladding attachment, 286
Class I standpipe hose connections, 133
Class III standpipes systems, 130
Classroom security function lockset, 158
CLT. *See* Cross laminated timber (CLT)
Code of Standard Practice for Cold-Formed Steel Structural Framing, 259
Column protection in light-frame construction, 82–83
Combustible projections, 86
    balconies and similar projections, 86
    bay and oriel windows, 86–87
Commercial cooking systems, 125
Communication equipment structures, 25
Concentrated business use areas, 141–142
Concentrated loads, 253
Concrete and reinforced masonry walls, anchorage, 328
Construction documents, 209
    general, 209
    roof rain load data, 211
    roof snow load data, 210
    special loads, 211
    wind design data, 210
Cooking appliances, 43
Cooking hood, 44
Cross laminated timber (CLT), 272

D

Dead-end smoke compartment, 38
Deck live load, 217
Delayed egress locking systems, 159–160
Delayed-action closer, 103
Delayed-action self-closing doors, 103–104
Domestic cooking systems, 127–128
    protection from fire, 127
        automatic fire-extinguishing system, 127
        ignition prevention, 127
Double fire walls, 92–93
    in light-frame construction, 92
    seismic design categories, 93
    structural stability, 92
Dry-chemical extinguishing systems, 125

E

Earthquake hazard reduction, 317–318
Earthquake loads, 227
    site class, 227
    site coefficients, 227, 228t
Egress through adjacent stories, 146
Egress travel
    common path, 170–171
    measurement, 170
Electrical systems, 45–46
Emergency elevator communication systems, 290–291
Emergency escape and rescue openings, 181–182
Emergency escape opening operation, 330–331
Enclosed parking garages ventilation, 35
Enclosure of atriums, 30–31
Existing structural elements, loading of, 322–323
Exit access doorway, 143t
Exit discharge illumination, 148–149
Exterior areas of assisted rescue, protection, 152
Exterior bearing walls, openings in, 280–281

F

Family/assisted-use toilet rooms, 186–187
Fasteners, 206
    in exterior applications/wet or damp locations, 258
    for preservative-treated wood, 268
    in treated wood, 268
Federal emergency management agency (FEMA), 234
Fire area concept, 69
Fireblocks and draftstops, 98
    Exception 1, 101
    Exception 2, 101
    Exception 3, 101
    Exception 4, 101
    Exception 5, 101
Fire partitions, 94–97
    continuity of, 96
    required extent, 96
    requirements, 96
Fire protection, 82–138
Fire protection of structural roof members, 74–75, 74t
Fire protection research foundation, 128
Fire pump and riser rom size, 113
    access, 113
    additional criterias, 113
    environment, 113
    heating equipments, 113
    lighting, 113
    marking on access doors, 113
Fire-resistance rating, 172
Fire-resistance-rated corridor, 292
Fire-retardant treated wood, 260–261
    in exterior walls, 78–79
Fire service access elevators, 294–295
Fire separation distance, 88
    allowable area, 88, 88t
    limitations on openings, 89
Fire sprinkler clearance, 243
Fire walls, 56–57
Fire watch during construction, 303–304
Flame spread testing, 109–110
    ASTM E 2404, 110
    ASTM E 2579, 110
    facings/wood veneers, 109
    laminated products factory-produced, 109
    NFPA 286, 110

Floor area allowances, 141t
Floor level exit sign location, 167
FM global, 126
Foam-water sprinkler system/foam-water spray systems, 125
Front approach, 190

G

Gaming, 188
    area, 188
    machines and gaming tables, 188
    machine type, 189
Gaming table type, 189
Gas cabinets, 51
Glass framing, deflection of, 212
Greenhouses, 2–3, 27
    assembly use, 13–14
Group A occupancies, 114, 135
    Group A-1 occupancy, 114
    Group A-2 occupancy, 115
    Group A-3 occupancy, 115
    Group A-4 occupancy, 115
    Group A-5 occupancy, 115
        spaces under grandstands/bleachers, 115
Group A-3, 161
    occupancy, 13–14, 161
Group B occupancy, 161
Group E occupancies, 117, 135
Group H high-hazard occupancy, 23
Group H occupancies, 29
Group H-3 storage facility, 24
Group I-1 assisted living housing units, 39–40
Group I-1 cooking facilities, 41–42
Group I-2 emergency illumination, 150
Group R-1 occupancies, 167
Group R-2 dormitory cooking facilities, 43
Group R-3 occupancies, 17, 138
Group R-3 fire separation distance, 76–77, 76t
Group R-4 facility, 138
Group R-4 fire alarm systems, 137–138
Group R-4 occupancies, 119, 138
Group R spaces, 143–145
Group S-1 moderate-hazard occupancy, 23
Group S-2 low-hazard occupancy, 23
Group S occupancy, 21
Group U occupancies, 25–27
Guard protection, 169

H

Header and girder spans, exterior walls, 276–279
Health care facilities code, 45–46
Heavy timber construction, 269
    columns, 269–270
    floor framing, 270
    floors, 270
        cross-laminated timber floors, 270
        sawn/glued-laminated plank floors, 271
    member's details, 269
    new locations, 273t
    partitions and walls, 270
        exterior walls, 270
        interior walls and partitions, 270
    roof decks, 271
        cross-laminated timber roofs, 271
        sawn/wood structural panel/glued-laminated plank roofs, 271
    roof framing, 270

Heavy timber exemption, 108
Higher education laboratories, 52
    application, 52–53
    conditions, 54
    design and number of laboratory suites per floor, 54t
    laboratory suite construction, 53
        automatic fire-extinguishing systems, 54
        floor assembly fire resistance, 53
        liquid tight floor, 54
        maximum number, 53
        means of egress, 53
        separation from other laboratory suites, 53
        separation from other nonlaboratory areas, 53
        standby or emergency power, 53
        ventilation, 53
    maximum allowable quantities of hazardous materials, percentage, 54
    primary considerations, 54
    scope, 52
High-rise buildings construction, 28–29
Hoistway opening protection, 292
Horizontal building separation allowance, 72–73
Hose stations, 130
Household model, 42

I

IBC. *See* International building code (IBC)
IBHS. *See* Insurance institute for business and home safety (IBHS)
ICC 500. *See* ICC/NSSA *Standard for design/construction of storm shelters*
ICC 500-compliant storm shelters, 214
ICC 600, 222
ICC-ES. *See* International code council evaluation service (ICC-ES)
ICC G1-2010. *See* ICC guideline for replicable buildings
ICC guideline for replicable buildings, 311
ICC/NSSA *Standard for design/construction of storm shelters*, 327
I-Codes. *See* International codes (I-Codes)
IEBC. *See* International existing building code (IEBC)
IFC. *See* International fire code (IFC)
IMC-compliant domestic hood, 42
Incidental uses, 70–71, 70t
Independent egress, 38
Insurance institute for business and home safety (IBHS), 205
Integrated fire protection system testing, 111
    damper control system, 111
    high-rise buildings, 111
    NFPA 4, 111
    smoke control system, 111
Interior exit stairway extension, 172
Interior wall and ceiling finish testing, 105–107
    acceptance criteria for NFPA 286, 105
    with ASTM E 84 or UL 723, 106
    dangers, 107
    with different requirements, 106
    textile wall coverings, 106
        acceptance criteria for NFPA 265, 106
        with ASTM E 84/UL 723, 107
        room corner test, 106
International building code (IBC), 1
International code council evaluation service (ICC-ES), 129
International codes (I-Codes), 311
International existing building code (IEBC), 315–346
International fire code (IFC), 4, 51
International mechanical code, 51
International plumbing code, 156
International residential code, 19

International staple, nail, and tool association (ISANTA), 262
ISANTA. *See* International staple, nail, and tool association (ISANTA)

L

Life safety code, 167, 175
Light-frame construction, 82–83
Live load reduction, 218–220
Live loads, 315–316
Loads with addition, changes, 344–346
Locking arrangements in educational occupancies, 157–158
Lodging houses, 18–20
Luminous egress path markings, 176

M

Masonry veneer systems
    adhered masonry, 196
    anchored masonry, 196
Mean recurrence intervals (MRI), 212
Means of egress, 141–182
Mechanically laminated decking, 264–265
Medical gas systems, 49–51
    general, 49
    interior supply location, 49
        gas cabinets, 51
        one-hour exterior room, 49
        one-hour interior room, 49–50
Membrane penetrations of shaft enclosures, 102
Mercantile group M occupancy, 15–16
Metal roof shingles, 199
Metal-plate connected wood trusses, 240
Mezzanine, 147
Mezzanine and equipment platform area limitations, 60–61
Minimum tire load, 221
Moderate-hazard occupancies, 29
Moderate-hazard storage, 23–24
Modular building institute, 300
Motor vehicle-related occupancies, 32–33
MRI. *See* Mean recurrence intervals (MRI)
Multiple garages, 34
Multiple occupancies, 213–214

N

Nails and staples, 262–263
National bureau of standards (NBS), 141
National design specification for wood construction (AWC NDS), 262, 265, 288
National earthquake hazard reduction program (NEHRP), 228
National electrical code (NFPA 70), 70
National electrical code®, 151
National flood insurance program (NFIP), 306
National institute of standards and technology (NIST), 141
NBS. *See* National bureau of standards (NBS)
NEHRP. *See* National earthquake hazard reduction program (NEHRP)
New York state energy research and development agency (NYSERDA), 287
NFIP. *See* National flood insurance program (NFIP)
NFPA 101. *See* Life safety code
NFPA 12. *See* Carbon dioxide extinguishing systems
NFPA 13. *See* Automatic sprinkler systems
NFPA 13R, 123
NFPA 14. *See* Standard for the installation of standpipe and hose systems

NFPA 16. *See* Foam-water sprinkler system/foam-water spray systems
NFPA 17. *See* Dry-chemical extinguishing systems
NFPA 17A. *See* Wet-chemical extinguishing systems
NFPA 2010. *See Standard for fixed aerosol fire extinguishing systems*
NFPA 750. *See* Automatic water mist systems; Water mist fire protection systems
NFPA 96, 125, 126
NFPA 99, 46
NIST. *See* National institute of standards and technology (NIST)
Nonseparated occupancies, 66–67
   group I-2, condition 2 occupancies, 66
   high-rise buildings, 66
Nonsprinklered buildings, 293
North American standard for cold-formed steel structural framing, 258
North American standard for seismic design of cold-formed steel structural systems, 258

O

Occupancy classification, 11–12
Occupancy-based code, 12
Occupant evacuation elevators, 296
   additional criteria, 297
   general, 296
   number, 296
   options, 297
   standby power load determination, 296
   use of, 297
Occupant load, 19–20, 177
Occupied roofs, 11, 58
   enclosures, 58
   exception 1, 59
   exception 2, 59
One-hour exterior room, 49
One-hour interior room, 49–50
Open-air assembly seating, 179–180

P

Party walls, 90–91
   construction, 91
   new allowance, 91
   new exception, 91
Penetrations, 174
   exit passageway, 174
   exit ramps, 175
   exit stairways, 174
Permeable floors and roofs
   supporting members, 274
   ventilation beneath balcony/elevated walking surfaces, 274
Plastic composites, 87
Play areas, 191
Plumbing/mechanical/electrical components, 243–244
Podium/pedestal buildings, 73
Polypropylene siding, 198
Power-driven metal caps, 201
Precast concrete diaphragms, seismic loads, 251–252
   plain and reinforced concrete, 251
Precast prestressed piles, 248–250
   axial load limit in seismic design categories C through F, 249
   seismic reinforcement in seismic design category C, 248
   seismic reinforcement in seismic design categories D through F, 248–249

Private garages
   and carports, 34
   regulation, 34
Projections, 84–85
   minimum distance, 84t
   modifications, 84–85
   provisions, 85
Prototypical construction documents, 308
Provisions update committee (PUC), 228
PUC. *See* Provisions update committee (PUC)

Q

Quick-response sprinklers, 40

R

Rack Manufacturers Institute's (RMI), 255
Rated corridors, 292–293
References, 7–8
Refuge area, 177–178
Relocatable buildings, 298
   compliance, 298
   general, 298
   inspection agencies, 299
   manufacturer's data plate, 299
   regulation, 300
   supplemental information, 299
Remote operation of locks, 157
Reorganization, 319–320
Repair garage, 4
Replicable buildings, 307–308
Replicable design, 308
   review and approval, 309
      approval, 310
      deficiencies, 310
      documentation, 310
      site-specific, 310–311
   site-specific application, 310
      architectural plans and specifications, 310
      general, 310
      submittal documents, 310
Replicable design submittal requirements, 308
   architectural plans and specifications, 308–309
   energy conservation requirements, 309
   general, 308
   structural plans/specifications/engineering details, 309
Residential Group R-2 occupancies, 17–18
Residential Group R-3 occupancies, 18
Retaining walls, 247
   design lateral soil loads, 247
   general, 247
Ring shank nails, 266–267
RMI. *See* Rack Manufacturers Institute's (RMI)
Roof sheathing ring shank (RSRS), 267
Rooftop equipment, fall arrest, 168–169
   guards, 168

S

Safety code for elevators and escalators, 290
SDC C. *See* Seismic design category C (SDC C)
Seismic design category C (SDC C), 249
Seismic force-resisting systems, 241
Seismic loads, 342–343
Seismic maps, 230–235
Seismic-force-resisting system, 332

Self-storage facilities, 23
Separated occupancies, 68–69, 68t
Shared spaces, 40
Single-exit buildings, 338–339
Site grading, 245–246
Size of doors, 154
Sleeping unit, 5–6
Smoke barriers, 36
    exit access travel distance, 37
    smoke compartment size, 36
Smoke control system, 31
Smoke-protected assembly seating, 180
Snow damage, 321
Sound transmission, engineering analysis of, 194–195
    air-borne sound, 194
    structural-borne sound, 194
Sprinklered building, 293
Sprinklered, one-story buildings, 64–65
Stairway door locks, 162–163
Stairway extensions, 172–173
Stairways in buildings, construction, 301–302
Standard for fixed aerosol fire extinguishing systems, 129
Standard for safety for household electric ranges, 128
Standard for the installation of standpipe and hose systems, 134
Stationary battery systems, 70
Steel joist institute (SJI) standard, 253
Storage Group S-1 occupancies, 23
Storage rooms, 21–22
Storm shelters, 326
    addition to Group E occupancy, 326
    additions to Group E facilities, 327
    loads, 215–216
    location, 327
    required occupant capacity, 326
Structural glass baluster panels, 282
Structural loads
    access to Risk Category IV, 333
    in historic Buildings, 335–336
    live loads, 332
    seismic loads, 332
    snow and wind loads, 332
Structural observations
    seismic resistance, 239
    structures, 238
    wind resistance, 239
Structural steel elements, 241–242
Suites, 6

T

Three arm waist-high turnstiles, 166
Truss designer, 240
Truss restraint, 240
Tsunami design geodatabase, 236

Tsunami design zone, 236
Tsunami loads, 236
    general, 237
    risk category, 236
Turnstiles, 164–166
    additional door, 165
    capacity, 164
    clear width, 164
    high turnstile, 165
    security access, 165
Type III and IV buildings, 78–79
Type VB greenhouses, 62–63

U

UL 1897. *See Uplift tests for roof covering systems*
UL 300, 125
UL 858. *See Standard for safety for household electric ranges*
Underlayment, 200–205, 202–204t
    ice barriers, 201
    scope, 200
United States geological survey (USGS), 233
    for California sources, 234
    for Central and Eastern US (CEUS) sources, 234
    for ground motion models, 235
    for Intermountain West and Pacific Northwest crustal sources, 234
Unreinforced masonry partitions, anchorage, 329, 341
Unreinforced masonry walls (URMs), 328
Uplift tests for roof covering systems, 208
URMs. *See* Unreinforced masonry walls (URMs)
USGS. *See* United States geological survey (USGS)

V

Ventilation control/fire protection of commercial cooking operations, 126
Ventilation, 35

W

Walk-in coolers and freezers, 184–185
Water mist fire protection systems, 126
Water course alteration, 306
Water-resistive barrier, 283–284
Weather covering minimum thickness, 196–197
Wet-chemical extinguishing systems, 125
Wind loads, 222–226
    applicability, 222–223
    basic design wind speed, 223–225
    notations, 222
Wind speeds, 226

Driving Growth and Affordability Through Innovation and Safety

An Overview of the International Code Council

The International Code Council is a member-focused association. It is dedicated to developing model codes and standards used in the design, build and compliance process to construct safe, sustainable, affordable and resilient structures. Most U.S. communities and many global markets choose the International Codes.

Services of the ICC

The organizations that comprise the International Code Council offer unmatched technical, educational and informational products and services in support of the International Codes, with more than 200 highly qualified staff members at 16 offices throughout the United States. Some of the products and services readily available to code users include:

- CODE APPLICATION ASSISTANCE
- EDUCATIONAL PROGRAMS
- PREFERRED PROVIDER PROGRAM FOR EDUCATORS
- CERTIFICATION PROGRAMS
- TECHNICAL HANDBOOKS AND WORKBOOKS
- PLAN REVIEW SERVICES
- DIGITAL PRODUCTS
- ONLINE MAGAZINES AND NEWSLETTERS
- PUBLICATION OF PROPOSED CODE CHANGES
- TRAINING AND INFORMATIONAL VIDEOS
- BUILDING DEPARTMENT ACCREDITATION PROGRAMS
- EVALUATION SERVICE FOR CODE COMPLIANCE
- EVALUATIONS UNDER GREEN CODES, STANDARDS AND RATING SYSTEMS

Additional Support for Professionals and Industry:

ICC EVALUATION SERVICE, LLC (ICC-ES)

ICC-ES is the industry leader in performing technical evaluations for code compliance, providing regulators and construction professionals with clear evidence that products comply with codes and standards.

INTERNATIONAL ACCREDITATION SERVICE (IAS)

IAS accredits testing and calibration laboratories, inspection agencies, building departments, fabricator inspection programs and IBC special inspection agencies.

SOLAR RATING & CERTIFICATION CORPORATION (ICC-SRCC)

The industry standard since 1980, ICC-SRCC fulfills the industry's need for a single, national program that allows manufacturers to rate and test the efficiency of solar equipment. In addition, ICC-SRCC now also includes the Small Wind Certification Council's services in its portfolio and can facilitate in independent certification of Wind Turbines.

S. K. GHOSH ASSOCIATES (SKGA)

The preeminent technical resource on structural codes and standards, SKGA supports Authorities Having Jurisdiction, design professionals and others through its specialized publications, seminars, research projects and services.

NEED MORE INFORMATION? CONTACT ICC TODAY!
1-888-ICC-SAFE (422-7233) | www.iccsafe.org

ES | ICC EVALUATION SERVICE

In Cooperation with **Innovation RESEARCH LABS**

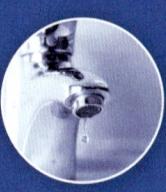

Accelerate your Building, Plumbing and Energy Products' Speed to Market with ICC-ES!

Your One-Stop Testing, Listing and Product Evaluation Service

- ICC-ES provides a one-stop shop for the evaluation, listing and now testing of innovative building products through our newly formed cooperation with Innovation Research Labs, a highly respected ISO 17025 accredited testing lab with over 50 years of experience.

- ICC-ES Evaluation Reports are the most widely accepted and trusted technical reports for code compliance. When you specify products or materials with an ICC-ES report, you avoid delays on projects and improve your bottom line.

- ICC-ES is a subsidiary of ICC, the publisher of the codes used throughout the U.S. and many global markets, so you can be confident in their code expertise.

- ICC-ES provides you with a free online directory of code compliant products at: **www.icc-es.org/Evaluation_Reports** and CEU courses that help you design with confidence.

WE CERTIFY & TEST:
- Air, Water & Vapor Barriers
- Cladding
- Doors
- Fasteners
- Roofing Materials & Accessories
- Wall Coverings/Systems
- Windows
- Floor/Deck Systems
- Flooring Materials
- Manufacturers Wood
- Plastic Lumber
- Plumbing Products

Look for the ICC-ES Marks of Conformity

WWW.ICC-ES.ORG | 800-423-6587 X3877

INTERNATIONAL CODE COUNCIL

People Helping People Build a Safer World

Take your career further, faster with ICC Membership

No other building code association offers more I-Code resources and related training to help you achieve your career goals than the International Code Council® (ICC®).

From discounts on top-notch training and educational programs to free code opinions from I-Code experts, ICC Members enjoy direct access to exclusive benefits including:

- Free I-Code book(s) or download to new Members*
- Posting resumes and job search openings through the ICC Career Center
- Mentoring programs and valuable networking opportunities at ICC events
- Participating in the development of codes and standards
- Corporate and Governmental Members: Your staff can receive free ICC benefits too* and more.

Plus, you'll save up to 25% on code books, training materials and more!

Join now. There's an ICC Membership category that's right for you. Visit **www.iccsafe.org/mem1** or call **1-888-ICC-SAFE** (422-7233), ext. 33804 to *join, renew or learn more.*

* *Some restrictions apply. Speak with an ICC Member Services Representative for details.*

ICC INTERNATIONAL CODE COUNCIL®

People Helping People Build a Safer World®

Have you seen ICC's Digital Library lately?

codes.iccsafe.org *offers convenience, choice, and comprehensive digital options*

Get FREE access 24/7

publicACCESS

Enjoy **FREE** access to the complete text of critical construction safety provisions including:

- International Codes®
- ICC Standards and Guidelines
- State and City Codes

premiumACCESS™

In addition to viewing the complete text online, **premiumACCESS** features make it easier than ever to save time and collaborate with colleagues.

Powerful features that work for you:

FREE Demo available!

- Advanced Search crosses your entire set of purchased products.
- Concurrent user functionality lets colleagues share access.
- Internal linking navigates between purchased books in your library.
- Print controls create a PDF of any section.
- Bookmarks can be added to any section or subsection.
- Highlighting with Annotation option keeps you organized.
- Color coding identifies changes.
- Tags can be added and filtered.

1-year and 3-year *premiumACCESS* subscriptions are available now for:
International Codes® | State Codes | Standards | Commentaries

Let codes.iccsafe.org start working for you today!